BIOPOLYMER RESEARCH TRENDS

Biopolymer Research Trends

Tamás S. Németh
Editor

Nova Science Publishers, Inc.
New York

For permission to use material from this book please contact us:
Telephone 631-231-7269; Fax 631-231-8175
Web Site: http://www.novapublishers.com

LIBRARY OF CONGRESS CATALOGING-IN-PUBLICATION DATA

Biopolymer research trends / Tamás S Németh (editor)
p.; cm
Includes bibliographical references and index.
ISBN-13: 978-1-60021-983-2 (hardcover)
ISBN-10: 1-60021-983-7 (hardcover)
1. Biopolymers. I Németh, Tamá S.
[DNLM: 1. Biocompatible Materials--chemistry. 2. Biopolmers--chemistry. 3. Spectrum Analysis--methods. QT 37.5.P7 B6158 2008]
QP801.B69B555 2008
572--dc22

2007030919

Published by Nova Science Publishers, Inc. ✢ New York

CONTENTS

PREFACE

Biopolymers are a special class of polymers produced by living organisms. Starch, proteins and peptides, DNA, and RNA are all examples of biopolymers, in which the monomer units, respectively, are sugars, amino acids, and nucleic acids. A major but defining difference between polymers and biopolymers can be found in their structures. Polymers, including biopolymers, are made of repetitive units called monomers. Biopolymers inherently have a well defined structure: The exact chemical composition and the sequence in which these units are arranged is called the primary structure. Many biopolymers spontaneously fold into characteristic compact shapes (see also "protein folding" as well as secondary structure and tertiary structure), which determine their biological functions and depend in a complicated way on their primary structures. Structural biology is the study of the structural properties of the biopolymers. In contrast most synthetic polymers have much simpler and more random (or statistic) structures. This new book presents leading-edge research from around the world in this dynamic field.

Expert Commentary - Unveiling sequence-stability and structure-stability relationships is a major goal of protein chemistry and structural biology. For globular proteins, despite the enormous amount of literature work in this field, no convincing solutions have been hitherto provided to these issues. Collagen represents an ideal system for such investigations due to its repetitive Gly-X-Y sequence and to its regular structure. The great abundance of collagen among vertebrates has made these analyses appealing also for their biological implications. The analysis of amino acid frequencies in collagen sequence and the use of host-guest model polypeptides have provided unambiguous indications on the sequence-stability relationships for this widespread protein. For several amino-acids clear hints on the structural bases of their stabilizing/destabilizing effects have been obtained by theoretical and experimental structural studies on collagen-like polypeptides. Paradoxically, there is no consensus on the structural determinants of triple helix stabilization by (4R,2S)-hydroxyproline, the residue that provides the strongest contribution to collagen stability. Proline hydroxylation has differentiate effects on triple helix stability depending on the position of proline in the Gly-X-Y sequence motif and on the diastereoisomer produced ((4R,2S)-hydroxyproline, Hyp or (4S,2S)-hydroxyproline, hyp)). In particular, replacement of Pro residues located at X or Y positions of Gly-Pro-Pro triplets with Hyp leads to destabilization or stabilization of the triple helix, respectively. On the other hand, the replacement of Pro residues with hyp has destabilizing effects in both X and Y positions. Over the years, a number of models have tried to explain the dependence of triple helix stability on proline hydroxylation. However, although significant advances have been achieved in the last decade, this subject is still open to debate. Intriguingly, even high resolution crystal structures of Hyp-containing polypeptide models have not provided a definite answer to this puzzling issue. In this commentary, a critical evaluation of the strengths and the drawbacks of the current hypotheses are presented. Future strategies needed to offer more insightful information on the structure and stability of this protein are also delineated.

Chapter 1 - Metallothioneins (MTs) are low molecular weight, cysteine-rich proteins with an exceptional heavy metal coordination capacity. Because of their ability to bind metals and to scavenge oxidant radicals, MTs are considered to play a role in metal homeostasis, metal detoxification and control of the oxidative stress. Although their high heterogeneity on the expression patterns, metal binding abilities and primary structure suggest very diverse functional specializations, the structural and functional studies have been mainly devoted to vertebrate and fungal MTs and their canonical cysteine-metal clusters.

This chapter will be focused on the new methodological procedures settled for the structural characterisation of some metallic MT aggregates. Five zinc complexes from the invertebrate and plant MTs, poorly described up to now in the literature, in addition to one Zn-MT complex from the well studied vertebrate MT family, have been analysed. The new discovered structural features of metal-MT clusters, in addition with the perspectives on MT research, have been also commented. In fact, much can be learnt about MT systems by using spectroscopic techniques such as Raman and IR spectroscopies, and Circular Dichroism, able to provide new structural information eventually related to the function of the metal binding. Despite the potentialities of these techniques, to our knowledge they have been scarcely used in MT conformational studies until now. Recently, the use of these spectroscopies has resulted to be very useful to approach unambiguously two basic structural points poorly described in MTs: the participation of chloride, sulfide ions and His residues to the metal-coordination sphere and the presence of secondary structure elements. In particular, ordered secondary structures, oppositely to what has been commonly accepted, are present in MTs from vertebrate, invertebrate and plant MTs, and could develop crucial roles in the determination of the functional properties of MTs.

The biosynthesis of intact metal-MT complexes, well corresponding to native forms, in sufficient quantity and purity for analytical spectrometric and spectroscopic characterization has been allowed by recombinant expression in *E.coli*. The spectroscopic analyses of the *in vivo*-synthesised metal-MTs have recently demonstrated the participation of extra-protein ligands, such as chloride and sulfide ions, in the metal-MT coordination environment of vertebrate, invertebrate and plant MTs. In conclusion, the coupling of analytical and spectroscopical techniques can be one of the most promising experimental strategies in the research on new hints on MTs.

Chapter 2 - In order to investigate in detail the influence of chemical modification on the internal structure of keratin fibers, which have a hierarchical structure, the authors have developed a new method for directly analyzing the structure of cross-sections at various depths of keratin fibers using Raman spectroscopy. This method involves embedding keratin fiber samples in an epoxy resin and microtoming the cured blocks on a microtome to 1-□m (white human hair) and 1.5-□m (black human hair) thickness, and then mounting the samples on a slide glass. The cross-sectional samples are then analyzed with a Raman microscope. Using this analytical technique, the Raman spectra of virgin black human hair, which had been impossible due to it's high melanin granule content, can be recorded. Also, the heterogeneous reaction between reducing agents (thioglycolic acid and L-cysteine), or a protein crosslinking agent (2-iminothiolane hydrochloride) and keratin fibers at the molecular level can be analyzed. Moreover, the secondary structure [the □-sheet and/or random coil (□/R) and the □-helix (□) contents] of cross-sections at various depths of keratin fibers changed by the chemical treatments (bleaching and permanent waving treatments), or chemical modification using 2-iminothiolane hydrochloride (2-IT) can be analyzed by amide I band analysis. Thus, the characterization of the cortex region, which consists of crystalline fibrous protein and the amorphous matrix is an effective method, since information about crystalline and amorphous protein structure can be obtained. Furthermore, the changes of the disulfide (-SS-) content, cysteic acid content, and random coil content show the level of damage on keratin fibers. It can be supposed that this method is a beneficial analytical tool to investigate more detailed internal structural changes due to the influence of not only external factors such as heating, permanent waving, bleaching treatments, and exposure to sunlight, but also internal factors such as aging and nutritional deficiencies on human hair, since the Raman spectra of virgin black human hair keratin fibers can be analyzed. Structural analysis of keratin fibers can be done to a much higher level of detail than previously as Raman spectroscopy can be used.

Chapter 3 - The investigation of molecular motions in biological polymers has been one of the basic trends in molecular biophysics for a long time. Many physical methods have been applied to studying biomolecular mobility. However, in spite of the large amount of experimental data there are still some methodological problems that are not yet completely resolved. One of the most essential ones is the ambiguity of transition from the first-hand experimental parameters to the parameters characterizing molecular motions. The most poorly defined characteristic of a motion is its geometry. There are almost no experimental techniques, except computer simulation, that provide direct and unambiguous information on the motional geometry models. At the same time, this information in many cases can be of high importance for revealing molecular mechanisms of the protein biological function. In this contribution the authors describe the experimental approaches that may solve this problem. These approaches are based on the complex experimental NMR study. One of the main advantages of NMR in respect to other physical methods is that it allows using different magnetic nuclei and different magnetic interactions (dipole-dipole and quadrupole couplings, chemical shift anisotropy) for probing the same kind of molecular mobility. The comparative quantitative analysis of different types of NMR data obtained on the same sample may allow the discrimination of different motional models directly from the experimental data. This complex approach is demonstrated by a study of molecular dynamics of a model system, homopolypeptide poly-L-lysine, and backbone dynamics of a protein barstar in solid state. Limitations as well as perspectives of the development of this approach are discussed in detail.

Chapter 4 - ZINDO quantum chemical calculations and vibronic theory of activation are used to study the effect of different distortions of the active center of carbonyl complexes of heme proteins and external electric fields on the magnitude of the C-O vibrational frequency and its relationship with the changes in the Fe-C frequency. It is shown that the experimentally observed negative linear correlation between these two frequencies stems from the variation of the electric field of the heme environment. Study of the effect of the electric field of the distal histidine on the C-O frequency allowed assigning a number of the CO infrared absorption sub bands of carboxymyoglobin to specific orientations and tautomeric states of the histidine. The results on the field dependence of the C-O vibrational frequency suggested that the width of this band should be sensitive to the large amplitude motion of the distal heme environment. The temperature dependencies of the C-O bands of carboxycomplexes of horseradish peroxidase, myoglobin and hemoglobin at different pH were quantitatively interpreted taking into account electrostatic coupling of the band to the motion of the heme environment. The analysis of the parameters of the fitting procedure showed that upon heating in the liquid solvent water molecule enters the heme pocket of the proteins with the capacious pocket (horseradish peroxidase and "open" conformation of myoglobin and hemoglobin), this water molecule mainly contributing into the temperature dependence of the band. In the glassy matrix the large amplitude motions of the pocket amino acids are arrested and the disordered water cannot enter or leave the pocket. Appearance of the water in the heme pocket causes transition of the protein to another conformational substate, at room temperature almost all protein molecules exist in this conformational substate. To the authors best knowledge this is the first observed example of almost full transition of a protein from one conformational substate to another, caused by the temperature change.

Chapter 5 - Although volatile general anesthetics are administered to over 20 million patients in the United States each year for a broad range of surgical procedures, their mechanisms of action remain poorly understood. Volatile general anesthetics are believed to exert their clinical effects by modulating the activity of neuronal plasma membrane ligand-gated ion channels, including the γ-aminobutyric acid type A receptor and the glycine receptor. The structures of these membrane proteins are currently unknown but their transmembrane domains are thought to be composed of the commonly occurring four-α-helix bundle protein fold, based upon homology modeling with the related nicotinic acetylcholine receptor Cys-loop ligand-gated ion channel. Site-directed mutagenesis studies on intact ligand-gated ion channels expressed in different cell types implicate the transmembrane domains of the protein as constituting volatile general anesthetic sites of action. A similar conclusion has been drawn using photoaffinity labeling of the nicotinic acetylcholine receptor with halothane, followed by microsequencing to identify volatile general anesthetic binding sites. Experimental studies with synthetic four-α-helix bundle proteins reveal that this folding motif is capable of binding several contemporary clinically used volatile general anesthetics with dissociation constants that correlate closely with their respective EC_{50} values (effective concentration in 50% of test subjects) in humans for maintaining the anesthetic state. Detailed biophysical studies on these synthetic four-α-helix bundle proteins provide insight into how volatile general anesthetic binding can lead to altered protein activity, by modulating the structure, flexibility and overall stability of the system. In addition, molecular dynamics simulations on both synthetic- and natural four-α-helix bundle protein domains provide further evidence for how biomolecular function can be modulated in the presence of bound volatile general anesthetic molecules. This chapter will present recent advances gained into the fundamental mechanisms of volatile general anesthetic action based upon studies with a number of different four-α-helix bundle motifs.

Chapter 6 - Using 6 samples of well-purified sodium-type gellan gums with different molar masses (Na-gellan, G1-G6, M_w =120x10^3–17x10^3 at 40°C), the effects of molar mass on the coil to double-helix transition in aqueous solutions with and without 25 mmol NaCl were studied by light scattering and circular dichroism measurements, viscometry, and differential scanning calorimetry.

In aqueous solutions with 25 mmol NaCl, the temperature dependences of M_w, molar ellipticity at 201nm $[\theta]_{201}$, intrinsic viscosity $[\eta]$, and DSC exothermic curves of G1-G6 samples were measured from 5 to 90°C. It was found that the coil to double-helix transitions for G1-G5 samples (M_w =120x10^3 –32x10^3) took place at almost the same temperature, and the coil to double-helix transition accelerated with increase of M_w. The G6 sample (M_w =17x10^3) did not form a double-helix at 25 °C suggesting that the lowest molar mass, below which no helix is formed, lies between M_w = 32x10^3 and M_w =17x10^3. The $[\eta]$ and M_w obtained in the temperature range from 40 to 25 °C can be explained by a simple coil/double-helix equilibrium model using the double-helix contents determined from circular dichroism data. The van't Hoff's transition enthalpy ΔH_{vH} of Na-gellans depended on M_w. It is concluded that the coil to double-helix transitions of Na-gellans are all-or-none type transitions, and accelerated with increasing M_w.

In aqueous solutions without NaCl, the coil to double-helix transitions of G1-G5 samples were investigated from the temperature dependences of viscosity number η_{sp}/c and $[\theta]_{201}$ within the temperature range from 5 to 90 °C. It was found that the coil to double-helix transition temperatures for G1-G5 samples are almost the same, irrespective of M_w, which increase with increasing polymer concentrations. ΔH_{vH} values of G1-G5 samples, which are nearly the same as the values obtained in aqueous 25 mmol NaCl solutions, depend markedly on M_w. These results suggest that, in the same way as the results obtained in aqueous solutions with 25 mmol NaCl, the coil to double-helix transition without NaCl accelerated with increasing M_w. From the results of viscometry and CD measurements in aqueous solutions without NaCl, the conformational behavior of G6 is considered to be different from those of G1-G5.

Chapter 7 - Oral cancers are a serious health problem in developing as well as developed countries, and more so in India and other south Asian countries. Survival rate of these cancers, despite advances in treatment modalities are one of the poorest. This can be attributed to lack of reliable screening and early detection. Optical spectroscopy methods which are sensitive to biomolecular compositions of systems can be potential alternatives/ adjuvant diagnosis/screening approaches. Due to high sensitivity and simplicity of instrumentation, so far, autofluorescence has been the most popular among the optical diagnostic methods. Despite its inherently weak nature, other attributes of Raman spectroscopy such as *in vivo* applicability, rich information content through molecular finger print, easy extraction of data, and most importantly, use of less harmful Near Infrared (NIR) radiation with larger penetration depths for excitation, make this spectroscopy as an ideal choice. Presently, several data mining methods are available to spectroscopists to achieve objective discrimination which is a major advantage of optical spectroscopy methods over conventional approaches. The authors have demonstrated the efficacy of Raman spectroscopic discrimination of healthy and pathological oral tissues based on spectral signatures analyzed by PCA.

In the present chapter, the authors will provide a brief overview of oral cancers, spectroscopic approach for oral cancer diagnosis and basics of Raman spectroscopy. The authors also share our experiences on Raman spectroscopic discrimination of normal and disease conditions in oral tissues

In: Biopolymer Research Trends
Editor: Tamas S. Nemeth, pp. 1-9

ISBN: 978-1-60021-983-2

Expert Commentary

UNDERSTANDING THE STRUCTURAL BASES OF COLLAGEN TRIPLE HELIX STABILITY

Alessia Ruggiero[a], Alfonso De Simone[a], Inesa Mesropyan[b], Luigi Vitagliano[a] and Rita Berisio[a]

[a]Istituto di Biostrutture e Bioimmagini, CNR, Via Mezzocannone 16.
I-80134 Napoli – Italy
[b]E. Andronikashvili Institute of Physics,
Department of Physics of Biological Systems, 6 Tamarashvili str.,
Tbilisi, Georgia, 0177

ABSTRACT

Unveiling sequence-stability and structure-stability relationships is a major goal of protein chemistry and structural biology. For globular proteins, despite the enormous amount of literature work in this field, no convincing solutions have been hitherto provided to these issues. Collagen represents an ideal system for such investigations due to its repetitive Gly-X-Y sequence and to its regular structure. The great abundance of collagen among vertebrates has made these analyses appealing also for their biological implications. The analysis of amino acid frequencies in collagen sequence and the use of host-guest model polypeptides have provided unambiguous indications on the sequence-stability relationships for this widespread protein. For several amino-acids clear hints on the structural bases of their stabilizing/destabilizing effects have been obtained by theoretical and experimental structural studies on collagen-like polypeptides. Paradoxically, there is no consensus on the structural determinants of triple helix stabilization by (4R,2S)-hydroxyproline, the residue that provides the strongest contribution to collagen stability. Proline hydroxylation has differentiate effects on triple helix stability depending on the position of proline in the Gly-X-Y sequence motif and on the diastereoisomer produced ((4R,2S)-hydroxyproline, Hyp or (4S,2S)-hydroxyproline, hyp)). In particular, replacement of Pro residues located at X or Y positions of Gly-Pro-Pro triplets with Hyp leads to destabilization or stabilization of the triple helix, respectively. On the other hand, the replacement of Pro residues with hyp has destabilizing effects in both X and Y positions. Over the years, a number of models have tried to explain the dependence of triple helix stability on proline hydroxylation.

However, although significant advances have been achieved in the last decade, this subject is still open to debate. Intriguingly, even high resolution crystal structures of Hyp-containing polypeptide models have not provided a definite answer to this puzzling issue. In this commentary, a critical evaluation of the strengths and the drawbacks of the current hypotheses are presented. Future strategies needed to offer more insightful information on the structure and stability of this protein are also delineated.

COMMENTARY

Collagen, an ubiquitous protein in multicellular organisms, is the main constituent of connective tissue and is the most abundant protein in higher vertebrates [1]. In these organisms, it accounts for one third of the total protein weight. Collagen-like molecules also have been found in lower eukaryotes such as mussels, worms, and sponges [2]. Also, collagen-like sequences have been identified from analyses of prokaryotic and viral genomes [3,4].

From a molecular point of view, collagen has three distinctive features: (a) a peculiar aminoacid composition, (b) a sequence made of repeated motifs, and (c) a triple helical structure. Collagen amino acid composition is characterized by an unusual abundance of iminoacids (proline and its hydroxylated derivatives) [5]. Indeed, collagen structure and function heavily depends on the hydroxylation, a post-translational process, of a fraction of proline residues present in its sequence. Incomplete hydroxylation of collagen polypeptide chains leads to the insurgence of severe diseases, such as scurvy. Other diseases related to malfunctioning prolyl-hydroxylases are also known [6]. Furthermore, the level of collagen prolyl-hydroxylation is related to the living temperature of the organism [7-9].

The analysis of collagen sequence clearly indicates that proline and hydroxyproline localization follows a rather rigid trend. Indeed, with very few exceptions, proline and hydroxyproline are located in the X and Y position, respectively, of collagen Gly-X-Y sequence motif. It is also worth mentioning that, among the possible hydroxyproline isomers, (4R,2S)-hydroxyproline (4RHyp) 4RHyp is the only one commonly present in collagen. Intriguingly, the closely related isomer (4S,2S)-hydroxyproline (4SHyp) is not present in collagen sequences, while the (3R,2S)-hydroxyproline (3RHyp) isomer has been sporadically detected [10,11].

More than fifty years ago it was suggested that collagen molecule is made of three distinct polypeptide chains wrapped around a common axis to form a triple helix structure [12,13]. Although subsequent structural investigations have confirmed this insightful prediction, the fine details of collagen triple helix motif are still highly debated [14-17].

Since the initial proposal of the triple helix model, many efforts have been devoted to unveil relationships between the abundance and the distribution of specific aminoacids within collagen sequence and specific properties of the triple helix. In this context, the main distinctive characteristic of collagen sequence, the presence of a Gly residue every third residue, could be immediately explained [13]. Indeed, glycine residues invariantly occupy an internal position in the triple helix that is not allowed for larger residues [18,19]. The identification of the structural bases for the observed frequency of the other aminoacids has proved very difficult. If the remarkable abundance of iminoacids could be explained by taking into account the high intrinsic preference of these residues for the conformation needed to build a triple helix structure, the basis of collagen stabilization by proline hydroxylation has hitherto remained elusive [20-22].

The complexity of collagen molecule and its fibrous nature prevents straightforward investigations on the full-length protein. To overcome this limitation a number of diversified strategies have been adopted. The use of peptide models embedding specific motifs has been quite successful due to the repetitive nature of collagen sequence/structure. This is especially true for sequence-stability relationships. Indeed, studies carried out by using host-guest model polypeptides have lead to the definition of a reliable scale of aminoacid/iminoacid propensities for collagen triple helix [23,24]. Furthermore, extensive analyses carried out on host-guest peptides containing proline derivatives have clearly demonstrated that the frequency and the location of specific diastereoisomer within collagen sequence is related to the role played by these iminoacids in triple helix stabilization. Indeed, 4RHyp has stabilizing or destabilizing effects when located in Y or X position of the Gly-X-Y triplets, respectively. On the other hand, 4SHyp has always destabilizing effects independent of its X or Y location [25,26].

Parallel investigations, performed by using both theoretical and experimental methodologies, have provided clear hints on the structural bases of the destabilizing/stabilizing effects on the triple helix exerted by some aminoacids (e.g. Met, Arg)[27-29]. For arginine, the specificity of this aminoacid for the Y position has also been explained [27,30]. Paradoxically, the definition of structure-stability relationships has proved to be extremely difficult for iminoacids. Indeed, there is no consensus on the structural determinants of triple helix stabilization by 4RHyp, the residue that plays the most important role in collagen function and stability. Apparently, the stabilizing effects of the Pro->4RHyp replacement in collagen should be easy to explain, since the structural consequence of this substitution is a simple insertion of an extra OH group in the rather rigid scaffold of the triple helix. However, all of the mechanisms proposed present severe drawbacks [1,20-22].

The first model [31], which was based on direct hydrogen-bonds between the OH of 4RHyp and backbone carbonyl groups, was rapidly discarded when the structure of collagen triple helix was proposed and validated [13]. Indeed, such direct interactions are not possible in the triple helix framework. Over the years, this initial suggestion was replaced by the idea that interactions between the OH group and main chain peptide groups could be mediate by water molecules. This hypothesis, initially proposed by Ramachandran [32] was corroborated by the first high resolution structures of collagen triple helix, that appeared to be highly hydrated in the crystalline state [16,33,34]. In this framework, however, the data on the inability of the peptides $(Gly\text{-}4RHyp\text{-}Pro)_{10}$, $(Gly\text{-}4SHyp\text{-}Pro)_{10}$, and $(Gly\text{-}Pro\text{-}4SHyp)_{10}$ to form triple helices [25,26], could be explained only by assuming that water-mediated networks of hydrogen bonds were highly stereospecific and compatible only with the presence of 4RHyp in the Y position. This model was seriously undermined by the discovery that some fluoroproline-containing peptides were more stable than their hydroxylated counterparts, despite the very low tendency of fluoroproline derivatives (Flp) to form hydrogen bonds [35]. The hyper-stability of the peptide $(Gly\text{-}Pro\text{-}4RFlp)_{10}$ was tentatively explained by invoking inductive effects of electron withdrawing groups such as F, OH in the 4R diasteroisomers of proline derivatives [35,36]. It was shown that these effects could increase the stability of the *trans* peptide bond state, the one observed in folded triple helices, over the *cis* one. The major drawback of this model is its inability to explain the data on $(Gly\text{-}4RHyp\text{-}Pro)_{10}$, which does not form triple helix [25]. Along this line, this model could not explain the destabilizing effect of Gly-4RHyp-Pro when inserted in either $(\text{Gly-Pro-4RHyp})_n$ and $(\text{Gly-Pro-Pro})_n$ host context [37,38].

In order to overcome this problem, a different mechanism, based on the distinct conformational properties of different diasteroisomers of proline derivatives, has been proposed. This mechanism, denoted as propensity-based model [39,40], takes into account the observation that proline rings exhibit similar propensities for both up and down conformations whereas 4RHyp (and 4RFlp) and 4SHyp (and likely 4SFlp) adopt preferentially the up and the down states, respectively. It is worth mentioning that up and down iminoacid states also display differences at backbone level [39,40]. Furthermore, high resolution structures of polypeptides with (Pro-Pro-Gly) repeats prevalently show alternating down-up states for proline residues located in X and Y positions [16,39,41-43]. All of these observations taken together provide an explanation for experimental data showing that, when located Y position, 4RHyp and 4SHyp have stabilizing and destabilizing effects on the triple helix, respectively (Table 1). For the same reason, intrinsic preferences explain why 4RHyp, which preferentially adopts the up state, have destabilizing effects when located in X [25,26]. Apparently, a conflict occurs for 4SHyp in X position (Table 1). This residue should be stabilizing, given its intrinsic preference for the down state. However, experimental data demonstrate that it is destabilizing. This discrepancy was solved by showing, using molecular modelling, that 4SHyp in X position generates severe steric clashes [39].

The propensity-based model has been recently supported by some recent structural investigations on the peptides:

P1 $(\text{Pro-Pro-Gly})_4$-Pro-4RHyp-Gly-$(\text{Pro-Pro-Gly})_4$ [38]
P2 $(\text{Pro-Pro-Gly})_4$-4RHyp-Pro-Gly-$(\text{Pro-Pro-Gly})_4$ [38]
P3 $(\text{Pro-Pro-Gly})_4$-Pro-4SHyp-Gly-$(\text{Pro-Pro-Gly})_4$ [38,44]

As expected, in P1 4RHyp adopts the up conformation, intrinsically favoured for this residue. Surprisingly, P2 4RHyp adopts the down conformation, intrinsically disfavoured for this residue, as a consequence of the triple helix constraints. This indicates that the destabilizing effects induced by 4RHyp located in X position are likely generated by the conflict between its intrinsic preference and the structural requirements of the triple helix. Similar considerations apply for data on the peptide P3, in which 4SHyp adopts the up conformation, intrinsically disfavoured for this residue [38,44].

Along this line, the tendencies (and the resulting effects) displayed by hydroxyproline-containing peptides are emphasized in the fluoro-containing derivatives of proline, due to the larger electronegativity of fluorine [35,45,46].

The agreement between the experimental data collected on Flp-containing peptides and the prediction of the propensity-based model are fairly good (Table 1). In this case, steric clashes which likely occur between 4SFlp in X and an adjacent chain do not lead to a full destabilization of the helix, as observed for 4SHyp, but they only limit the stabilizing effects produced by the intrinsic propensity of the imonoacid. Indeed, the replacement Pro->4SFlp in X position (Tm 31.4->54.5 °C) has smaller stabilizing effects when compared to the replacement Pro->4RFlp in Y (Tm 31.4-> 77 °C) [46].

The different behavior of 4SFlp and 4SHyp in X position may be ascribed (a) to differences local geometries of the residues and/or (b) to the known differences between C-F and C-OH bond distances.

Table 1. Stability of some polypeptides made of repeating Gly-X-Y sequence motifs. Values of Tm are from [46]. The prediction of the propensity of the propensity-based model is also reported

Peptide	Tm (°C)	Prediction
(Pro-Pro-Gly)10	31.4	Reference peptide
(Pro-4RHyp-Gly)10	62.2	Stabilization
(Pro-4SHyp-Gly)10	No helix	Destablization
(4RHyp-Pro-Gly)10	No helix	Destabilization
(4SHyp-Pro-Gly)10	No helix	Stabilization[a]
(Pro-4RFlp-Gly)10	77.0	Hyper-Stabilization
(Pro-4SFlp-Gly)10	No helix	Destabilization
(4RFlp-Pro-Gly)10	No helix	Destabilization
(4SFlp-Pro-Gly)10	54.5	Hyper-Stabilization[b]

[a]The destabilization of 4SHyp in the X position has been attributed to repulsive effects of its OH group with the atoms of an adjacent polypeptide chain.

[b]The reduced stabilization of 4SFlp in X position may attributed to some repulsive effects of its -F group with the atoms of an adjacent polypeptide chain.

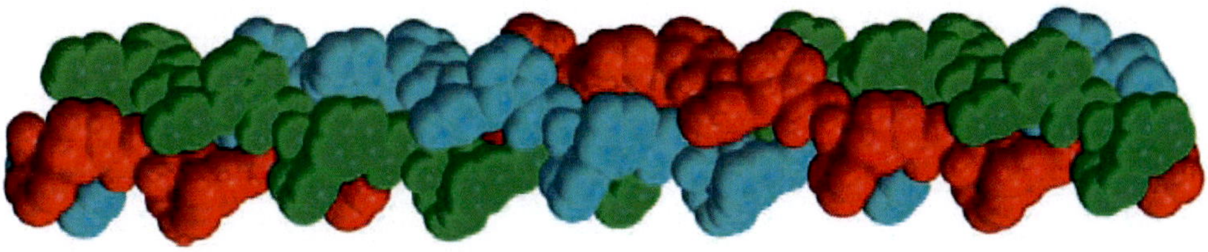

Figure 1. A model of collagen triple helix. Different colors have been used for the three polypeptide chains.

An indirect support to this idea comes from the observation that a single Gly-Pro-4RFlp triplet has destabilizing effects if embedded in a Gly-Pro-4RHyp context [37]. A definitive answer to this issue will be likely identified when structural data on Flp-containing peptides will become available.

Although the propensity-based model is able to provide a reasonable explanation for the stability of peptides containing only a single type of proline derivative [39], the picture is much more complicated for polypeptides containing residues at X and Y positions that can mutually interact [47,48]. Particularly impressive are the stabilizing effects exerted by 4RHyp at the X position when the residue located at the Y position is either 4RHyp or Thr [37,46,49-53]. This clearly indicates that stabilization/destabilization induced by a specific residue is context-dependent. A new challenge in collagen research field is the dissection of the role of different energetic contributions. Attempts to explain the "unexpected" high stability of peptides embedding (4RHyp-Thr-Gly) and (4RHyp-4RHyp-Gly) triplets have been based on hypothetical hydrogen bonding interactions, either direct or water-mediated, involving the -OH groups of these residues [50,52,54]. Even crystal structure determinations of peptides containing 4RHyp-4RHyp-Gly or 4RHyp-Thr-Gly triplets have not provided conclusive answer to this issue [46,54] [PDB code 2D3H]. The high resolution structure of (Gly-4RHyp-4RHyp)$_{10}$ has indicated that water mediated interactions favor a polyproline II conformation of the individual chains [54]. In other words, each chain is pre-organized to adopt the conformation required for the triple helix. However, this idea cannot explain the inability of the peptide (Gly-4RHyp-Pro)$_{10}$ to fold in a triple helical structure [26]. More recent investigations have been focused on the role played by dipole-dipole interactions in the stabilization of this class of peptides [55].

Altogether, these considerations clearly indicate how difficult is the mechanistic interpretation of the data on the stability of collagen triple helix, even when high resolution structural information is available. Therefore, it is not surprising that, in spite of the efforts devoted, the understanding of structural bases of thermostability of globular proteins, which exhibit a higher structural variability, is hitherto very limited. In this context, analyses on repetitive models of fibrous proteins may be important to pinpoint the impact and the role of specific interactions, which may also be involved in the stabilization of globular protein.

Finally, it should be noticed that the great majority of thermodynamic and structural studies have been conducted on peptide models with an over-represented iminoacid content, if compared to the real iminoacid/aminoacid distribution in collagen. Indeed, with rather few exceptions, the triplets of these models contained iminoacids in both X and Y position. On the other hand, the percentage of triplets with iminoacids in both X and Y is only 13% in real collagen. Furthermore, iminiacid containing triplets are rarely consecutive in the collagen sequence, which presents a large abundance of triplets with aminoacids in both X and Y (41%) or with iminoacid in a single position (46%). These simple considerations indicate that the currently available data and hypothesis, essentially derived using iminoacid rich sequences, may be heavily biased. In the near future, focused efforts aimed at collecting information on models with low/moderate content of iminoacids should be undertaken.

ACKNOWLEDGEMENTS

We thank Regione Campania (L.R. 5 2003) for financial support. I.M. thanks Boehringer-Ingelheim. for travel support.

REFERENCES

[1] Brodsky B, Persikov AV. Molecular structure of the collagen triple helix. *Adv. Protein Chem.* 2005;70:301-339.

[2] Engel J. Versatile collagens in invertebrates. *Science* 1997;277 (5333):1785-1786.

[3] Smith MC, Burns N, Sayers JR, Sorrell JA, Casjens SR, Hendrix RW. Bacteriophage collagen. *Science* 1998;279 (5358):1834.

[4] Rasmussen M, Jacobsson M, Bjorck L. Genome-based identification and analysis of collagen-related structural motifs in bacterial and viral proteins. *J. Biol. Chem.* 2003;278 (34):32313-32316.

[5] Ramshaw JA, Shah NK, Brodsky B. Gly-X-Y tripeptide frequencies in collagen: a context for host-guest triple-helical peptides. *J. Struct. Biol.* 1998;122 (1-2):86-91.

[6] Morello R, Bertin TK, Chen Y, Hicks J, Tonachini L, Monticone M, Castagnola P, Rauch F, Glorieux FH, Vranka J, Bachinger HP, Pace JM, Schwarze U, Byers PH,

Weis M, Fernandes RJ, Eyre DR, Yao Z, Boyce BF, Lee B. CRTAP is required for prolyl 3- hydroxylation and mutations cause recessive osteogenesis imperfecta. *Cell* 2006;127 (2):291-304.

[7] Burjanadze TV. New analysis of the phylogenetic change of collagen thermostability. Biopolymers 2000;53 (6):523-528.

[8] Burjanadze TV. Hydroxyproline content and location in relation to collagen thermal stability. *Biopolymers* 1979;18 (4):931-938.

[9] Privalov PL. Stability of proteins. Proteins which do not present a single cooperative system. *Adv. Protein Chem.* 1982;35:1-104.

[10] Schumacher MA, Mizuno K, Bachinger HP. The crystal structure of a collagen-like polypeptide with 3(S)-hydroxyproline residues in the Xaa position forms a standard 7/2 collagen triple helix. *J. Biol. Chem.* 2006;281 (37):27566-27574.

[11] Rhodes RK, Miller EJ. Physicochemical characterization and molecular organization of the collagen A and B chains. *Biochemistry* 1978;17 (17):3442-3448.

[12] Rich A, Crick FHC. Structure of collagen. *Nature*, 1955. pp. 915-916.

[13] Ramachandran GN, Kartha G. Structure of collagen. *Nature,* 1954. pp. 269-270.

[14] Okuyama K, Xu X, Iguchi M, Noguchi K. Revision of collagen molecular structure. *Biopolymers* 2006;84 (2):181-191.

[15] Okuyama K, Okuyama K, Arnott S, Takayanagi M, Kakudo M. Crystal and molecular structure of a collagen-like polypeptide (Pro-Pro-Gly)10. *J. Mol. Biol.*, 1981. pp. 427-443.

[16] Kramer RZ, Vitagliano L, Bella J, Berisio R, Mazzarella L, Brodsky B, Zagari A, Berman HM. X-ray crystallographic determination of a collagen-like peptide with the repeating sequence (Pro-Pro-Gly). *J. Mol. Biol.* 1998;280 (4):623-638.

[17] Kramer RZ, Bella J, Mayville P, Brodsky B, Berman HM. Sequence dependent conformational variations of collagen triple-helical structure. *Nat. Struct. Biol.* 1999;6 (5):454-457.

[18] Ramachandran GN, Kartha G. Structure of collagen. *Nature* 1955;176:593-595.

[19] Rich A. Structure of collagen. *Nat. Struct. Biol.* 1998;5 (10):858-859.

[20] Berisio R, Vitagliano L, Mazzarella L, Zagari A. Recent progress on collagen triple helix structure, stability and assembly. *Protein Pept. Lett.* 2002;9 (2):107-116.

[21] Jenkins CL, Raines RT. Insights on the conformational stability of collagen. *Nat. Prod. Rep.* 2002;19 (1):49-59.

[22] Engel J, Prockop DJ. Does bound water contribute to the stability of collagen? *Matrix Biol.* 1998;17 (8-9):679-680.

[23] Persikov AV, Ramshaw JA, Brodsky B. Collagen model peptides: Sequence dependence of triple-helix stability. *Biopolymers* 2000;55 (6):436-450.

[24] Persikov AV, Ramshaw JAM, Kirkpatrick A, Brodsky B. Amino acid propensities for the collagen triple-helix. *Biochemistry*, 2000. pp. 14960-14967.

[25] Inouye K, Sakakibara S, Prockop DJ. Effects of the stereo-configuration of the hydroxyl group in 4- hydroxyproline on the triple-helical structures formed by homogenous peptides resembling collagen. *Biochim. Biophys. Acta* 1976;420 (1):133-141.

[26] Inouye K, Kobayashi Y, Kyogoku Y, Kishida Y, Sakakibara S, Prockop DJ. Synthesis and physical properties of (hydroxyproline-proline-glycine)10: hydroxyproline in the

X-position decreases the melting temperature of the collagen triple helix. *Arch. Biochem. Biophys*. 1982;219:198-203.

[27] Vitagliano L, Nemethy G, Zagari A, Scheraga HA. Stabilization of the triple-helical structure of natural collagen by side-chain interactions. *Biochemistry* 1993;32 (29):7354-7359.

[28] Yang W, Chan VC, Kirkpatrick A, Ramshaw JA, Brodsky B. Gly-Pro-Arg confers stability similar to Gly-Pro-Hyp in the collagen triple-helix of host-guest peptides. *Journal of Biological Chemistry*. United States, 1997. pp. 28837-28840.

[29] Shah NK, Ramshaw JAM, Kirkpatrick A, Shah C, Brodsky B. A Host-Guest Set of Triple-Helical Peptides: Stability of Gly-X-Y Triplets Containing Common Nonpolar Residues. *Biochemistry,* 1996. pp. 10262-10268.

[30] Kramer RZ, Bella J, Brodsky B, Berman HM. The crystal and molecular structure of a collagen-like peptide with a biologically relevant sequence. *J. Mol. Biol.* 2001;311 (1):131-147.

[31] Gustavson KH. The function of hydroxyproline in collagens. *Nature* 1955;175:70-74.

[32] Ramachandran GN, Bansal M, Bhatnagar RS. A hypothesis on the role of hydroxyproline in stabilizing collagen structure. *Biochim. Biophys. Acta* 1973;322 (1):166-171.

[33] Bella J, Eaton M, Brodsky B, Berman HM. Crystal and molecular structure of a collagen-like peptide at 1.9 A resolution [see comments]. Science 1994;266 (5182):75-81.

[34] Bella J, Brodsky B, Berman HM. Hydration structure of a collagen peptide. *Structure* 1995;3 (9):893-906.

[35] Holmgren SK, Taylor KM, Bretscher LE, Raines RT. Code for collagen's stability deciphered. *Nature* 1998;392 (6677):666-667.

[36] Holmgren SK, Bretscher LE, Taylor KM, Raines RT. A hyperstable collagen mimic. *Chem. Biol.* 1999;6 (2):63-70.

[37] Persikov AV, Ramshaw JA, Kirkpatrick A, Brodsky B. Triple-helix propensity of hydroxyproline and fluoroproline: comparison of host-guest and repeating tripeptide collagen models. *J. Am. Chem. Soc.* 2003;125 (38):11500-11501.

[38] Jiravanichanun N, Hongo C, Wu G, Noguchi K, Okuyama K, Nishino N, Silva T. Unexpected puckering of hydroxyproline in the guest triplets, hyp-pro-gly and pro-allohyp-gly sandwiched between pro-pro-gly sequence. *Chembiochem* 2005;6 (7):1184-1187.

[39] Vitagliano L, Berisio R, Mazzarella L, Zagari A. Structural bases of collagen stabilization induced by proline hydroxylation. *Biopolymers* 2001;58:459-464.

[40] Vitagliano L, Berisio R, Mastrangelo A, Mazzarella L, Zagari A. Preferred proline puckerings in cis andtrans peptide groups: Implications for collagen stability. *Protein Sci*. 2001;10 (12):2627-2632.

[41] Okuyama K, Hongo C, Fukushima R, Wu G, Narita H, Noguchi K, Tanaka Y, Nishino N. Crystal structures of collagen model peptides with Pro-Hyp-Gly repeating sequence at 1.26 A resolution: implications for proline ring puckering. *Biopolymers* 2004;76 (5):367-377.

[42] Berisio R, Vitagliano L, Mazzarella L, Zagari A. Crystal structure of the collagen triple helix model [(Pro-Pro-Gly)(10)](3). *Protein Sci.* 2002;11 (2):262-270.

[43] Okuyama K. Structural study of collagen based on single crystal analyses of model peptides. *Pept. Sci.,* 2001. pp. 263-264.

[44] Jiravanichanun N, Nishino N, Okuyama K. Conformation of alloHyp in the Y position in the host-guest peptide with the pro-pro-gly sequence: implication of the destabilization of (Pro-alloHyp-Gly)10. *Biopolymers* 2006;81 (3):225-233.

[45] Hodges JA, Raines RT. Stereoelectronic and steric effects in the collagen triple helix: toward a code for strand association. *J. Am. Chem. Soc.* 2005;127 (45):15923-15932.

[46] Nishi Y, Uchiyama S, Doi M, Nishiuchi Y, Nakazawa T, Ohkubo T, Kobayashi Y. Different effects of 4-hydroxyproline and 4-fluoroproline on the stability of collagen triple helix. *Biochemistry* 2005;44 (16):6034-6042.

[47] Persikov AV, Ramshaw JA, Brodsky B. Prediction of collagen stability from amino acid sequence. *J. Biol. Chem.* 2005;280 (19):19343-19349.

[48] Persikov AV, Ramshaw JA, Kirkpatrick A, Brodsky B. Electrostatic interactions involving lysine make major contributions to collagen triple-helix stability. *Biochemistry* 2005;44 (5):1414-1422.

[49] Bann JG, Bachinger HP. Glycosylation/Hydroxylation-induced stabilization of the collagen triple helix. 4-trans-hydroxyproline in the Xaa position can stabilize the triple helix. *J. Biol. Chem.* 2000;275 (32):24466-24469.

[50] Mizuno K, Hayashi T, Bachinger HP. Hydroxylation-induced stabilization of the collagen triple helix. Further characterization of peptides with 4(R)-hydroxyproline in the Xaa position. *J. Biol. Chem.* 2003;278 (34):32373-32379.

[51] Mizuno K, Hayashi T, Peyton DH, Bachinger HP. Hydroxylation-induced stabilization of the collagen triple helix. Acetyl-(glycyl-4(R)-hydroxyprolyl-4(R)-hydroxyprolyl)(10)-NH(2) forms a highly stable triple helix. *J. Biol. Chem.* 2004;279 (36):38072-38078.

[52] Berisio R, Granata V, Vitagliano L, Zagari A. Imino acids and collagen triple helix stability: characterization of collagen-like polypeptides containing Hyp-Hyp-Gly sequence repeats. *J. Am. Chem. Soc.* 2004;126 (37):11402-11403.

[53] Doi M, Nishi Y, Uchiyama S, Nishiuchi Y, Nishio H, Nakazawa T, Ohkubo T, Kobayashi Y. Collagen-like triple helix formation of synthetic (Pro-Pro-Gly)10 analogues: (4(S)-hydroxyprolyl-4(R)-hydroxyprolyl-Gly)10, (4(R)-hydroxyprolyl-4(R)-hydroxyprolyl-Gly)10 and (4(S)-fluoroprolyl-4(R)-fluoroprolyl-Gly)10. *J. Pept. Sci.* 2005;11 (10):609-616.

[54] Schumacher M, Mizuno K, Bachinger HP. The crystal structure of the collagen-like polypeptide (glycyl-4(R)-hydroxyprolyl-4(R)-hydroxyprolyl)9 at 1.55 A resolution shows up-puckering of the proline ring in the Xaa position. *J. Biol. Chem.* 2005;280 (21):20397-20403.

[55] Improta R, Berisio R, Vitagliano L. Contribution of dipole-dipole interactions to the stability of collagen triple helix. 2007. Submitted.

In: Biopolymer Research Trends
Editor: Tamas S. Nemeth, pp. 11-48

ISBN: 978-1-60021-983-2

Chapter 1

RESEARCH PROGRESS ON METALLOTHIONEINS: INSIGHTS INTO STRUCTURE, METAL BINDING PROPERTIES AND MOLECULAR FUNCTION BY SPECTROSCOPIC INVESTIGATIONS

Jordi Domènech[1,3], Anna Tinti[2] and Armida Torreggiani[3*]

[1] Departament de Genètica, Facultat de Biologia, Universitat de Barcelona, Av. Diagonal 645, 08028 Barcelona (Spain);

[2] Biochemistry Department, University of Bologna, Via Belmeloro 8/2, 40126 Bologna (Italy)

[3] Istituto per la Sintesi Organica e la Fotoreattività, Consiglio Nazionale delle Ricerche, Via P. Gobetti 101, 40129 Bologna (Italy)

ABSTRACT

Metallothioneins (MTs) are low molecular weight, cysteine-rich proteins with an exceptional heavy metal coordination capacity. Because of their ability to bind metals and to scavenge oxidant radicals, MTs are considered to play a role in metal homeostasis, metal detoxification and control of the oxidative stress. Although their high heterogeneity on the expression patterns, metal binding abilities and primary structure suggest very diverse functional specializations, the structural and functional studies have been mainly devoted to vertebrate and fungal MTs and their canonical cysteine-metal clusters.

This chapter will be focused on the new methodological procedures settled for the structural characterisation of some metallic MT aggregates. Five zinc complexes from the invertebrate and plant MTs, poorly described up to now in the literature, in addition to one Zn-MT complex from the well studied vertebrate MT family, have been analysed. The new discovered structural features of metal-MT clusters, in addition with the perspectives on MT research, have been also commented. In fact, much can be learnt about MT systems by using spectroscopic techniques such as Raman and IR

* For correspondence: Dr. Armida Torreggiani, Istituto ISOF (CNR), Via P. Gobetti n° 101, 40129 Bologna (Italy). Tel: +39 051 6399787; Fax: +39 051 6399844; e-mail: torreggiani@isof.cnr.it

spectroscopies, and Circular Dichroism, able to provide new structural information eventually related to the function of the metal binding. Despite the potentialities of these techniques, to our knowledge they have been scarcely used in MT conformational studies until now. Recently, the use of these spectroscopies has resulted to be very useful to approach unambiguously two basic structural points poorly described in MTs: the participation of chloride, sulfide ions and His residues to the metal-coordination sphere and the presence of secondary structure elements. In particular, ordered secondary structures, oppositely to what has been commonly accepted, are present in MTs from vertebrate, invertebrate and plant MTs, and could develop crucial roles in the determination of the functional properties of MTs.

The biosynthesis of intact metal-MT complexes, well corresponding to native forms, in sufficient quantity and purity for analytical spectrometric and spectroscopic characterization has been allowed by recombinant expression in *E.coli*. The spectroscopic analyses of the *in vivo*-synthesised metal-MTs have recently demonstrated the participation of extra-protein ligands, such as chloride and sulfide ions, in the metal-MT coordination environment of vertebrate, invertebrate and plant MTs. In conclusion, the coupling of analytical and spectroscopical techniques can be one of the most promising experimental strategies in the research on new hints on MTs.

ABBREVIATIONS

AA	amino acids
CD	Circular Dichoism
ESI-MS	Electron Spray Ionisation – Mass Spectrometry
FT	Fourier-Transform
GC-FPD	Flame Photometric Detector Gas Chromatography
ICP-AES	Inductively-Coupled Plasma Atomic Emission Spectroscopy
IR	Infrared
M	metal (generalised)
MP	Metalloprotein
MT	Metallothionein
R	peptidic chain

1. INTRODUCTION

Metals, Metalloproteins and the Particular Case of Metallothioneins

Metal ions (Zn, Cu, Co, Fe, Mg, ..) are essential oligoelements that develop important roles in the main biological processes as structural agents or enzymatic cofactors. As structural agents, metal ions can determine the protein structure and modulate regulatory interactions, like in the well-known *zinc-finger* proteins (Vallee and Falchuk, 1993; Auld, 2001; Maret, 2006; Pierrel *et al.*, 2007). As enzymatic cofactors, due to their particular redox properties, metal ions play key roles in the catalytic centers of enzymes participating in fundamental redox processes as respiration, nitrogen fixation and photosynthesis (Degtyarenko, 2000). The crucial participation of metal ions in the structure and functionality of metalloprotein (MP) systems has been described in more than 800 MPs enclosing the paradigmatic cases of clorofila, ferredoxin, hemocyanin, lactoferrin, transferrin and porfirins, which structural diversity is related to functional specificities (Degtyarenko, 2000; Auld, 2001; Kulkarni *et al.*, 2006).

Metal ions can be bound to MP by two main types of ligands, endogenous and/or exogenous. The first type is constituted by the amino acid side chains taking part in metal coordination by their functional groups, i.e. the thiol (-SH), azo (-NH), and carboxylate groups ($-COO^-$) of Cys, His and Glu (or Asp), respectively. In the second case, inorganic anions, such as phosphate, chloride, sulfide ions, or water molecules can stabilize the metal in the metalloproteic aggregate. The combination of different ligands determines both the metal binding preferences of MP and its functional properties (Degtyarenko, 2000; Auld, 2001).

Under conditions of metal overload, toxic effect can be observed since metals can develop interferences in the normal cell metabolism, interfering in the enzyme activity and enhancing apoptosis processes (Bertin and Averbeck, 2006). Recently, an increasing number of human diseases are thought to be related to disturbances in metal ion homeostasis, including metal ion overload and deficiency disorders (i.e., anemia, haemochromatosis, Menke's disease, Wilson's disease) and neurodegenerative diseases (i.e., Alzheimer's, Friedreich's ataxia and Parkinson's diseases). Therefore, one of the challenges faced by every cell as well as by whole organism is to maintain appropriate concentrations of essential nutrient metals while excluding xenobiotic toxic metals. Toward that end, all organisms have developed mechanisms for metal homeostasis and detoxification to maintain metal levels within physiological limits.

Metal homeostasis, being a critical point in biological systems, is regulated at three main levels, by means of metal chelators, metal transporters and protein sensors. The first level acts by scavenging metals and decreasing their biological availability. Phytochelatins and ferritins are examples of this type of systems (Clemens, 2006; Rouault, 2006). The second level is observed in the case of metal transporters, specific metal-binding proteins which participate in the redistribution of metal ions by protein channels and transporters, or transferring the metal to metalloproteins being synthesised, as in the case of transferrin and albumin (Sarkar, 1987). Protein sensors constitute the third level, and, in general manner, their metal-binding activity trigger cellular or systemic responses to metal levels. This activity is classically described in the case of the Zn-finger proteins, as the Zn(II) availability in the media determines their DNA-binding activity, which determines the expression levels of the genes regulated by themselves (Maret, 2005). In this sense, protein sensors *detect* metal concentration and initiate cellular and systemic responses.

Metallothioneins (MTs) are a unique class of low-molecular weight metalloproteins, characterised by a high content in Cys (almost a third of the total amino acids), a sulfur-containing amino-acid, from there the name (thio means sulfur). MTs are present in a huge range of living organisms and have the ability to bind metals mainly through the thiol group of the Cys residues, forming tetrahedral metallic clusters (Kägi, 1993).

MTs constitute an exceptional case among MPs, as they can act in the three metal homeostatic levels. MTs, binding metals, are able to protect the cellular mechanisms from the toxic effects of metals (Vasak, 1991; Amiard *et al.*, 2006) and can act as metal-transporter, transferring metals to other MPs or other intracellular compartments (Maret *et al.*, 1997; Jacob *et al.*, 1998). Petering *et al.* have effectively summarized the functional aspects of MT related to metal ion homeostasis and sequestration of toxic metals: "Because of this unusual kinetic lability as well as the thermodynamic stability [owing to soft acid-soft base interactions] of the metallothionein species that are formed, metallothionein acts as a sink for the binding of a variety of essential and toxic metal ions which enter cells." (Petering *et al.*, 1992).

It has been proposed that MTs can also develop metal-sensor roles by a simple mechanism: when metal ions, such as Cu(I) or Cd(II), are present in the media, they displace the Zn(II) ions bound to the MT and the consequent rising Zn(II) concentration induce MT synthesis through the activation of expression factors as MTF1 (Haq *et al.*, 2003; Amiard *et al.*, 2006).

In a similar way, the exposure of Zn-MT to oxidative stress can also cause the liberation of Zn(II) ions by oxidation of the Cys residues. The free Zn(II) ions trigger the *de novo*-synthesis of MTs, which collaborate in the control of oxidative stress by scavenging and neutralizing oxidant radicals as hydroxyl radical (Kumari *et al.*, 1998). So, the redox-active Zn-Cys coordination environment of MTs has been considered to be a central node in cellular signaling, interconverting redox and zinc signals (Maret, 2003). Moreover, by binding metal ions as Cu and Fe, MT would avoid them to participate in reactions critical for oxidative stress, as the Fenton reaction (Meneghini, 1997; Viarengo *et al.*, 2000).

MTs are widely distributed among the living organisms, and their primary structure show extreme heterogeneity, enclosing MTs from 35 to 150 aa length and very diverse Cys-motifs (Cys-Cys, Cys-X-Cys, Cys-Cys-Cys, Cys-X-X-Cys,...) (Binz and Kägi, 2001). Binz and Kagi proposed in 1999 (Binz and Kägi, 1999) a classification of MTs (available in digital format from Expasy server (Binz, 2007)) based both on primary structure (amino acid content and Cys patterns) and phylogenetic parameters, which resulted in 15 main families. The main families, their principal structural features and taxonomic relationships are shown in Figure 1.

Three of these families (Vertebrate, Crustacea and Echinoderm) present MTs which show sequence homology. Their tertiary structures display functional analogy, as they are formed by two independent domains (*bidominial* folding model) when binding divalent ions like Zn(II) or Cd(II) (Figure 1). These domains can be of two types: type α domains coordinate 4 divalent metals bound to 11 Cys and type β domains coordinate three divalent metals bound to 9 Cys (Armitage *et al.*, 1982; Riek *et al.*, 1999).

On the contrary, no homology is observed between the rest of the invertebrate (nematode, mollusk, insect,…), fungi, protozoan, prokariotic and plant MT families, which present higher sequence heterogeneity (Fowler *et al.*, 1987) and whose secondary and tertiary structures are yet unknown. The only exceptions to this general lack of knowledge in these MTs are constituted by prokariote (SmtA) and fungi (CUP1 and *N.crassa* MT), whose structures show particular one-domain *monodominial* folding models when coordinate Zn(II) or Cu(I), respectively, (Bertini *et al.*, 2000; Blindauer *et al.*, 2001; Cobine *et al.*, 2004; Calderone *et al.*, 2005). In fact, structural studies have been traditionally centered in only 6 of the 15 MT families, and mainly refer to vertebrate, crustacean, fungal, echinoderm and prokaryote MTs (Pande *et al.*, 1986; Wagner *et al.*, 1987; Narula *et al.*, 1995; Wang *et al.*, 1996; Riek *et al.*, 1999; Bertini *et al.*, 2000; Blindauer *et al.*, 2001; Oz *et al.*, 2001; Munoz *et al.*, 2002; Wang *et al.*, 2006), but structural data on MTs from other families are traditionally poor. Moreover, data on secondary structure in MTs are at present quite poor respect to the huge amount of data regarding primary and tertiary structures.

In addition, it has been recently shown that metal coordination in MTs could be more complex than the paradigmatic Cys-metal clusters. Participation of either endogenous (His, Asp, Glu) (Blindauer *et al.*, 2001; Villarreal *et al.*, 2006; Domenech Ph.D Thesis 2007;

Domenech *et al.*, 2007b; Leszczyszyn *et al.*, 2007) or exogenous ligands (sulfides or chlorides), recently evidenced (Maret *et al.*, 2002; Domenech *et al.*, 2003; Capdevila *et al.*, 2005; Villarreal *et al.*, 2005) open new perspectives and methodological hints in the study of the functional properties of MTs.

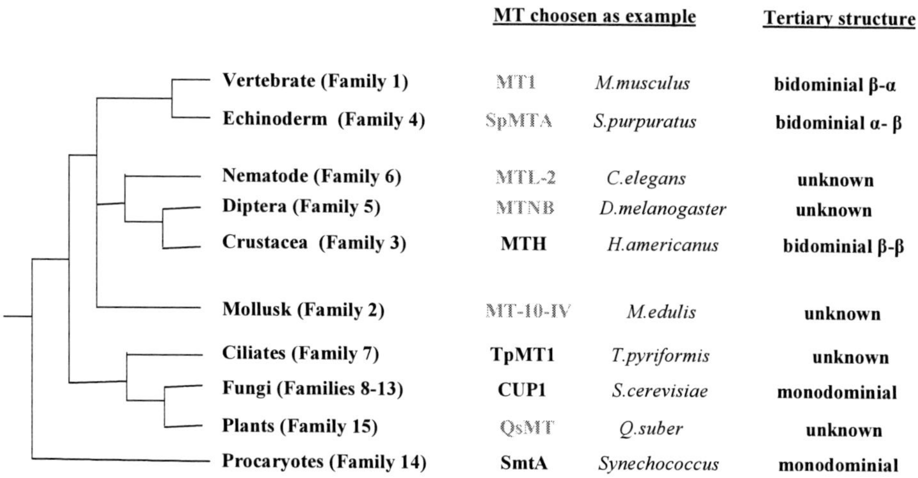

Figure 1. Taxonomic relationships between the 15 MT families proposed by Binz and Kägi (Binz and Kägi, 1999), and an example of each MT family. The length of the branches are not representative of phylogenetic distances between the taxonomic groups or the MT sequences. The proposed tertiary structure model, when known, is noted on the right. The six MTs whose Raman and IR studies are described in this chapter are noted in grey.

In this chapter we will describe the capacities of some spectroscopic techniques, till now not exploited to the outmost, on the characterization of the metal-binding properties and secondary and tertiary structure of MTs. Six *in vivo*-synthesized Zn-complexed MTs representative of different MT families will be analysed, enclosing the well-known vertebrate (family 1) and echinoderm (family 4), and the poorly understood nematode (family 6), diptera (family 5), molluscan (family 2) and plant (family 15) MTs (Figure 1).

Zinc complexes are very important since zinc is the second most abundant transition metal found *in vivo*. Biological complexes of this element contain zinc only in its divalent oxidation state; hence, the coordination chemistry of Zn(II) complexes is dominated by the tetrahedral geometry for this ion. Because of the single stable oxidation state for zinc solution, this metal does not play a redox active role in biological processes, but neverless partecipates in a vast range of enzymatic reactions as Lewis acid or as a structural factor. MTs are postulated to act as storage and/or transport proteins for Zn(II) (Stillman *et al.*, 1992).

A short description of the main structural characteristics available till now on the six MTs that will be analysed, is reported below:

- *MT1 metallothionein* belongs to the mammal *Mus musculus* (mouse, vertebrate). Vertebrate MT forms present homology between them and almost identical structural features, as shown by NMR and X-ray diffraction studies (Furey *et al.*, 1987; Wagner *et al.*, 1987; Robbins *et al.*, 1991). Vertebrate MTs are considered the paradigm of MTs, and they are the focus of most functional and structural studies on MTs, including Raman, IR and CD studies (Pande *et al.*, 1986). MT1 contains 20 Cys residues arranged in Cys-Cys or Cys-X-Cys patterns (Table 1) which form two metallic clusters of 4 and 3 metals, respectively, when coordinating divalent metal ions (Zn(II) or Cd(II)).

- *SpMTA metallothionein* is one of the MTs of the sea urchin (echinoderm, family 4) *Strongylocentrotus purpuratus*, and has been studied by traditional NMR techniques on Cd-coordinated forms (Riek *et al.*, 1999; Zangger *et al.*, 1999); here vibrational and CD spectroscopic studies on its *in vivo*-synthesised Zn-MT aggregates will be described. Its Cys residues are arranged in patterns similar to those of MT1, and form analogue metal clusters, but with an inverted directionality along the MT sequence respect to MT1. It contains one aromatic amino acid, unusual in paradigmatic MTs (Table 1).
- *MTL-2 metallothionein* is part of the MT system of the nematode-worm *Caenorhabditis elegans*, the only organism representative of family 6. MTL-2 contains 18 Cys and an His residue at its C-terminal edge, susceptible to participate in metal coordination (Table 1) (Imagawa *et al.*, 1990; Slice *et al.*, 1990). The only data available on its folding and structure, at our knowledge, come from Raman, IR and CD studies of its Zn-coordinated forms (Domenech, Ph.D Thesis, 2007).
- *MTNB metallothionein* from family 5, is part of the fly *Drosophila melanogaster* MT system, which has been thoroughly studied from the functional point of view (Valls *et al.*, 2000; Domenech *et al.*, 2003; Egli *et al.*, 2006). In spite of the interest suggested by its exceptionally short sequence (Table 1) (Mokdad *et al.*, 1987) and particular metal coordination features (Domenech *et al.*, 2003), its structural features are poorly understood.

Table 1. Amino acid sequences of MT1, SpMTA, MTL-2, MTNB, MT-10-IV and QsMT (Cys residues are indicated in bold while aromatic and acid amino acids are in italic). The total amino acid content and the number of Cys, aromatic and acidic residues is also reported

MT	Amino acid sequence	Total A.A.	Cys	Arom. A.A.	Acidic A.A.
MT1	MDPNCSCSTGGSCTCTSSCACKNCKCT SCKKSCCSCCPVGCSKCAQGCVCKGA ADKCTCCA	61	20	0	2
SpMTA	MPDVKCVCCKEGKECACFGQDCCKT GECCKDGTCCGICTNAACKCANGCKC GSGCSCTEGNCAC	64	20	1 Phe	7
MTL-2	MVCKCDCKNQNCSCNTGTKDCDCSD AKCCEQYCCPTASEKKCCKSGCAGGC KCANCECAQAAH	63	18	1 His 1 Tyr	7
MTNB	MVCKGCGTNCQCSAQKCGDNCACNK DCQCVCKNGPKDQCCSNK	43	12	0	3
MT-10-IV	MPAPCNCIETNVCICDTGCSGEGCRCG DACKCSGADCKCSGCKVVCKCSGSCA CEGGCTGPSTCKCAPGCSCK	73	20	0	6
QsMT	MSCCGGNCGCGTGCKCGSGCGGCKM FPDISSEKTTTETLIVGVAPQKTHFEGSE MGVGAENGCKCGSNCTCDPCNCK	77	14	1 His 2 Phe	7

- *MT-10-IV metallothionein* from the marine mollusk *Mytilus edulis*, is one of the invertebrate MTs which present a higher sequence similarity with the well-known vertebrate MTs (Mackay *et al.*, 1993). It belongs to family 2, and its use as a biomarker in marine pollution studies has made this family one of the best known from the physiological point of view.
- *QsMT metallothionein* belongs to the plant *Quercus suber* (from family 15) and present the particular distribution of Cys residues typical of this MT family: a *spacer* region devoid of Cys residues is flanked by two Cys-rich regions with 6 and 8 Cys residues, respectively (Table 1). It presents an His residue in its spacer region. Genetic engineering procedures have allowed to individuate the effect of each domain on the QsMT functionality when binding Zn(II), Cd(II) or Cu(I) ions, supporting for this MT a *codominial* folding model: the two Cys-rich domains of QsMT would bind metals together in a unique metallic cluster (Domenech *et al.*, 2006; Domenech *et al.*, 2007a). This folding model had been previously proposed for other MTs of the same family, like *Pea sativum* MT (PsMT) (Kille *et al.*, 1991). The structural study of QsMT in its *in vivo*-synthesized Zn- and Cd-complexed forms is very informative of the particular structural and functional properties of plant MTs.

2. Experimental Approaches for the Study of MTs

2.1. *In Vivo*-Synthesis and Determination of the Metallopeptide Composition

The study of MPs from the structural point of view requires a huge amount of protein. For this reason, to obtain homogenous samples, MPs are usually obtained in their demetallated form (Apo) from native tissues or *in vitro* synthesis, and subsequently remetallated *in vitro*. Such procedures facilitate also the obtention of the MP with a chosen metallic cation useful for structural studies purposes (i.e. ^{113}Cd for NMR studies), but metal substitution could also imply structural changes respect to the *in vivo* folding. The synthesis of metallic aggregates of a metalloprotein for its structural study should be performed in physiological conditions to be certain that the obtained aggregates are representative of those formed in biological systems.

To obtain homogenous sample preparations of *in vivo*-synthesized MPs, a simple procedure of heterologous synthesis is available, by genetic engineering and biochemical procedures (Capdevila *et al.*, 1997). The pGEX system is a plasmid vector which allows the cloning of the cDNA coding for the protein to study. When the cDNA coding for a MP is cloned in pGEX, the resulting plasmid (pGEX-MP) can be introduced by transformation to bacterian *E.coli* BL21 cells. pGEX plasmid allows to time-regulate the expression of the cloned cDNA by the exposure to a synthesis-inducing factor called IPTG. The addition of this factor together with metal ions in the media where the bacteria are growing will induce the synthesis of the metallic aggregates (M-MP) inside the cell. The metal bound to the aggregates will depend on which metal is supplemented to the media.

By sonication, centrifugation and chromatographic procedures, the M-MP aggregates can be purified in a chosen buffer. As a result, the M-MP aggregates synthesized in an *in vivo* environment, well representative of those formed in the biologic systems, will be obtained in purity and amounts enough for spectroscopic structural studies.

The first step to characterize the M-MP aggregates is to quantify the metal content of the resulting M-MP aggregates. An analytical technique, Inductively-Coupled Plasma Atomic-Emission Spectroscopy (ICP-AES), allows to measure the metal and sulfur concentration in the sample containing the pure M-MP aggregates and to calculate a mean stoichiometry M/MP by relating the M content to the MP concentration. The MP concentration is calculated from the S content in the sample, assuming that it exclusively comes from Met and Cys residues of the peptide (Bongers *et al.*, 1988; Cols *et al.*, 1997; Domenech *et al.*, 2007b).

2.2. Structural Study of MTs

Many different chemico-physical techniques are utilized to study the structural properties of polypeptides, proteins free or bound to metals. The known structural chemistry of MTs comes from a variety of sources but currently it is heavily based on the detailed results from both NMR (^{111,113}Cd and ^{1}H) and X-ray diffraction (Braun *et al.*, 1992; Robbins and Stout, 1992). For example, the metal-thiolate connectivities, especially the critically important differentiation between terminal thiolate (one sulfur binds to one metal ion) and bridging thiolates (one sulfur binds to two metal ions) have been deduced by NMR measurements (Braun *et al.*, 1992). Also the overall structure of the two domains in the mammalian proteins comes from the results of the X-ray diffraction measurements (Robbins and Stout, 1992), as well as the important metal-sulfur bond length information that have allowed to propose the three-dimensional structure of rabbit liver Zn_7-MT 2A (Kägi, 1993).

The disadvantage of these two techniques is that they require large amounts of sample. Due to the technical difficulties in obtaining homogenous samples in the necessary amounts for applying these techniques, the structural studies have been often performed on metal-reconstituted samples, where the *in vivo*-bound metal is substituted by other metals by reconstitution of the aggregates *in vitro* (i.e. ^{113}Cd). Moreover, cystallization is not a trivial and successful endeavour. In view of the fact that so far the structures of only a small part of all discovered MTs are known, it becomes clear that complementary methods for structure analysis are needed to obtain the necessary information for understanding the relationship between structure, function and dynamics of MTs.

To infer information on the structure of *in vivo*-synthesised metallopeptides, two other spectroscopic techniques, circular dicroism (CD) and vibrational spectroscopy (Raman and IR), can be very useful, since they require lesser sample amounts and their performances are not restricted to a limited number of metals, but are useful to study MTs containing each type of metal.

In fact, the absence of aromatic amino acids in MTs is an important and characteristic property of these peptides. It is spectroscopically significant because optical spectroscopy is able to measure absorption and circular dicroism (CD) spectra for the thiolate to metal charge-transfer transitions occurring in the 200-400 nm range. This region would be normally completely masked by the presence of aromatic groups and, therefore, unavailable for the study by optical methods.

The capability of CD in the study of metallic aggregates is well described in the literature (Rupp and Weser, 1978; Stillman *et al.*, 1991). In addition to details of the metal binding chemistry, available by the accessibility to the charge-transfer bands due to the metal-ligand binding (200-400 nm), also information on changes in the constitution, degree of folding and coordination geometries of the metallic clusters can be provided from changes in intensity and wavelength of these bands. The association of the absorbances to different types of metal-coordination environments allows to identify the formation of particular metal-ligand interactions and to study their evolution under different experimental conditions, i.e. temperature, metal concentration or pH. In addition, CD is a technique habitually used to describe the secondary structure content of a protein. However, in the case of MTs, the signals due to metal-ligand bonds can affect some of the signals associated to secondary structure elements.

Investigation of protein structure by vibrational spectroscopy (Raman and IR) has been practised for more than three decades, during which specific band assignments, signatures of secondary structure and Raman markers of side chain environments have been established. (Miura and Thomas, 1995; Thomas, 1999; Tuma and Thomas, 2002; Tuma, 2005). The more recent technical devices increased the achievable signal-to-noise ratio in Raman spectroscopy. This has made difference spectroscopy applicable in many different problems and difference spectroscopy has emerged as an indispensable tool for detecting minute structural changes (Callender and Deng, 1994; Tuma *et al.*, 2001; Benevides *et al.*, 2002).

Vibrational spectroscopies can imply lyophilization procedures. In this case, the structural features of the aggregates should be controlled before and after lyophilization, for example by CD analysis after resolubilisation of the lyophilized sample or by measuring its metal content. Changes in these parameters would indicate degradation of the sample during the lyophilization process; in this case Raman and IR results would no more be physiologically representative.

Vibrational spectroscopies are generally considered low-resolution techniques which provide global insight into the overall secondary structure of proteins without being able to establish the precise three-dimensional location of individual elements. Moreover, they have been widely applied for obtaining information on the presence of some functional moiety, such as -SH, S-S, etc., the microenvironment of some amino acid residues as Tyr, Trp, and the metal coordination sites (i.e. His, COO^-) (Carey, 1982; Tu, 1982; Parker, 1983).

Vibrational spectroscopy has been successfully used by the authors for the structural characterization of different proteins and the identification of the coordination sites in some metal-ligand systems (Torreggiani and Fini, 1998; Torreggiani and Fini, 1999; Torreggiani *et al.*, 2005; Torreggiani *et al.*, 2006a; Torreggiani *et al.*, 2006b; Torreggiani *et al.*, 2007)

In the case of heme proteins, spectroscopic techniques have also been able to identify the presence of Fe^{2+} or Fe^{3+}, giving results about the oxidation state of the metal, the metal-ligand bonds present in the protein and clarifying the structure of the complex (Tu, 1982; Parker, 1983).

The peptidic group of polypeptides and proteins gives rise to many representative vibrational bands, (amide A, B and amide I-VII), not all of these bands are observable in both IR and Raman spectra. For the study of protein conformation, some of them are particularly useful. In particular, the amide A band, at about 3280 cm^{-1}, due to N-H stretching, is intense in the IR spectrum but weak in Raman. Its wavenumber gives information about the H-bond strength (Spassov *et al.*, 2006).

The amide I band, observed both in the IR and Raman spectra in the range 1695-1630 cm^{-1} (Figure 2), mainly related to C=O stretching mode, is particularly useful for the protein secondary structure determination.

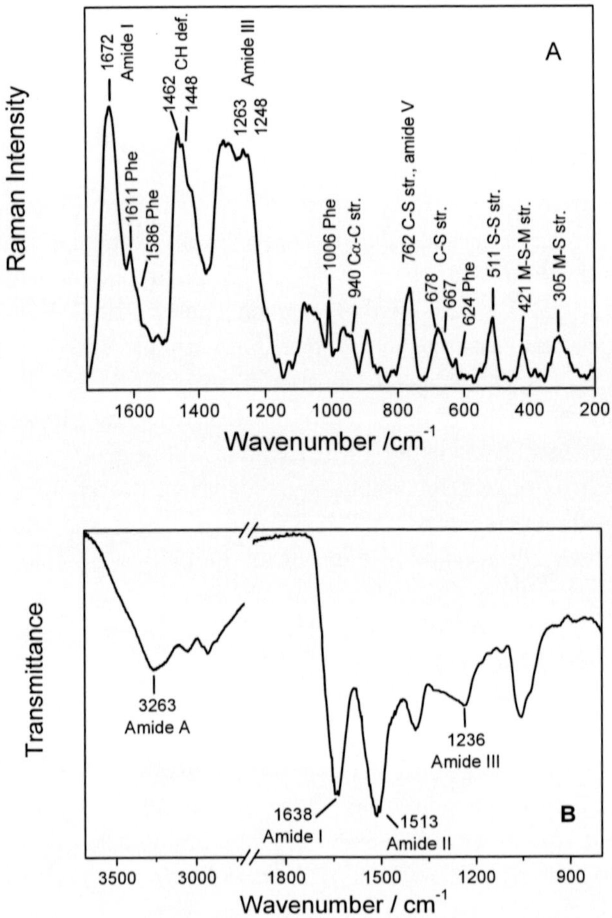

Figure 2. (A) Raman and (B) IR spectrum of the echinoderm Zn-SpMTA metallothionein in the 200-1750 and 800-3700 cm^{-1} range, respectively. Str = stretching vibration; def = deformation vibration.

The amide II band, mainly due to the NH deformation, active only in IR, is observed in the range 1560-1510 cm^{-1}, but alone does not give important information (Figure 2B). The presence of the above mentioned three bands in the spectrum of an "unknown" sample indicate the presence of amidic groups. Important for the determination of the secondary structure of a protein is also the amide III band, at about 1200-1300 cm^{-1}, resulting from coupled C-N stretching and N-H bending motions, IR and Raman active (Figure 2). Traditionally, the position and the intensities of amide bands have been used for empirical estimation of secondary structure (Tu, 1982; Parker, 1983). Later on, the broad amide bands have been analysed using the spectral deconvolution of the envelop into a few component bands (Sane *et al.*, 1999; Pelton and McLean, 2000). The decomposed bands were then related to a set of reference spectra obtained for proteins with known three-dimensional structure and the fraction of secondary structure were computed (Williams and Dunker, 1981; Williams, 1983; Williams, 1986; Berjot *et al.*, 1987). The accuracy of these methods is generally comparable to similar analyses of CD spectra, being the estimated errors in general few percentages. The same methods have then been applied to calculate secondary structure percentages of proteins with unknown structure, because of difficulties to obtain them in a crystalline form.

Table 2 summarises the Raman and IR frequency range of amide I and III bands, generally diagnostic of specific protein secondary structures. Other amide bands have been observed in the spectra, but normally their features do not have a so strictly "diagnostic" character.

Beside the amide bands, there are other bands sensitive to protein conformation. As an example, Raman spectra can show a band in the 890-945 cm^{-1} range, attributable to a skeletal C-C stretching vibration, which is typical of alfa-helix. This band disappears or become weaker upon conversion to beta-sheet or random coil. For beta-sheet conformation, a characteristic band lies in the 1020-1060 cm^{-1} region.

As regards the side-chain amino acid residues, the aromatic ring gives rise to characteristic bands as that observed for Phe at 1005 cm^{-1} (ring breathing), intense in the Raman spectrum; this band indicates the presence of Phe residues but it is not sensible to the amino acid environment (Figure 2A). Other bands typical of aromatic amino acids lie in the 1600 cm^{-1} region (1605 Phe, $\approx$1585 cm^{-1} Phe, Tyr and Trp, 1618 cm^{-1} Tyr) and also at lower wavenumbers (Figure 2). In the case of Tyr, intense bands observed in the Raman spectrum lie at about 1255, 1210, 1176, 850, 830 and 646 cm^{-1} (Siamwiza *et al.*, 1975), most of which are due to the aromatic ring of this amino acid.

The frequency and the relative intensity of the Tyr Fermi doublet at about 830 and 850 cm^{-1} is indicative of presence/absence of H-bond and gives information about Tyr environment (buried/exposed) (Tu, 1982; Parker, 1983). In the case of Trp, an intense Raman band at about 1363 cm^{-1} can be correlated to Trp environment (buried/exposed).

Table 2. Wavenumbers of the principal Raman and IR bands characteristic of protein secondary structure

Structure	Raman bands (cm^{-1})	
	Amide I	Amide III
beta-turn	1632-1648 1680-1697	1260-1290
alfa-helix	1650-1657	1268-1309
unordered structure	1660-1666	1240-1256
beta-sheet	1667-1680	1227-1240
	IR bands (cm^{-1})	
	Amide I	Amide III
beta-turn	1655-1675 1680-1696	1270-1290
alfa-helix	1652-1660	1290-1310
unordered structure	1640-1655	1255-1259
beta-sheet	1610-1644	1207-1245

If other aromatic residues are absent or present in a low percentage, it is possible to identify the weak band due to the C=C stretching vibration of His residues, whose frequency is strongly dependent by the tautomeric form of His (tautomer I or II, also referred as N_τ–H or N_π–H) and its involvement in metal ion chelation (Miura *et al.*, 1998; Takeuchi, 2003; Torreggiani *et al.*, 2003) (Figure 3). The presence of the I or II tautomers is also indicated by the bands at 1282 or 1260 cm^{-1}, respectively.

As regards Cys, the presence of a well-resolved Raman band in the 2500-2600 cm^{-1} region, where no other group displays bands, indicates the presence of the -SH group. The wavenumber of this band is indicative of the presence and strength of hydrogen bonds (Raso *et al.*, 2001). Other bands, due to C-S-S-C vibrations, give information about the conformation of C-S-S-C groups. Raman bands, attributable to C-S stretching, can be observed in the 630-670 (*gauche* conformation) and 700-745 cm^{-1} (*trans* conformation) region, and those attributable to the S-S stretching in the 500-550 cm^{-1} range (Akhtar and Edwards, 1997; Nakamura *et al.*, 1997). Also Met residues give rise to a C-S stretching band at 718-728 cm^{-1} (Nogami *et al.*, 1975; Torreggiani *et al.*, 2006a).

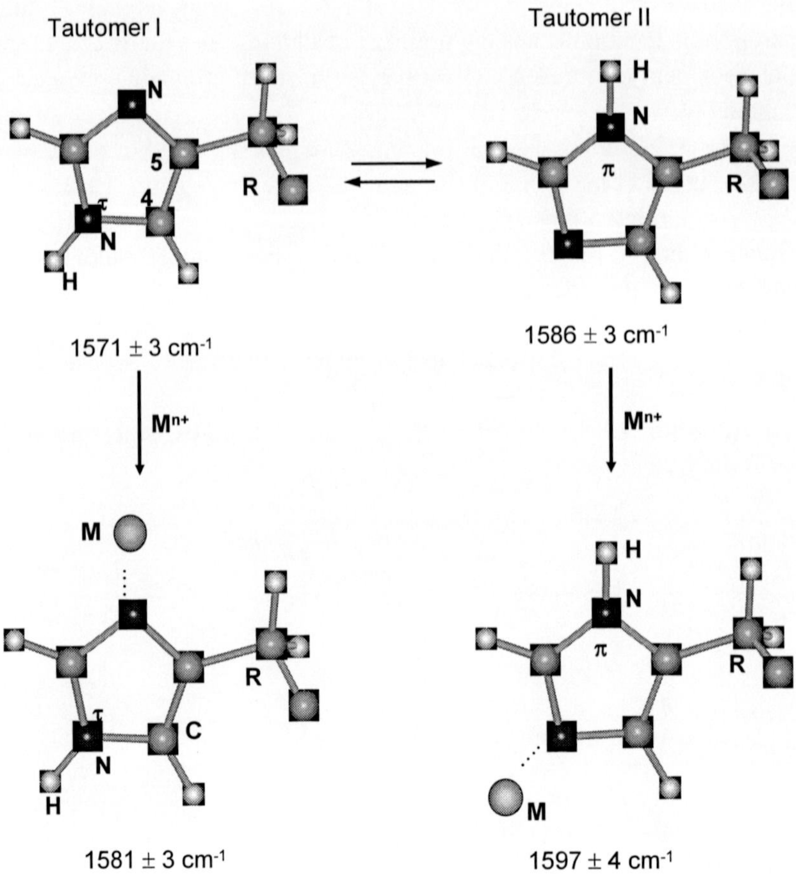

Figure 3. The tautomeric equilibrium of the imidazole moiety of His, the metal-coordinated tautomeric forms, and the corresponding wavenumbers of one Raman marker band (C_4=C_5 stretching). M = metal, R = peptidic chain.

Other bands characteristic of sulphur-containing groups have been observed in the 410-430 cm^{-1} region and at lower wavenumbers, but some attributions are still unclear even if some of these components have been attributed to -S- bridging atoms bonded to metal ions (Jensen, 2003). Bands attributable to a metal bonded to Cys sulfur can be observed in the 430-280 cm^{-1} region. (Quail *et al.*, 1996; Miura *et al.*, 1998; Broderich *et al.*, 2000). Raman bands at 270, 302 and 343 $cm^{-1}$are attributed to Fe-$S_{bridging}$ stretching while those observed at 378 and 417 cm^{-1} are related to Fe-$S_{terminal}$ stretching (Broderich *et al.*, 2000). The presence of bands at fixed frequencies, i.e. at about 290 cm^{-1}, indicate a cubane-type cluster (Lover *et al.*, 1997), whereas binuclear centres are indicated by Raman bands at 282, 327, 340, 367, 395, and 426 cm^{-1} in [2Fe-2S] ferredoxin (Rotsaert *et al.*, 2003).

In spite of its wide application to the study of protein structure and metal-ligand interactions, till now vibrational spectroscopy has been scarcely used in the study of MTs (Pande *et al.*, 1986; Shi *et al.*, 2002).

As example of "diagnostic" Raman bands for the study of protein structure, Figure 2A reports the Raman spectrum of the *echinoderm* Zn-SpMTA in the 200-1800 cm^{-1} region, together with the more significant attributions. The 2500-2600 cm^{-1} region is not reported in the figure, since this protein, containing 20 Cys residues, does not display any band in this region. Moreover, S-S and C-S stretching bands are intense in the spectrum, while the bands typical of M-S bonds are less intense, indicating that a discrete quantity of sulfur atoms are in an oxidised state. Other important bands, very intense in the spectrum, are amide I and III bands, which maxima are in a range typical of beta-sheet and disordered structure/alfa-helix conformations, respectively. The typical bands of Phe residue are also indicated. No other aromatic aminoacid is present in this MT. For a comparison, Figure 2B reports the IR spectrum of the same MT in the 800-3700cm^{-1} region. It can be observed that the bands typical of the peptidic bond are well evident also in the IR spectrum, while the bands due to M-S bonds are better evident in the Raman spectrum because of their low wavenumber and high polarizability of the bonds.

3. Information on MTs Structure Obtained by CD, Raman and IR Spectroscopies

In this section seven cases of *in vivo*-synthesized metal-MT aggregates and the information about their structure obtained from Raman, IR and CD spectroscopic analyses, will be reviewed. The metal-MT aggregates, taken as models, belong to different MT families (see Section 1). The *in-vivo* synthesized Zn-coordinated forms have been used for the Raman, IR and CD analysis. In order to show the influence of the metal bound, an analysis of Cd-QsMT in comparison with Zn-QsMT is also included. The recombinant metallopeptides spectroscopically analysed contain variable amounts of metal and sulfide ions, quantitatively evaluated by acid ICP-AES and GC-FPD measurements. As shown in Figure 4, the metal content ranges from 3.5 (Zn-QsMT) to 7.6 (Zn-MT-10-V), whereas that of sulfide ions (S^{2-}) varies from 0.6 to 3 sulfides per protein.

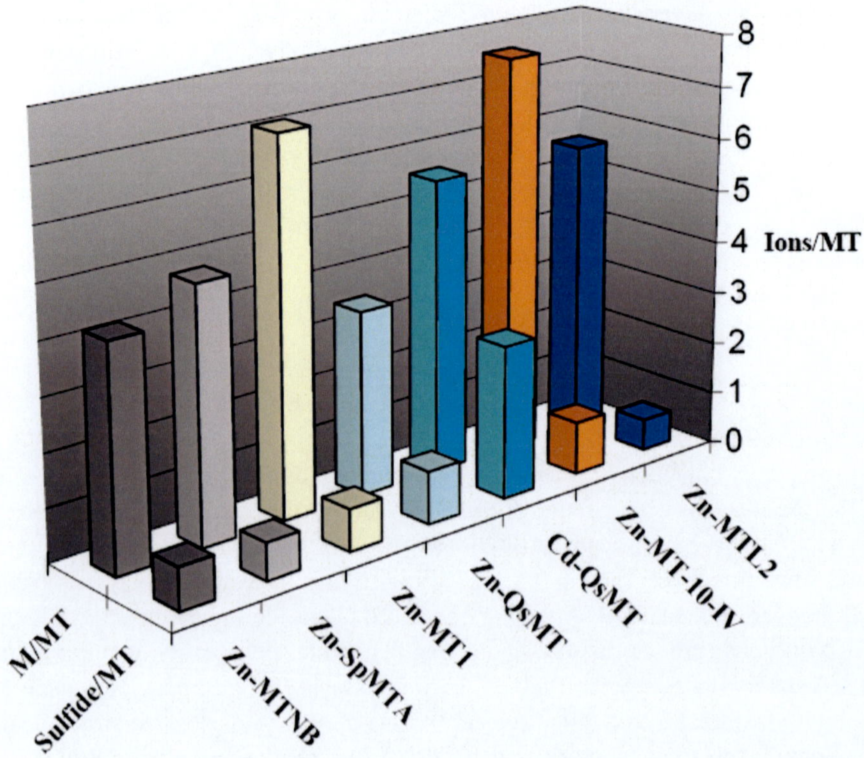

Figure 4. Block diagram of mean metal and sulfide ion content present in the MT aggregates. The M / MT molar ratios have been calculated by acid ICP-AES measurements and an error of ± 0.1 has to be considered. Sulfide (S^{2-}) /MT ratio has been measured by GC-FPD.

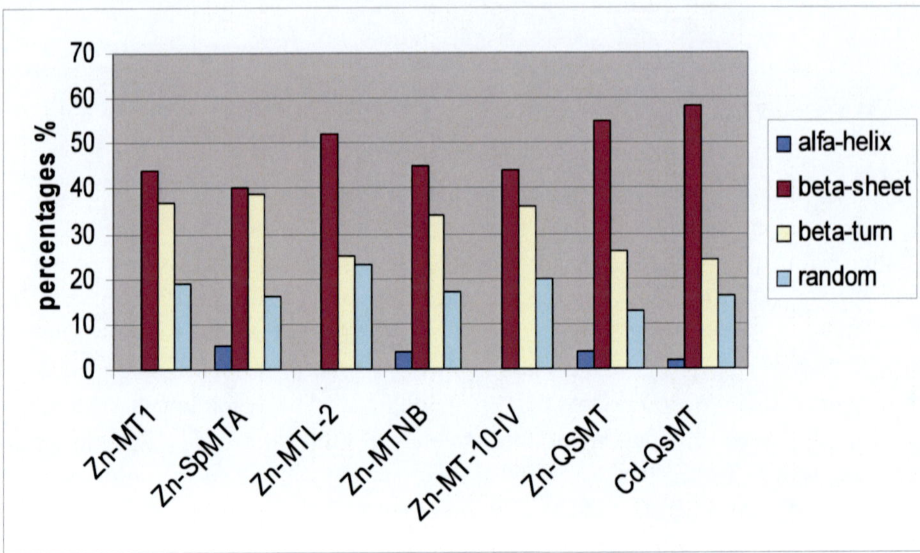

Figure 5. Block diagram of percentages of the secondary structures found in the MT aggregates calculated by the analysis of the Raman amide I and IR amide III regions.

3.1. Secondary Structure

Although secondary structure elements, such as beta-sheet, beta-turn or alpha helix, have a high functional importance in MPs like *zinc fingers* (Miura *et al.*, 1998), in MTs for a long time it has been generally assumed that secondary structure elements are poor or functionally insignificant (Cobine *et al.*, 2004; Calderone *et al.*, 2005). Most of the literature reviews dealing with MT structure were focused mainly on the primary and tertiary structure, conceding scarce interest to secondary structure. Nowadays, some authors claim for a crucial participation of secondary structure elements in the functional properties of MTs (Rigby and Stillman, 2004; Vergani *et al.*, 2005; Rigby *et al.*, 2006) and proofs of their existence and functional importance in MTs are growing, as shown in Table 3. IR and Raman spectroscopies can shed light on this important aspect of MTs structure.

To quantify the contribution of a distinct secondary structure motif to the overall structure of the metal-MT complexes, the method proposed by Alix was used on the amide I Raman bands appearing between 1600 and 1700 cm^{-1} (Alix *et al.*, 1988). This methodology is based on an equation obtained through a multiparametric analysis of the correlations between X-ray structural and spectroscopic Raman data from a large set of reference proteins, and calculates the percentages of structural elements in a protein as a linear function of calculated parameters of the amide I Raman band (Alix *et al.*, 1988; Alix and Pedanou, 1994; Fagnano *et al.*, 1995; Torreggiani and Fini, 1998). The calculated percentages of secondary structure of M-MT complexes are shown in Figure 5.

To confirm the results obtained by Raman amide I band, a quantitative evaluation of the secondary structure by IR spectra can also be obtained. The most studied infrared band of proteins is the amide I although a water vibration almost coincides with the amide I band, making studies in protonic aqueous solution difficult. The amide III band, although less intense than amide I, can facilitate the structural analysis of proteins since no water absorption occurs in this region (Figure 2B) (Byler and Susi, 1986; Arrondo *et al.*, 1993; Hollosi *et al.*, 1994; Wi *et al.*, 1998).

The analysis of the broad IR amide III band has been carried out on the Zn-QsMT and Cd-QsMT spectra by means of a gaussian curve fitting method based on a least-square fit of several curves to an experimental profile (Domenech *et al.*, 2007b). An indication of the number and position of the fitting bands has been obtained by the fourth-derivative spectra (Maddams, 1982) and the secondary structure contents have been calculated from the integrated intensities of the individually assigned bands and their fraction of the total intensity (Parker, 1983; Cai and Singh, 1999). As example, the curve fitting on both the amide I and III region of the IR spectrum of Cd-QsMT is reported in Figure 6. The IR spectrum of the Cd-QsMT exhibits eight and ten components in the 1205-1320 and 1720-1580 cm^{-1} regions, respectively, resulting from the contribution of different secondary structure elements.

In all the six MT isoforms a relevant contribution of β-sheet and β-turn segments has been found, whereas the α-helix content has resulted to be almost negligible (Figure 5). The two higher beta-sheet contents have been found in Zn-MTL-2 and Zn-QsMT, as well as the lowest beta-turn percentages, whereas the random percentage values have resulted to be very similar in all the M-MT aggregates here described. Thus, sensitive differences in secondary structure have been found in MTs from different families. Interestingly, the secondary structure percentages of Zn-MT1 (from vertebrate) and Zn-MT-10-IV (from invertebrate), which show the higher sequence similarity among the considered MTs, resulted to be almost equal (44% β-sheet, ≈ 37% β -turn, ≈ 20% random, 0% α-helix). The large content of β-turns and the lack of α-helix segments evidenced in the Zn-MT1 structure is in agreement with the data obtained by different techniques (Boulanger *et al.*, 1983; Qing *et al.*, 1995).

Table 3. Reported data on secondary structures of MTs, derived from theoretical calculations or experimental measurements by different techniques

MT Family	MT	Secondary structure elements	Technique	Reference
1 Vertebrate	MT1	Two short stretches of 3-10 helix and three half-turns	NMR	Zangger *et al.*, 1999
	MT1	β-turns	Raman, IR	Pande *et al.*, 1986; Qing *et al.*, 1995
	MT2	Several β-bends (10 or 11)	Theor. study	Boulanger *et al.*, 1983
	MT1, MT2	Half-turns 27%, 3-10 helix 13-18 %, Random 45-49%, β sheet 10-12%	FT-IR, FT-Raman	Shi *et al.*, 2002
	MT2	Two 3-10 helix; numerous half-turns	NMR	Wagner *et al.*, 1986
	MT2	Helix 3-10 and half-turns	NMR	Wagner *et al.*, 1987
	MT2	Several half-turns and 3-10-helical segments	NMR	Arseniev *et al.*, 1988
	MT2	Short helix in the α-domain	X-ray	Robbins *et al.*, 1991
	MT3	one α-helix	NMR, CD	Wang *et al.*, 2006; Hasler *et al.*, 1998
	MT1, MT2, MT3	One α-helix, helix 3-10 and half-turns	NMR	Oz *et al.*, 2001
	Fish MT	Two α-helix, one 3_{10} helix	Theor. study	Scudiero *et al.*, 2005
3 Crustacean	MTH (lobster)	Three β-turns, one short 3-10 helix, one α-helix segment	NMR	Munoz *et al.*, 2002
	MT crab	One short segment of α-helix, several type I β turns	NMR	Narula *et al.*, 1995
4 Echinoderms	SpMTA	Helix 3-10 and half-turns	NMR	Riek *et al.*, 1999
12 Fungal 5	Yeast CUP1	One α-helix	NMR	Bertini *et al.*, 2000
14 Procaryota	Bacterian SmtA	Several α-helix and β-sheet	X-ray	Cook *et al.*, 1998; Blindauer *et al.*, 2001
15 Plant	Plant MT	β-sheet	Theor. study, molec. dinamics	Zhu, 2000; Bilecen *et al.*, 2005

As far as QsMT is concerned, the Zn(II) coordination resulted to slightly favour major content of β-turns and minor β-sheet content in relation to Cd-QsMT, whereas it did not affect the percentages of disordered conformation (Domenech *et al.*, 2007b).

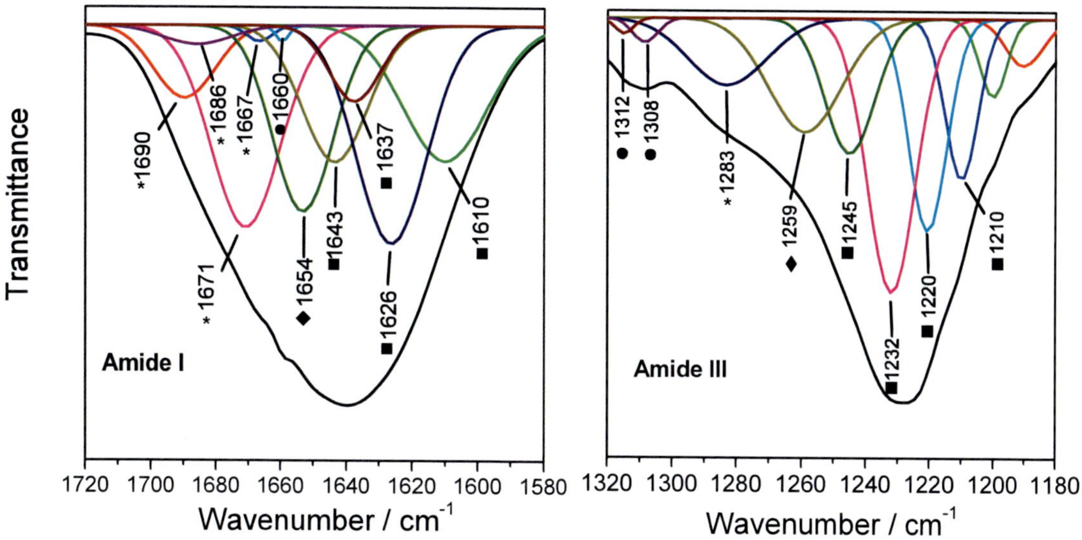

Figure 6. Curve fitting of the IR Amide I and III bands of Cd-QsMT. The component assignments to the different secondary structure elements are indicated as follows: ■ β-sheet; ◆ = Random; ● = α-helix; * = β-turn.

3.2. Aromatic Aminoacids

Almost all the MTs considered are completely devoid of aromatic residues. Only three MTs contains one or two aromatic residues: Phe residue is present in the sequence of SpMTA and QsMT, Tyr in MTL-2, and His in MTL-2 and QsMT (Table 1). In the Raman spectra of these MTs the typical bands of these aromatic residues can be observed (see section 2.2 and Figure 2A).

As regards Tyr residues, the phenolic moiety generally exhibits an intense Raman doublet in the spectral interval 820-860 cm^{-1}, whose intensity ratio is diagnostic of specific donor or acceptor roles of the phenoxyl OH group. In the Raman spectrum of MTL-2 the Tyr residue gives rise to an anomalous singlet (≈ 855 cm^{-1}) rather than a doublet in this spectral region. This finding is indicative of non-hydrogen-bonded state of the Tyr phenoxyl group (Arp *et al.*, 2001). The absence of the phenoxyl group hydrogen bonding in MTL-2 may be attributed to the location of the Tyr residue within a highly hydrophobic region, and then tightly packed domain. This result can be relevant in elucidation of the MTL-2 tertiary structure. In fact, it has been assumed that MTL-2, when binds divalent ions, would fold into a bi-dominial structure with 9 Cys residues in each domain (You *et al.*, 1999). In such structure the Tyr residue, situated between the two putative domains, would be exposed to the solvent, as it would not be inside one of the folded-domains. On the contrary, the buried position of this Tyr residue supports its inclusion in a folded-domain rather than its inter-dominial free position, this arguing against a bidominial 9 Cys+9 Cys structure for MTL-2.

The marker bands of His residues will be examined in section 3.3.1 where the metal binding involving endogenous ligands is considered.

3.3. Metal Coordination Sites

Recently, it has been shown that ligands other than Cys can participate in the coordination sphere of metals in MTs. Two main types have been identified: endogenous ligands such as imidazole moiety of His residues (Blindauer *et al.*, 2001; Tucker *et al.*, 2004; Villarreal *et al.*, 2006) and/or exogenous ligands such as inorganic ions (i.e. sulfide or chloride ions)(Maret *et al.*, 2002; Domenech *et al.*, 2003; Capdevila *et al.*, 2005; Villarreal *et al.*, 2005; Tio *et al.*, 2006; Domenech *et al.*, 2007a). This section will be focused on the information that Raman, IR and CD spectroscopies can give about the metal-coordination sites of MTs.

3.3.1. Endogenous Ligands

Cysteine

Cys residues contain thiol groups(-SH) able to deprotonate and originate thiolate groups ($-S^-$) which can coordinate a wide diversity of metals. The binding of several Cys to a metal ion can generate a metallic cluster with diverse coordination geometries depending on the bound metal. In the constitution of such aggregates, the sulfur atom of one Cys can bind to one or two metals: in the first case, Cys is indicated as "*terminal*", whereas in the second as "*bridging*". It has been extensively described that both kind of M-Cys bonds are present in M-MT aggregates (Furey *et al.*, 1987).

One of the main features of the M-Cys binding is the striking combination of high thermodynamic but low kinetic stability, thus, MTs are able to bind metals very tightly, but also promptly exchange the bound metal to other proteins (Maret, 2004). Metallic substitution follows the order characteristic of tiolate group affinity, it is $Hg^{2+} > Ag^{1+} = Cu^{1+} > Cd^{2+} > Zn^{2+}$. For this reason Zn can be substituted in MT by toxic xenobiotic metals such as Cd ions.

The oxidation of Cys (R-SH) can result on the liberation of the bound metal and the formation of Cystine (R-S-S-R) (Kumari *et al.*, 1998; Maret, 2003). Thus, in MTs Cys can be present in three main states: free, coordinated to a metal, or oxidised with the setting up of a disulfide bridge (Cystine). All the three states of Cys can be evidenced by Raman spectroscopy in the spectral region typical of sulfur-containing moiety vibrations. Moreover, metal-coordinated Cys can give rise to different vibrational bands depending on the structural environment of the M-Cys clusters.

As regards free Cys, their presence in MT can be evidenced by the -SH stretching band appearing in the 2500-2600 cm^{-1} Raman region. This band is diagnostic for the presence of the thiol moiety since no other group gives rise to bands in this wavenumber range. Since this band has not been found in all the examined MTs, it can be concluded that these MTs do not contain free Cys residues, as it could be expected from the high facility of Cys to be oxidised at neutral pH or its ability to chelate metal ions.

The 500-550 cm^{-1} Raman region bears information on the disulfide bridges, allowing a qualitative evaluation of the overall oxidation state of a MT (Figure 2A). In all the MTs considered in this chapter only a weak contribution of the disulfide vibration was found in the Raman spectra, indicating that almost all the cysteinic sulfurs are involved in metal coordination. However, the presence of different cystine amounts has been revealed in Zn-

MTs, by considering the peak heights of the disulfide bands obtained after normalising the spectra on the intensity of the Raman band at about 1450 cm^{-1} (CH_2 in-plane deformation) which is assumed not to be sensitively affected by structural changes. In the case of Zn-SpMTA and Zn-MT1 the higher intensity of the S-S bands visible in the Raman spectrum of Zn-SpMTA (at 512 and 523 cm^{-1}) has clearly revealed the formation of a larger amount of dimerised cystine in this metal aggregate than in that of MT1, which contains an identical number of the Cys residues in its sequence (Figure 7). The intensity of the S-S bands has followed this decreasing order: Zn-SpMTA> Zn-MTL-2 > Zn-QsMT ≅ Zn-MT-10-IV > Zn-MTNB ≅ Zn-MT1> Cd-QsMT. Since among protein side-chain interactions, the disulfide bond is particularly important because it gives additional stability to the folding, the highest amounts of S-S bonds found in Zn-SpMTA and Zn-MTL-2 could play a structural role or be indicative of instability of the metal clusters in these metal aggregates.

The exact position of the S-S bands is determined to some extent by the conformation of the residues contributing the disulfide bridges and hence the protein's three-dimensional structure. In the case of Zn-SpMTA a doublet at 512 and 523 cm^{-1} was visible (Figure 7), corresponding to the stretching modes of disulfides in two different conformations (*gauche-gauche-gauche* and *gauche-gauche-trans*, respectively). On the contrary for Zn-MT1, the possibility of a mixture of conformations for the very few disulfides formed should be excluded since only the weak component at 522 cm^{-1} was visible.

As regards Cys bound to metal ions, Raman spectra allow to differentiate the vibrational frequencies of M-S bonds for the seven MTs reviewed in this chapter. The Raman region from 200 to 800 cm^{-1} provides useful information about the bonds in which sulphur atoms participate, also thanks to the absence of Trp residues, which originate intense bands in this region.

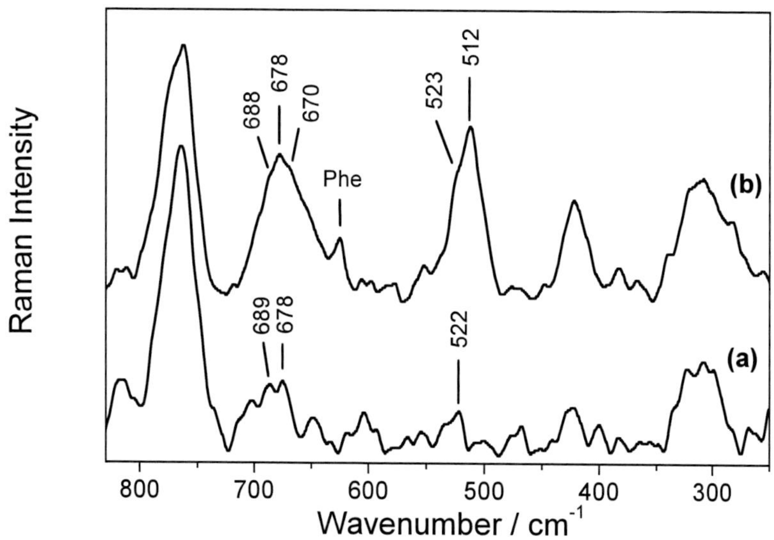

Figure 7. Raman Spectra of (a) Zn-MT1 and (b) Zn-SpMTA in the 800-250 cm^{-1} region where the bands due to the vibrations of disulfide bridges (νS-S and νC-S) are detectable. The spectra have been normalised on the 1450 cm^{-1} band.

The involvement of the cysteinic sulfur atoms in Zn^{2+} binding has been clearly visualised in the Raman spectra through several bands attributable to the M-S stretching modes at low wavenumbers (< 500 cm^{-1}) (Figure 8). (Adams and Cornell, 1968; Han *et al.*, 1989; Boldyrev and Simons, 1997; Miura *et al.*, 1998) Contributions of the sulfur bridging and terminal ligands to the MT Raman spectra have been qualitatively identified for Zn- and Cd-QsMT (Domenech *et al.*, 2007b).

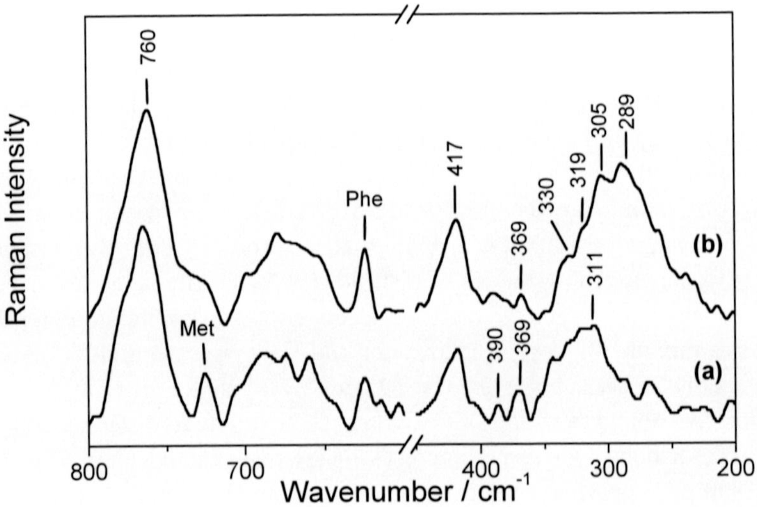

Figure 8. Raman Spectra of (a) Zn-QsMT and (b) Cd-QsMT in the 800-200 cm^{-1} region where the M-S bands (νM-S and νC-S) are detectable. The spectra have been normalised on the 1450 cm^{-1} band.

The highest wavenumber bands (395-430 cm^{-1}) are essentially due to metal-S bridging vibrations, whereas the lowest wavenumber modes (250-370 cm^{-1}) are contributed by both S-terminal and S-bridging ligands (Figure 8). The high number of the M-S stretching bands as well as their broadening has also suggested the formation of metal clusters with different geometry in the two metallated QsMT, well in agreement with the differences in size, electrostatic and covalent bonding forces of zinc and cadmium ions (Domenech *et al.*, 2007b).

The C-S stretching modes originated from M-Cys bonds give rise to a band at ca. 760 cm^{-1}, peculiar of the M-MT complexes (Figure 8) (Pande *et al.*, 1986). In the case of QsMT metal aggregates the curve fitting analysis of this band has allowed to evidence a minor variability in the $S-CH_2$ bond geometry in the Zn(II) coordination environment in respect to that in Cd-QsMT, in concordance with the predominant species detected by ESI-MS measurements (Domenech *et al.*, 2007b).

The metal-binding environment can be also analyzed by CD spectroscopy. In fact, the metal-ligand charge transfer bands are visible in CD spectra at wavelengths correlated to the metal-binding environment. In all the Zn-MT aggregates a signal at about 240 nm, attributable to Cys-Zn binding, is generally observed (Figure 9). However, some spectral differences of the diverse Zn-MT aggregates have suggested differences in the metallic clusters. The Zn-MT1 and Zn-SpMTA CD spectra are mainly conformed by an *exciton coupling* centered at 240 nm (Figure 9).

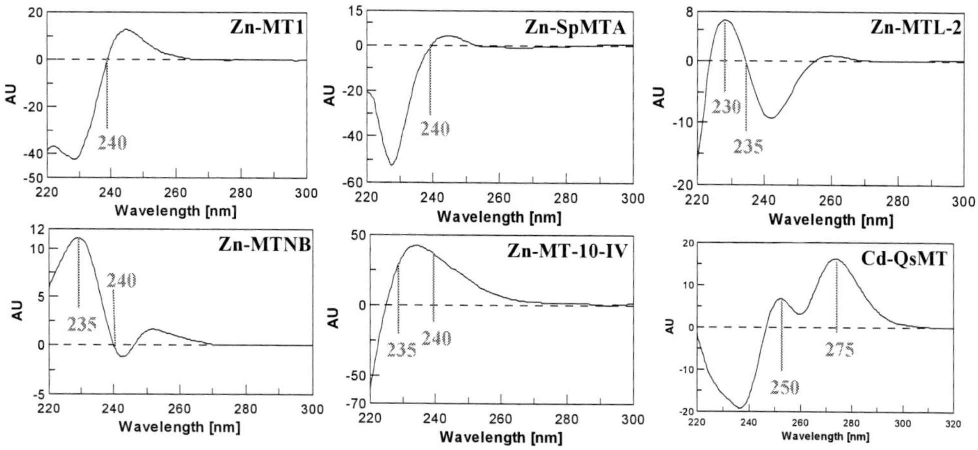

Figure 9. Circular Dichroism spectra of some representative Zn- and Cd-MT aggregates. The main features of the CD spectra appear marked in grey.

Since the signal profile and its intensity are a measure of the structuration degree of the metallic aggregate, the lower symmetry of the Zn-SpMTA CD spectrum is indicative of a lower degree of structure of its metal clusters in comparison with those of Zn-MT1, in agreement with the lower metal/MT content present in Zn-SpMTA (4.8 istead of 7.1) (Figure 4). The high chirality of Zn-MT-10-IV at 240 nm argues in favour of exceptionally structured Cys-Zn metallic clusters. In spite of all the five Zn-MT aggregates show signals near to 240 nm, both MT1 and MT-10-IV show the highest intensities at this wavelength. This spectral feature indicates a higher degree of Zn-Cys structuration for these aggregates than those of other Zn-MT complexes.

The Cd-QsMT CD spectrum shows a high chirality profile (as usually observed for Cd-MT aggregates) with two main absorptions: a gaussian band at *ca.* 275 nm and a shoulder at about 250 nm. The first informs about the participation of sulfide ligands in the Cd-MT complexes (Capdevila *et al.,* 2005), whereas the second has been attributed to the involvement of His in the otherwise tetrahedral $Cd(SCys)_4$ chromophores (Romero-Isart *et al.*, 1999; Villarreal *et al.*, 2006).

In conclusion, the use of CD and Raman spectroscopies can be useful for evaluating differences in the Cys-metal binding environments and detecting the presence of disulfide bridges. The analysis of inorganic sulfide-related bands will be discussed in section 3.2.2.

Histidine

The two nitrogen atoms (N_τ and N_π) of the His residue are potential donors for transition metal ions and its coordination can be detected by using Raman marker bands such as the C_4=C_5 stretching vibration. In fact, this band appears at different wavenumbers depending if N_τ or N_π are bonded to the metal ions (Figure 3) (Miura *et al.*, 1998; Torreggiani *et al.*, 2000a; Torreggiani *et al.*, 2000b; Domenech *et al.*, 2007b). Thus, although His Raman bands are weak compared to those of aromatic amino acids, their identification is still possible in proteins with a low content of Tyr or Phe, such as MTs. QsMT and MTL-2 contain one His residue and, respectively, two Phe or one Tyr (Table 1). The eventual His involvement in the metal binding has been evaluated by carrying out the curve fitting analysis of the 1630-1565 cm^{-1} spectral range (Figure 10) that allows to distinguish also the contribution of overlapped weak bands.

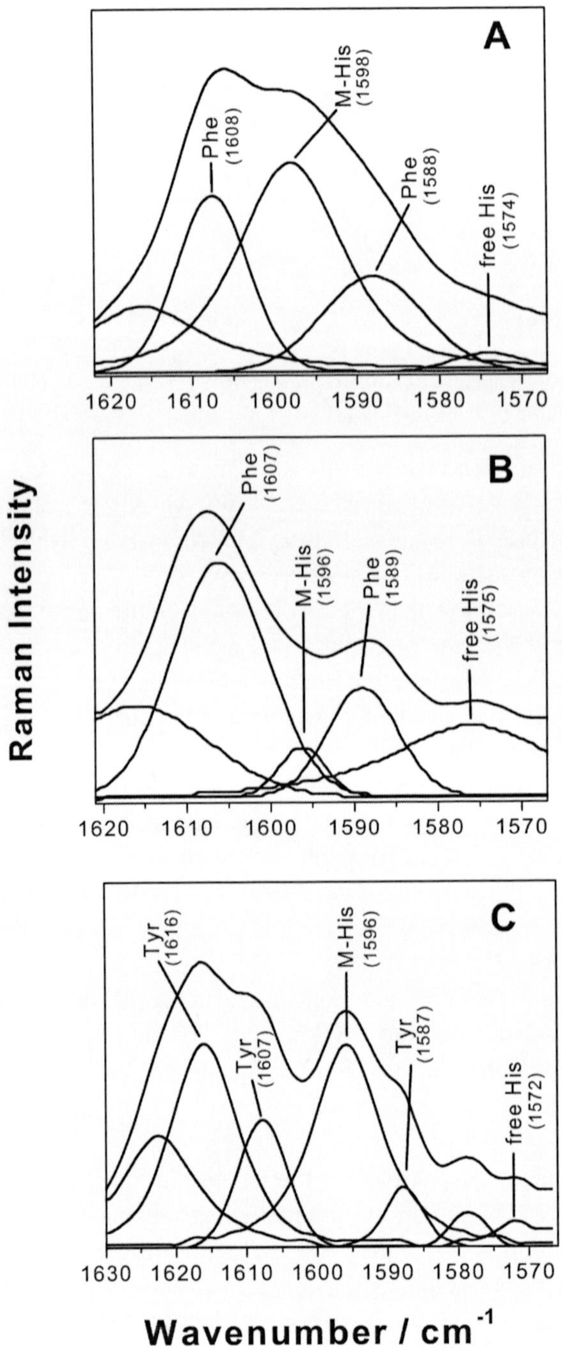

Figure 10. Curve fitting analysis of the 1630-1565 cm^{-1} Raman region of His-containing MTs: (A) Cd-QsMT, (B) Zn-QsMT and (C) Zn-MTL-2. The components due to free His and metal-His system are visible at about 1575 and 1600 cm^{-1}, respectively.

Free His can give rise to a band at about 1570 and 1585 cm^{-1}, depending on the tautomeric form assumed by the imidazole ring of His (Figure 3). Generally, His is present as tautomer I since this is the most stable form (Ashikawa and Itoh, 1979); thus, the expected band of free His should be at $\approx$1575 cm^{-1}. By the curve fitting of the Raman spectrum of Cd-QsMT, the two components due to Phe are well visible at 1608 and 1588 cm^{-1} as well as two bands at 1598 and 1574 cm^{-1}, attributable to coordinated (as tautomer II) and free His (in the I tautomer form), respectively. By considering the integrated intensity of these bands, it can be concluded that His residues are mainly coordinated to Cd^{2+} ions ($\approx$ 80 %). By using the same procedure on the Raman spectrum of Zn-QsMT, again these four components have been evidenced, but in this case His resulted to be mainly present as free tautomer ($\approx$ 90%). Analogously, the curve fitting analysis of the Raman spectrum of Zn-MTL-2 has shown three components attributable to Tyr residue (1616, 1607 and 1587 cm^{-1}), and two peaks due to coordinated and free His. Also in such aggregates about 80% of the N_τ–imidazole of His takes part in the metal binding. The same tautomeric form of His has been found to be involved in metal binding in MPs such as hemoglobin (Dickerson, 1983) and in *Clostridium pasteurianum* iron hydrogenases (Adams, 1990; Peters *et al.*, 1998).

The identification of His participation in metal-MT aggregates by Circular Dichroism is less clear, but its participation in the CD spectra is quite evident.

In MT1-derived peptides genetically engineered to substitute Cys by His, the participation of His in metal binding has implied a blue-shift of the CD bands and the appearance of particular shoulders at 250 nm in Cd-aggregates (Romero-Isart *et al.*, 1999). Such effect has also been observed in the CD of Cd-QsMT, where a pronunciated shoulder at 250 nm is evident (Figure 9), and attributable to the participation of His to Cd-coordination, in agreement with the Raman data. In fact, the lowering of pH to 4.5, which implies the protonation of His, causes a decrease of this band in Cd-QsMT CD spectrum (Domenech *et al.*, 2007a), so indicating that this signal can be associated to the participation of His residues to metal coordination.

Blue-shift effects can be also observed in Zn-MTL-2 (which *exciton-coupling* shape is centered at 235 nm instead of the expectable signals observed for the other Zn-MTs at 240 nm) and can be attributed to the metal coordination of His residue (Figure 9).

These observations demonstrates that the parallel use of Raman and CD spectroscopies can constitute a suitable tool for the detection of the His participation in metal coordination in MTs. In fact, the Raman analysis allows the unambiguous detection of the metal-His binding, while CD can be useful for following the evolution of the coordinated His along acidic or metallic titrations of MTs. Since His residues have been found in MTs from several MT families, (including vertebrate (family 1), nematode (family 6), fungi (family 8), plants (family 15) and bacteria (family 14)) (Winge *et al.*, 1985; Blindauer *et al.*, 2001; Tucker *et al.*, 2004; Villarreal *et al.*, 2006), their functional role could be investigated by approaches equivalent to those used for QsMT and MTL-2.

Carboxylate Groups

Carboxylate groups from Asp or Glu side chains or Carboxy-terminal edges of proteins can take part in metal coordination (Degtyarenko, 2000; Auld, 2001) and play key roles as supporting ligands in a diverse array of metalloprotein active sites (Holm *et al.* 1996). Such carboxylates are notable for the facility with which they adopt different binding modes

(Rardin *et al.* 1991), in particular during the catalytic cycles of dioxygen-activating mono- and di-iron enzymes (Wallar and Lipscomb, 1996). However, it has been assumed that COO^- groups do not play a role in metal coordination in MTs, although acidic residues are quite abundant in MT sequences and are often situated near to metal-coordinating Cys (Table 1).

The carboxyl group exhibits the COO^- symmetric stretch vibration in the 1400-1420 cm^{-1} region, giving rise to a band almost intense in Raman spectra but very weak in IR. By analysing the second derivative spectra of the six MTs considered, in order to better distinguish also weak adjacent peaks, two components, attributable to the COO^- vibrations, were detected at about 1420 and 1410 cm^{-1} in the spectra of Zn-SpMTA, Zn-MTL-2 and Zn-MT10-IV. In particular, the presence of the lower wavenumber component suggests the possible involvement of some COO^- groups in the metal chelation. Thus, the COO^- groups from MT lateral chains could play a role in the stabilization of the MT structure not only by ionic interactions but also metal coordination. This participation would not be a structural feature restricted to one MT family, but a more general behaviour of MTs, since the spectral results concerning the carboxylate groups were obtained for MTs from organisms of different MT-families (Figure 1).

3.2.2. Exogenous Metal Ligand Sites

Inorganic Sulfide Ions

Sulfide ions (S^{2-}) play key functional roles in metalloenzymes like ferredoxins, proteins characterised by the presence of polymetallic systems (iron-sulfur clusters) containing iron ions with variable oxidation state (Johnson *et al.*, 2005). Biological [Fe-S] clusters are characterized by the presence of multiple iron ions bridged by sulfide ions and coordinated to the protein, generally via Cys residues. The most common types of [Fe-S] clusters, found in the widest variety of proteins and enzymes, are the [2Fe-2S] and [4Fe-4S] clusters, which contain the indicated number of iron and sulfide ions and are typically bound to the protein by four cysteines (Figure 11A). These clusters have diverse roles in biology, acting as catalytic centers, structural elements, and sensors in regulating gene expression (Broderick, 2007).

For a long time the functional role of sulfide ions was considered to be limited to their enzymatic redox properties; however, the discovering of *crystallites* changed this vision. Crystallites are structures constituted by metal ions and phytochelatins, enzimatically synthesized proteins produced in higher plants and some fungi upon exposure to heavy metals. These structures could contain sulfide anions, that can increase the metal detoxification potential of the aggregates (Hayashi *et al.*, 1986; Winge *et al.*, 1992; Hall, 2002) (Figure 11B).

MTs are proteins similar to phytochelatins in many way, including the high number of Cys residues in the protein and the fact that both are responsible for the detoxification of heavy metals. Only recently the presence of sulfide ions has been discovered also in *in vivo*-synthesized M-MT aggregates, probably since most of the data referring to MT structure available to date comes from non-biological synthesis of M-MT complexes (González-Duarte, 2003). The presence of the acid-labile S^{2-} ligands has been determined both qualitatively and quantitativey by analytical, spectroscopic, and spectrometric techniques and the features of the recovered Zn(II)- and Cd(II)-MT complexes correlate well with those reported for plants and yeast phytochelatins, therefore bridging the behaviour gap between both types of metal-binding molecules.

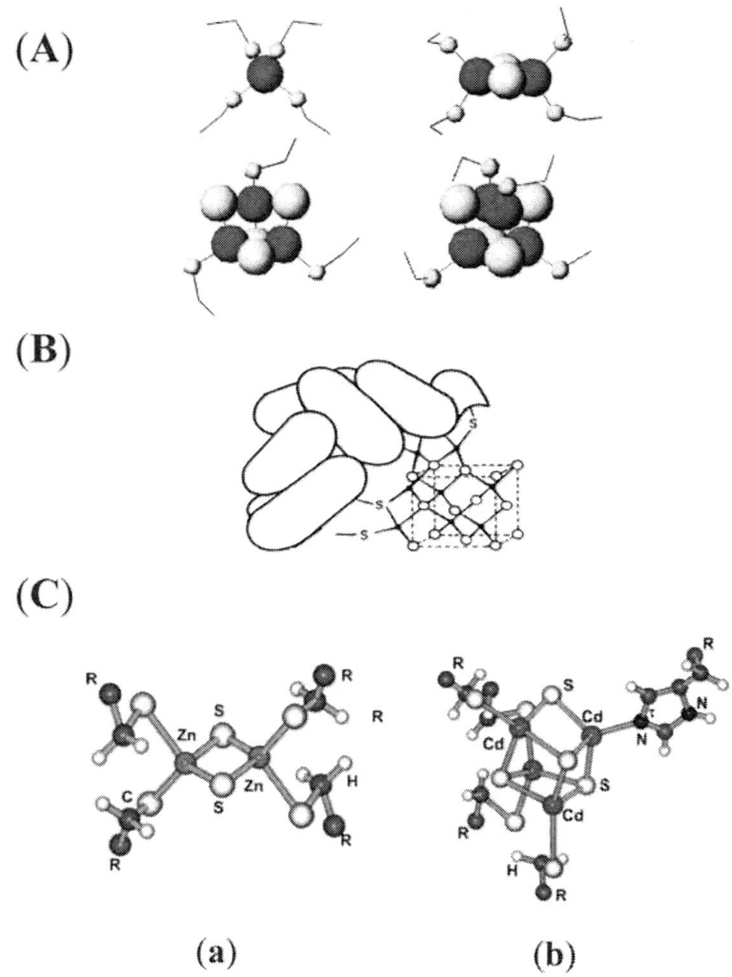

Figure 11. Structure of the metallic clusters proposed for (A) iron-sulfur proteins; mono-, bi-, tri-, and tetranuclear active-site structure type, present in rubredoxin, [2Fe-2S] ferredoxin, [3Fe-4S] aconitase, and [4Fe-4S] ferrodoxin, respectively; (B) phytochelatins (Winge *et al.*, 1992); (C) plant MTs: (a) Zn-QsMT and (b) Cd-QsMT (Domenech *et al.*, 2007b).

The presence of S^{2-} ions was hypothesised on the basis of the unusually low stoichiometry of Cd-MT complexes determined by ICP-AES in comparison with the values expected from the corresponding Zn-MT complexes and the number of Cys residues available for metal coordination. By using acid ICP-AES, that involves acidification of the sample prior to the conventional method to favour loss of acid-labile ligands such as H_2S, the protein concentration was correctly determined and new M/MT ratios consistent with the expected stoichiometries were obtained. In addition to CD and ESI-MS measurements, the S^{2-} quantity was finally quantified by GC-FPD, considered as the most suitable methodology for the sulfide direct detection at low concentrations (Capdevila *et al.*, 2005).

The acid-labile sulfide have been found in nearly all the recombinat Zn(II)-MT and Cd(II)-MT complexes (Capdevila *et al.*, 2005), thus they are present in species formed *in vivo*, that is, in a physiological, although heterologous, environment. The amount of S^{2-}

depends on the MT and the coordinated metal, but generally (oppositely to what described for phytochelatins) its presence does not increase the chelating potential of the MT, as the divalent metal content of the aggregates remains the same in spite of the sulfide content is usually higher in Cd(II)-containing aggregates than in Zn(II)-containing ones. As reported above, the recombinantly synthesized MTs contain variable amounts of sulfide ions (Figure 4) (Capdevila *et al.*, 2005; Egli *et al.*, 2006; Tio *et al.*, 2006; Domenech *et al.*, 2007a; Pagani *et al.*, 2007), determined by the protocol reported in Capdevila *et al.*, 2005.

Evidences of the participation of sulfide ions to metal coordination have been found also by the analysis of the Raman spectra of the M-MT aggregates in the 400-440 cm^{-1} region (νS-M), although the presence of both bridging sulfide anions and bridging Cys did not allowed the definitive assignments of the band components. A broad band, visible at about 417 cm^{-1} in the Raman spectra of Cd(II)- and Zn(II)-QsMT (Figure 8), was assigned to the M-S_b-M bond vibrations (S_b standing for bridging sulfur, and M for metal) on the basis of the spectral similarities with the 410-420 cm^{-1} bands of ferredoxins. This assignment was in well agreement also with the different sulfide content revealed in the two samples (Figure 4): the higher band intensity visible in the Cd-QsMT spectrum would be consistent with a more important participation of the sulfur bridging ligands in the former, in accordance with the analytical data reported in Figure 4. On the basis of other spectral similarities of sulfur-metal band in the 250-370cm^{-1} region of M-QsMT aggregates and ferredoxins, the formation of different metal cluster geometries in the two metallated QsMT has been proposed (Domenech *et al.*, 2007b): a cubane-type cluster could be formed in the presence of Cd(II) ions and a binuclear centre could be present in Zn-QsMT (Figure 11C).

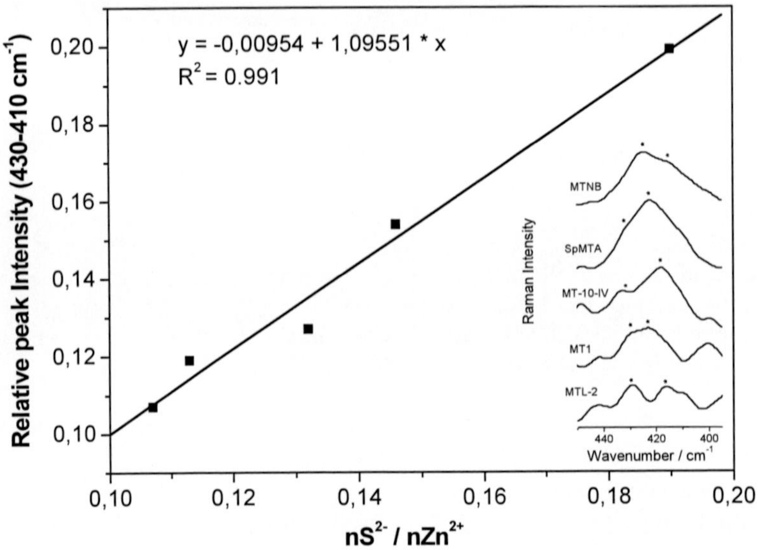

Figure 12. Linear correlation between the sulfide/zinc molar ratio and the relative intensity of the Raman bands in the 410-430 cm^{-1} region. The values reported in the plot are obtained for the following Zn-MT aggregates (increasing order): Zn-MTL-2, Zn-MT1, Zn-MT-10-IV, Zn-SpMTA, Zn-MTNB. Inset of the figure: Raman spectral region of the five Zn-MT complexes where the metal-S^{2-} stretching vibrations give a relevant contribution.

As regards the other five MTs reviewed in this chapter, by reporting the relative peak intensity of the two main bands in the 415-430 cm^{-1} range as a fuction of the sulfide/zinc molar ratio, a very good linear correlation was found (Figure 12) (Domenech Ph.D Thesis 2007). The linear relationship between the Raman intensity of the two bands and the relative sulfide content clearly indicates the relevant contribution of the metal-S^{2-} stretching vibrations to these bands. Thus, this Raman region could be considered a marker of sulfur-atom bridging ligands.

As regards CD spectroscopy, also this technique can give information on the participation of sulfide to the metallic clusters. For example, in the case of Cd-QsMT a strong positive-gaussian signal is observed at 275 nm (Figure 9). Such signal does not correspond to $Cd(SCys)_4$ chromophores, absorbing at approximately 250 nm, but can be attributed to sulfide ions involved in metal coordination, on the basis of similar 260-280 nm transitions observed in the CD spectra of Cd-phytochelatins containing sulfide anions. The study of other Cd-MT forms has shown that a high diversity of CD shapes can be obtained from Cd-sulfide-containing forms, probably related to structural differences (Capdevila *et al.*, 2005). As regards Zn-MT aggregates, no clear evidences on sulfide detection by CD measurements have been published, in spite of the low-intensity absorptions that have been detected in the CD spectra of some sulfide-containing Zn-MT aggregates and tentatively attributed to sulfide-zinc signals (Domenech *et al.*, 2006; Domenech, 2007).

Chlorides

It is now commonly accepted that non-proteic ligands contribute to the structure and stability of M-MT species, although this contribution may differ substantially depending on the MT and the metal ions involved. Since chloride (Cl^-) has been related to ATP-MT1 interaction (Maret *et al.*, 2002), the capacity of a MT to establish this association, or not, may be of crucial biological relevance. NMR studies on mammalian metallothionein MT2 showed that Cl^- ions are able to participate to metal binding (Maret *et al.*, 2002). This evidence has been later confirmed also by CD studies on MTs from three different families: *Drosophila* Zn-MTNB, mollusk Zn-MT-10-IV and mammalian Cd-MT1 (Domenech *et al.*, 2003; Villarreal *et al.*, 2005; Domenech Ph.D Thesis 2007). In these aggregates, particular CD absorptions partially due to Cl^- were observed: in Zn-containing MT aggregates, the signal due to the Cl^--M interaction appeared near to 230 nm, while it appeared at 240 nm in Cd-containing MT aggregates. However, these absorptions resulted not to be always indicative of the Cl^- participation, and strictly, the Cl^- ions were not detectable by mass spectrometry analysis (Domenech *et al.*, 2003; Villarreal *et al.*, 2005). In the case of Zn-MTL-2, for example, a positive signal was visible at $\approx$ 230 nm (Figure 9), but it could be due both to the Cl^--M interaction or to the positive edge of the negative exciton-coupling centered at 235 nm, attributed to His participation in metal-coordination.

Raman spectroscopy can be an helpful tool also for evaluating the capacity of a MT to establish the association with Cl^- ions. In fact, the Zn-Cl^- stretching vibration gives rise to a band at about 290 cm^{-1}. This band was clearly visible in two of the six Zn-MT complexes reviewed, Zn-MTNB and Zn-MT-10-IV, thus supporting the participation of this non-proteic ligand to the stabilisation of their structure. This result was further confirmed by the CD data, also suggesting the chloride participation in these metal complexes (Domenech, 2007).

4. The Growing Field of the Functional Properties of MTs

Metal-ligand interactions are critical components of metalloprotein assembly, folding, stability, electrochemistry, and catalytic function. In fact, metalloprotein engineering involves controlling the delicate interplay between the forces involved in protein folding and the geometric and electronic requirements of the bound metal ion.

As explained above, the metal-binding preferences and functional properties of MTs depend on the composition and structure of the metallic centres. Of the naturally encoded amino acids used to bind metal ions, Cys residues are the main ligands involved in the metal coordination in MTs. Nature employs thiolate-rich metalloprotein active sites from archaea to higher organisms. This observed ubiquity within biological systems emphasizes the operational importance of cysteine-rich metalloprotein active sites, which perform functions ranging from gene expression to enzymatic catalysis. Perhaps the most straightforward function of a thiolate-rich metal-ion active site is to utilize the metal ligand binding thermodynamics to structurally stabilize the protein fold.

The diversity of Cys disposition patterns along the MT sequence would account for the observed differences in the functional properties. However, it has been stated that MTs with identical Cys-patterns present diverse functional properties. This is the case, for example, for MT1 and MT4 of *M.musculus* from the vertebrate MT family (Tio *et al.*, 2004), Cd-MT and Cu-MT from the molluscan *H.pomatia* (family 2) (Winge and Miklossy, 1982) or MTNB, MTNC and MTND from *D. melanogaster* metallothionein system (Egli *et al.*, 2006). This argues in favour of a crucial participation of non-Cys aminoacids in the determination of the functional properties.

In this chapter we have reviewed some elements that can play active roles in the determination of the functional properties of an MT, i.e. the presence of the secondary structure elements and the participation of non-sulfur-containing amino acids in metal coordination.

As regards the secondary structure, the few literature data indicate in many cases the importance of secondary structure elements at several levels for MT functionality(Wagner *et al.*, 1987; Riek *et al.*, 1999; Bertini *et al.*, 2000; Blindauer *et al.*, 2001; Oz *et al.*, 2001; Munoz *et al.*, 2002; Scudiero *et al.*, 2005;Domenech *et al.*, 2007b). In prokaryote SmtA and mammalian MT3 metallothioneins, the MT regions presenting secondary structure elements play key roles in the physiological functionality (Cook *et al.*, 1998; Blindauer *et al.*, 2001; Oz *et al.*, 2001; Wang *et al.*, 2006). Additionally, it has been recently demonstrated that in plant QsMT metallothionein, the loss of a Cys-devoid region - with an attributed beta-sheet conformation (Domenech *et al.*, 2007b) - implies a decrease of its *in vivo* Cd- and Cu-detoxification abilities (Domenech *et al.*, 2006; Domenech *et al.*, 2007a).

As far as the metal ligands are concerned, the His participation in M-MT aggregates had been demonstrated years ago, but it was considered rare in MTs. In this chapter we have underlined that this participation is not restricted to one MT family and can be clearly detected by some methodologies.

Carboxylate ligands are usually involved in Fe coordination, but they are quite unusual in Zn or Cd coordination. However, the Raman data have indicated a participation of such ligands in metal stabilization in some of the examined M-MTs. As most of MT functional studies and classifications are centred in the Cys patterns, the abundance of Glu and Asp residues makes necessary to take into account also the possible participation of their COO^- groups as ligands. In fact, Asp/Glu side chains have been assigned various roles in metal binding and selectivity of metalloproteins, based mainly on their charge rather than on their identity. In example, due to the negative charge of the carboxylate group, interactions with the metal cation in a buried protein cavity are not only thermodynamically favorable but also generally more favorable than those of other neutral ligands; thus, Asp/Glu of metalloproteins are thought to be mainly responsible for sequestering the metal cation from physiological fluids.

Both secondary structure elements and participation of non-sulfur-containg residues in metal coordination should be elements to take into account to describe factors controlling the metal-binding behaviour of MT.

Moreover, the eventual participation of sulfide and chloride ions expands the functional possibilities of MTs. In particular, the recovery of sulfide-containing complexes has striking theoretical implications since S^{2-} ion is an universal cell component and its presence can be related with physiological events also proposed for MT function candidates (i.e. neurotransmission, neuromodulation) (Kimura, 2000; Lowicka and Beltowski, 2007). On the other hand, sulfide ions can be also toxic (Milby and Baselt, 1999). By considering that the general ability of MTs to bond sulfide ions in ferredoxin or phytochelatin-like structures has just been demonstrated (Capdevila *et al.*, 2005; Domenech, 2007; Domenech *et al.*, 2007b), the ability of MTs to perform a protective sulfide-transporter role for MTs could be postulated.

New data on native MT forms could shed light into the physiological significance of sulfides in MTs. Moreover, the description of such properties can be of high interest for the structural, biomedical, and biotechnological applications of MTs.

In a general view, the study of MT families different from the well-known vertebrate MT is a key point to achieve a better comprehension of the functional and biological role of MTs.

5. Conclusion

Metallobiomolecules are highly elaborated coordination complexes, and their fundamental metal-ligand interactions are critical components of metalloprotein folding, assembly, stability, electrochemistry, and catalytic function.

Herein, we have decribed the benefits in using some traditional coordination chemistry methods to define the metal-ion binding properties of MTs toward metal ions such as Zn(II). In particular, the coupling of CD and vibrational spectroscopy constitutes a very informative experimental strategy for the analysis of *in vivo*-synthesized M-MT aggregates. In fact, the vibrational techniques can shed light on secondary structures eventually present in MTs and the ligands involved in metal coordination. The oxidation state of Cys residues and their participation in the metal chelation can be clearly defined, as well as the eventual involvement of His residues and carboxylate groups. As regards exogenous metal ligands, Raman spectroscopy in particular allows to identify vibrational bands which can be considered markers of sulphide bridging ligands, since their intensity is linearly correlated with sulphide/metal molar ratio.

The synthesis of M-MTs aggregates in an *in vivo* environment is another important step in improving the knowledge on MT structure since it allows to obtain representative models of the biological systems in purity and amount enough for carrying on spectroscopic structural studies. Thus, the structural study of M-MT aggregates *in vivo*-synthesized instead of metal-reconstituted ones seems to be important to achieve physiologically-representative results.

In conclusion, many advantages, such as the possibility of investigating aggregates synthesized *in vivo* without limitations in the coordinated metal, the capability of detecting several structural features at the same time, and the requirement of low sample amount, make to propose the coupling of analytical and spectroscopical techniques as one of the most promising experimental strategies in the research on new hints on MTs.

Acknowledgements

We thank Professor Giancarlo Fini for the critical reading of this chapter and the useful suggestions.

References

Adams DM, Cornell JB (1968) Metal sulphur vibrations. Part II. Dithiolate and dithiophosphate complexes and correlations for g(M-S). *J. Chem. Soc.*(A):1299-1303.

Adams MW (1990) The structure and mechanism of iron-hydrogenases. *Biochim. Biophys. Acta* 1020:115-145.

Akhtar W, Edwards HGM (1997) Fourier-transform Raman spectroscopy of mammalian and avian keratotic biopolymers. *Spectrochim. Acta* [A]. 53:81-90.

Alix AJP, Pedanou G (1994) In: *Proceeding of the Fourteenth International Conference on Raman Spectroscopy*, 22-24 August 1994, Hong Kong. (Yu NT, Li X, eds), p 122. Chichester: Wiley.

Alix AJP, Pedanou G, Berijot M (1988) Fast Determination of the Quantitative Secondary Structure of Proteins by using Some Parameters of the Raman Amide I Band. *J. Mol.Struct.* 174:159-164.

Amiard JC, Amiard-Triquet C, Barka S, Pellerin J, Rainbow PS (2006) Metallothioneins in aquatic invertebrates: their role in metal detoxification and their use as biomarkers. *Aquat. Toxicol.* 76:160-202.

Armitage IM, Otvos JD, Briggs RW, Boulanger Y (1982) Structure elucidation of the metal-binding sites in metallothionein by 113Cd NMR. *Fed. Proc.* 41:2974-2980.

Arp Z, Autrey D, Laane J, Overman SA, Tomas GJ (2001) Tyrosine Raman signatures of filamentous Virus Ff are diagnostic of non-hydrogen-bonded phenoxyls: demonstration by Raman and Infrared spectroscopy of p-cresol vapor. *Biochemistry* 40:2522-2529.

Arrondo JL, Muga A, Castresana J, Goni FM (1993) Quantitative studies of the structure of proteins in solution by Fourier-transform infrared spectroscopy. *Prog. Biophys. Mol. Biol.* 59:23-56.

Arseniev A, Schultze P, Worgotter E, Braun W, Wagner G, Vasak M, Kagi JH, Wuthrich K (1988) Three-dimensional structure of rabbit liver [Cd7]metallothionein-2a in aqueous solution determined by nuclear magnetic resonance. *J. Mol. Biol.* 201:637-657.

Ashikawa I, Itoh K (1979) Raman spectra of polypeptides containing L-histidine residues and tautomerism of imidazole side chain. *Biopolymers* 18:1859-1876.

Auld DS (2001) Zinc coordination sphere in biochemical zinc sites. *Biometals* 14:271-313.

Benevides JM, Juuti JT, Tuma R, Bamford DH, Thomas GJ, Jr. (2002) Characterization of subunit-specific interactions in a double-stranded RNA virus: Raman difference spectroscopy of the phi6 procapsid. *Biochemistry* 41:11946-11953.

Berjot M, Marx J, Alix AJP (1987) Determination of the secondary structure of proteins from the Raman amide I band: The reference intensity profiles method. *J. Raman Spectrosc.* 18:289-300.

Bertin G, Averbeck D (2006) Cadmium: cellular effects, modifications of biomolecules, modulation of DNA repair and genotoxic consequences (a review). *Biochimie* 88:1549-1559.

Bertini I, Hartmann HJ, Klein T, Liu G, Luchinat C, Weser U (2000) High resolution solution structure of the protein part of Cu7 metallothionein. *Eur. J. Biochem.* 267:1008-1018.

Bilecen K, Ozturk UH, Duru AD, Sutlu T, Petoukhov MV, Svergun DI, Koch MH, Sezerman UO, Cakmak I, Sayers Z (2005) *Triticum durum* metallothionein. Isolation of the gene and structural characterization of the protein using solution scattering and molecular modeling. *J. Biol. Chem.* 280:13701-13711.

Binz P-A (2007) *Metallothioneins: classification and list of entries.* In: http://expasy.org/cgi-bin/lists?metallo.txt.

Binz AP, Kägi JHR (1999) Metallothionein: Molecular evolution and classification. In: *Metallothionein IV* (Klaassen CD, ed), pp 7-14. Basel: Birkhäuser.

Binz PA, Kägi, J.H.R. (2001) http://www.biochem.unizh.ch/mtpage/MT.html. In.

Blindauer CA, Harrison MD, Parkinson JA, Robinson AK, Cavet JS, Robinson NJ, Sadler PJ (2001) A metallothionein containing a zinc finger within a four-metal cluster protects a bacterium from zinc toxicity. *Proc. Natl. Acad. Sci. USA* 98:9593-9598.

Boldyrev AI, Simons J (1997) Ab initio study of the bonding of zinc atoms to first- and second-row main group atoms. *Molecular Physics* 92:365-379.

Bongers J, Walton CD, Richardson DE, Bell JU (1988) Micromolar protein concentrations and metalloprotein stoichiometries obtained by inductively coupled plasma atomic emission spectrometric determination of sulfur. *Anal. Chem.* 60:2683-2686.

Boulanger Y, Goodman CM, Forte CP, Fesik SW, Armitage IM (1983) Model for mammalian metallothionein structure. *Proc. Natl. Acad. Sci. USA* 80:1501-1505.

Braun W, Vasak M, Robbins AH, Stout CD, Wagner G, Kagi JH, Wuthrich K (1992) Comparison of the NMR solution structure and the x-ray crystal structure of rat metallothionein-2. *Proc. Natl. Acad. Sci. USA* 89:10124-10128.

Broderick JB (2007) Assembling iron-sulfur clusters in the cytosol. *Nature Chemical Biology* 3:243-244.

Broderich JB, Henshaw TF, Wojtuszewski K, Smith SR, Trojan MR, McGhan R, Kopf A, Kibbey M, Broderick WE (2000) Pyruvate formate-lyase-activating enzyme: strictly anaerobic isolation yelds active enzyme containing a $[3Fe\text{-}4S]^{+}$ cluster. *Biochem. Biophys. Res. Comm.* 269:451-456.

Byler DM, Susi H (1986) Examination of the secondary structure of proteins by deconvolved FTIR spectra. *Biopolymers* 25:469-487.

Cai S, Singh BR (1999) Identification of beta-turn and random coil amide III infrared bands for secondary structure estimation of proteins. *Biophys. Chem.* 80:7-20.

Calderone V, Dolderer B, Hartmann HJ, Echner H, Luchinat C, Del Bianco C, Mangani S, Weser U (2005) The crystal structure of yeast copper thionein: the solution of a long-lasting enigma. *Proc. Natl. Acad. Sci. USA* 102:51-56.

Callender R, Deng H (1994) Nonresonance Raman difference spectroscopy: a general probe of protein structure, ligand binding, enzymatic catalysis, and the structures of other biomacromolecules. *Annu. Rev. Biophys. Biomol. Struct.* 23:215-245.

Capdevila M, Cols N, Romero-Isart N, Gonzalez-Duarte R, Atrian S, Gonzalez-Duarte P (1997) Recombinant synthesis of mouse Zn3-beta and Zn4-alpha metallothionein 1 domains and characterization of their cadmium(II) binding capacity. *Cell Mol. Life Sci.*53:681-688.

Capdevila M, Domenech J, Pagani A, Tio L, Villarreal L, Atrian S (2005) Zn- and Cd-metallothionein recombinant species from the most diverse phyla may contain sulphide; (S^{2-}) ligands. *Angew. Chem. Int. Ed. Engl.* 44:4618-4622.

Carey PR (1982) *Biochemical applications of Raman and Resonance Raman Spectroscopies*. New York: Academic Press.

Clemens S (2006) Evolution and function of phytochelatin synthases. *J. Plant Physiol.* 163:319-332.

Cobine PA, McKay RT, Zangger K, Dameron CT, Armitage IM (2004) Solution structure of Cu6 metallothionein from the fungus *Neurospora crassa*. *Eur. J. Biochem.* 271:4213-4221.

Cols N, Romero-Isart N, Capdevila M, Oliva B, Gonzalez-Duarte P, Gonzalez-Duarte R, Atrian S (1997) Binding of excess cadmium(II) to Cd7-metallothionein from recombinant mouse Zn7-metallothionein 1. UV-VIS absorption and circular dichroism studies and theoretical location approach by surface accessibility analysis. *J. Inorg. Biochem.* 68:157-166.

Cook WJ, Kar SR, Taylor KB, Hall LM (1998) Crystal structure of the cyanobacterial metallothionein repressor SmtB: a model for metalloregulatory proteins. *J. Mol. Biol.* 275:337-346.

Degtyarenko K (2000) Bioinorganic motifs: towards functional classification of metalloproteins. *Bioinformatics* 16:851-864.

Dickerson RE, Gei I. (1983) *Hemoglobin: Benjamin/Cummings*, Menlo Park, California.

Domenech J (2007) PhD Thesis: Study of the structure-function relationship in Metallothioneins of Invertebrates, Plants and Protozoans. In: *Genetics*. Universitat de Barcelona.

Domenech J, Palacios O, Villarreal L, Gonzalez-Duarte P, Capdevila M, Atrian S (2003) MTO: the second member of a *Drosophila* dual copper-thionein system. *FEBS Lett.* 533:72-78.

Domenech J, Mir G, Huguet G, Capdevila M, Molinas M, Atrian S (2006) Plant metallothionein domains: functional insight into physiological metal binding and protein folding. *Biochimie* 88:583-593.

Domenech J, Orihuela R, Mir G, Molinas M, Atrian S, Capdevila M (2007a) The Cd-binding abilities of recombinant *Quercus suber* metallothionein, QsMT: bridging the gap between phytochelatins and metallothioneins. *J. Biol. Inorg. Chem.* 12:867-82.

Domenech J, Tinti A, Capdevila M, Atrian S, Torreggiani A (2007b) FT-Raman and FT-IR approach to the structure of the Zn- and Cd-complexes of a type 2 plant (*Quercus suber*) metallothionein. *Biopolymers* 86:240-248.

Egli D, Domenech J, Selvaraj A, Balamurugan K, Hua H, Capdevila M, Georgiev O, Schaffer W, Atrian S (2006) The four members of the *Drosophila* Metallothionein family exhibit distinct yet overlapping in heavy metal homeostasis and detoxification. *Genes to Cells* 11:647-658.

Fagnano C, Fini G, Torreggiani A (1995) Raman spectroscopic study of the avidin-biotin complex. *J. Raman Spectrosc.* 26:991-995.

Fowler BA, Hildebrand CE, Kojima Y, Webb M (1987) Nomenclature of metallothionein. Experientia Suppl 52:19-22.

Furey WF, Robbins AH, Clancy LL, Winge DR, Wang BC, Stout CD (1987) Crystal structure of Cd, Zn metallothionein. *Experientia Suppl.* 52:139-148.

González-Duarte P (2003) In: *Comprehensive Coord. Chem II* (McCleverty JA, Meyer TJ, eds), vol 8, pp 213-228. Amsterdam: Elsevier-Pergamon.

Hall JL (2002) Cellular mechanisms for heavy metal detoxification and tolerance. *J. Exp. Bot.* 53:1-11.

Han S, Czernuszewicz RR, Kimura T, Adams MWW, Spiro TG (1989) Fe_2S_2 Protein Resonance Raman spectra revisited: structural variation among Adrenodoxin, Ferredoxin, and red paramagnetic protein. *J. Am. Chem. Soc.* 111:3505-3511.

Haq F, Mahoney M, Koropatnick J (2003) Signaling events for metallothionein induction. *Mutat. Res.* 533:211-226.

Hasler DW, Faller P, Vasak M (1998) Metal-thiolate clusters in the C-terminal domain of human neuronal growth inhibitory factor (GIF). *Biochemistry* 37:14966-14973.

Hayashi Y, Nakagawa CW, Murasugi A (1986) Unique properties of Cd-binding peptides induced in fission yeast, *Schizosaccharomyces pombe*. *Environ. Health Perspect.* 65:13-19.

Hollosi M, Majer Z, Ronai AZ, Magyar A, Medzihradszky K, Holly S, Perczel A, Fasman GD (1994) CD and Fourier transform ir spectroscopic studies of peptides. II. Detection of beta-turns in linear peptides. *Biopolymers* 34:177-185.

Holm RH, Kennepohl P, Solomon E I. (1996) Structural and Functional Aspects of Metal Sites in Biology, *Chem. Rev.* 96: 2239-2314

Imagawa M, Onozawa T, Okumura K, Osada S, Nishihara T, Kondo M (1990) Characterization of metallothionein cDNAs induced by cadmium in the nematode *Caenorhabditis elegans*. *Biochem. J.* 268:237-240.

Jacob C, Maret W, Vallee BL (1998) Control of zinc transfer between thionein, metallothionein, and zinc proteins. *Proc. Natl. Acad. Sci. USA* 95:3489-3494.

Jensen JO (2003) Vibrational frequencies and structural determination of Al_4S_6. *J. Mol. Struct. (Theochem.)* 664-665:37-45.

Johnson DC, Dean DR, Smith AD, Johnson MK (2005) Structure, function, and formation of biological iron-sulfur clusters. *Annu. Rev. Biochem.* 74:247-281.

Kägi JHR (1993) Evolution, structure and chemical activity of class I metallothioneins: An overview. In: *Metallothionein III, Biological Roles and Implications*. (Suzuki KT INaKM, ed), pp 29-55: Birkhauser Verlag, Basel.

Kille P, Winge DR, Harwood JL, Kay J (1991) A plant metallothionein produced in *E. coli*. *FEBS Lett*. 295:171-175.

Kimura H (2000) Hydrogen sulfide induces cyclic AMP and modulates the NMDA receptor. *Biochem. Biophys. Res. Commun*. 267:129-133.

Kulkarni PP, She YM, Smith SD, Roberts EA, Sarkar B (2006) Proteomics of metal transport and metal-associated diseases. *Chemistry* 12:2410-2422.

Kumari MV, Hiramatsu M, Ebadi M (1998) Free radical scavenging actions of metallothionein isoforms I and II. *Free Radic. Res*. 29:93-101.

Leszczyszyn OI, Schmid R, Blindauer CA (2007) Toward a property/function relationship for metallothioneins: Histidine coordination and unusual cluster composition in a zinc-metallothionein from plants. *Proteins* 68:922-35.

Lover T, Bowmaker GA, Seakins JM, Cooney RP (1997) Vibrational spectroscopic study of thiophenolate-capped nanoclusters of CdS and of Cadmium tiophenolate complexes. *Chem. Mater*. 9:967-975.

Lowicka E, Beltowski J (2007) Hydrogen sulfide (H_2S) - the third gas of interest for pharmacologists. *Pharmacol. Rep*. 59:4-24.

Mackay EA, Overnell J, Dunbar B, Davidson I, Hunziker PE, Kagi JH, Fothergill JE (1993) Complete amino acid sequences of five dimeric and four monomeric forms of metallothionein from the edible mussel *Mytilus edulis*. *Eur. J. Biochem*. 218:183-194.

Maddams W SM (1982) The measurement of derivative IR spectra. III: the effect of band width and band shape on resolution enhancement by derivative spectroscopy. *Spectrochim. Acta* [A] 38:459-466.

Maret W (2003) Cellular zinc and redox states converge in the metallothionein/thionein pair. *J. Nutr*. 133:1460S-1462S.

Maret W (2004) Zinc and sulfur: a critical biological partnership. *Biochemistry* 43:3301-3309.

Maret W (2005) Zinc coordination environments in proteins determine zinc functions. *J. Trace Elem. Med. Biol.* 19:7-12.

Maret W (2006) Zinc coordination environments in proteins as redox sensors and signal transducers. *Antioxid. Redox Signal.* 8:1419-1441.

Maret W, Larsen KS, Vallee BL (1997) Coordination dynamics of biological zinc "clusters" in metallothioneins and in the DNA-binding domain of the transcription factor Gal4. *Proc. Natl. Acad. Sci. USA* 94:2233-2237.

Maret W, Heffron G, Hill HA, Djuricic D, Jiang LJ, Vallee BL (2002) The ATP/metallothionein interaction: NMR and STM. *Biochemistry* 41:1689-1694.

Meneghini R (1997) Iron homeostasis, oxidative stress, and DNA damage. *Free Radical. Biol. Med.* 23:783-792.

Milby TH, Baselt RC (1999) Hydrogen sulfide poisoning: clarification of some controversial issues. *Am. J. Ind. Med.* 35:192-195.

Miura T, Thomas GJJ (1995) Proteins: Structure, Function and Engineering. In: *In Subcellular Biochemistry.* (Biswas BB, Roy S, eds), p 55. New York: Plenum Press.

Miura T, Satoh T, Takeuchi H (1998) Role of metal-ligand coordination in the folding pathway of zinc finger peptides. *Biochim. Biophys. Acta* 1384:171-179.

Mokdad R, Debec A, Wegnez M (1987) Metallothionein genes in *Drosophila melanogaster* constitute a dual system. *Proc. Natl. Acad. Sci. USA* 84:2658-2662.

Munoz A, Forsterling FH, Shaw CF, 3rd, Petering DH (2002) Structure of the (113)Cd(3)beta domains from *Homarus americanus* metallothionein-1: hydrogen bonding and solvent accessibility of sulfur atoms. *J. Biol. Inorg. Chem.* 7:713-724.

Nakamura K, Era S, Ozaki Y, Sogami M, Hayashi T, Murakami M (1997) Conformational changes in seventeen cystine disulfide bridges of bovine serum albumin proved by Raman spectroscopy. *FEBS Lett.* 417:375-378.

Narula SS, Brouwer M, Hua Y, Armitage IM (1995) Three-dimensional solution structure of *Callinectes sapidus* metallothionein-1 determined by homonuclear and heteronuclear magnetic resonance spectroscopy. *Biochemistry* 34:620-631.

Nogami N, Sugeta H, Miyazawa T, (1975) C-S stretching vibrations and molecular conformations of methyl propyl sulphide and related alkyl sulfides. *Chem. Lett:*147.

Oz G, Zangger K, Armitage IM (2001) Three-dimensional structure and dynamics of a brain specific growth inhibitory factor: metallothionein-3. *Biochemistry* 40:11433-11441.

Pagani A, Villarreal L, Capdevila M, Atrian S (2007) The *Saccharomyces cerevisiae* Crs5 Metallothionein metal-binding abilities and its role in the response to zinc overload. *Mol. Microbiol.* 63:256-269.

Pande J, Pande C, Gilg D, Vasak M, Callender R, Kagi JH (1986) Raman, infrared, and circular dichroism spectroscopic studies on metallothionein: a predominantly "turn"-containing protein. *Biochemistry* 25:5526-5532.

Parker FS (1983) *Application of Infrared, Raman and Resonance Raman Spectroscopy* in *Biochemistry*. New York: Plenum Press.

Pelton JT, McLean LR (2000) Spectroscopic methods for analysis of protein secondary structure. *Anal. Biochem.* 277:167-176.

Peters JW, Lanzilotta WN, Lemon BJ, Seefeldt LC (1998) X-ray crystal structure of the Fe-only hydrogenase (CpI) from *Clostridium pasteurianum* to 1.8 angstrom resolution. *Science* 282:1853-1858.

Petering DH, Krezoski S, Chen P, Pattanaik A, Shaw CF (1992) In *Metallothioneins*; Stillman M J, Shaw C F, III, Suzuki K T, Eds; New York: VCH Publishers, Chapter 7.

Pierrel F, Cobine PA, Winge DR (2007) Metal Ion availability in mitochondria. *Biometals* 20:675-82.

Qing H, Baolin Z, Fenling S, Zuyi T (1995) Raman and ATR Infrared studies of the conformation of metallothionein in solution. *Spectrosc. Lett.* 28:829-838.

Quail MA, Jordan P, Grogan JM, Butt JN, Lutz M, Thomson AJ, Andrews SC, Guest JR (1996) Spectroscopic and voltammetric characterization of the bacterioferritin-associated ferredoxin of *Escherichia coli*. *Biochem. Biophys. Res. Comm.* 229:635-642.

Rardin RL, Tolman WB, Lippard S (1991) *New J. Chem.*, 15, 417-430.

Raso SW, Clark PL, Haase-Pettingell C, King J, Thomas GJ (2001) Distinct cysteine sulfhydryl environments detected by analysis of Raman S-H markers of Cys-Ser mutant proteins. *J. Mol. Biol.* 307:899-911.

Riek R, Precheur B, Wang Y, Mackay EA, Wider G, Guntert P, Liu A, Kagi JH, Wuthrich K (1999) NMR structure of the sea urchin (*Strongylocentrotus purpuratus*) metallothionein MTA. *J. Mol. Biol.* 291:417-428.

Rigby KE, Chan J, Mackie J, Stillman MJ (2006) Molecular dynamics study on the folding and metallation of the individual domains of metallothionein. *Proteins* 62:159-172.

Rigby KE, Stillman MJ (2004) Structural studies of metal-free metallothionein. *Biochem. Biophys. Res. Commun.* 325:1271-1278.

Robbins AH, McRee DE, Williamson M, Collett SA, Xuong NH, Furey WF, Wang BC, Stout CD (1991) Refined crystal structure of Cd, Zn metallothionein at 2.0 A resolution. *J. Mol. Biol.* 221:1269-1293.

Robbins AH, Stout CD (1992) Crystal structure of metallothioneins. In: *Metallothioneins* (Stillman MJ, Shaw CF, Suzuki KT, eds), pp 31-54. New York: VCH Publishers.

Romero-Isart N, Cols N, Termansen MK, Gelpi JL, Gonzalez-Duarte R, Atrian S, Capdevila M, Gonzalez-Duarte P (1999) Replacement of terminal cysteine with histidine in the metallothionein alpha and beta domains maintains its binding capacity. *Eur. J. Biochem.* 259:519-527.

Rotsaert FJ, Pikus JD, Fox BG, Markley JL, Sanders-Loehr J (2003) N-isotope effects on the Raman spectra of Fe(2)S(2) ferredoxin and Rieske ferredoxin: evidence for structural rigidity of metal sites. *J. Biol. Inorg. Chem.* 8:318-326.

Rouault TA (2006) The role of iron regulatory proteins in mammalian iron homeostasis and disease. *Nat. Chem. Biol.* 2:406-414.

Rupp H, Weser U (1978) Circular dichroism of metallothioneins. A structural approach. *Biochim. Biophys. Acta* 533:209-226.

Sane SU, Cramer SN, Przybycien TM (1999) A holistic approach to protein secondary structure characterization using amide I band Raman spectroscopy. *Anal. Biochem.* 269:255-272.

Sarkar B (1987) Metal protein interactions. *Prog. Food Nutr. Sci.* 11:363-400.

Scudiero R, Temussi PA, Parisi E (2005) Fish and mammalian metallothioneins: a comparative study. *Gene* 345:21-26.

Shi YB, Fang JL, Liu XY, Du L, Tang WX (2002) Fourier transform IR and Fourier transform Raman spectroscopy studies of metallothionein-III: amide I band assignments and secondary structural comparison with metallothioneins-I and -II. *Biopolymers* 65:81-88.

Siamwiza MN, Lord RC, Chen MC, Takamatsu T, Harada I, Matsuura H, Shimanouchi T (1975) Interpretation of the doublet at 850 and 830 cm^{-1} in the Raman spectra of tyrosyl residues in proteins and certain model compounds. *Biochemistry* 14:4870-4876.

Slice LW, Freedman JH, Rubin CS (1990) Purification, characterization, and cDNA cloning of a novel metallothionein-like, cadmium-binding protein from *Caenorhabditis elegans*. *J. Biol. Chem.* 265:256-263.

Spassov S, Beekes M, Naumann D (2006) Structural differences between TSEs strains investigated by FT-IR spectroscopy. *Biochim. Biophys. Acta* 1760:1138-1149.

Stillman MJ, Shaw CF, Suzuki KT (1991) Metallothioneins-Synthesis, Structure and Properties of Metallothioneins, Phytochelatins and Metal-Thiolate Complexes. New York: VCH Publishers.

Stillman MJ, Shaw CF, Suzuki KT (eds. 1992) Metallothioneins. New York: VCH Publishers.

Takeuchi H (2003) Raman structural markers of tryptophan and histidine side chains in proteins. *Biopolymers* 72:305-317.

Thomas GJ, Jr. (1999) Raman spectroscopy of protein and nucleic acid assemblies. *Annu Rev. Biophys. Biomol. Struct.* 28:1-27.

Tio L, Villarreal L, Atrian S, Capdevila M (2004) Functional differentiation in the mammalian metallothionein gene family: metal binding features of mouse MT4 and comparison with its paralog MT1. *J. Biol. Chem.* 279:24403-24413.

Tio L, Villarreal L, Atrian S, Capdevila M (2006) The Zn- and Cd-clusters of recombinant mammalian MT1 and MT4 metallothionein domains include sulfide ligands. *Exp. Biol. Med. (Maywood)* 231:1522-1527.

Torreggiani A, Fini G (1998) The binding of biotin analogues by streptavidin: a Raman spectroscopic study. *Biospectroscopy* 4:197-208.

Torreggiani A, Fini G (1999) Drug-antiserum molecular interactions: a Raman spectroscopic study. *J. Raman Spectrosc.* 30:295-300.

Torreggiani A, Bonora S, Fini G (2000a) Raman and IR spectroscopic investigation of zinc(II)-carnosine complexes. *Biopolymers* 57:352-364.

Torreggiani A, Tamba M, Bonora S, Fini G (2000b) Binding of Cu (II) to carnosine: a Raman and IR spectroscopic study. *Biopolymers* 57:149-159.

Torreggiani A, Tamba M, Bonora S, Fini G (2003) Raman and IR study on copper binding of histamine. *Biopolymers* 72:290-298.

Torreggiani A, Tamba M, Manco I, Faraone-Mennella MR, Ferreri C, Chatgilialoglu C (2005) Radiation damage of lysozyme in a biomimetic model: some insights by Raman spectroscopy. *J. Mol. Struct.* 767:744-747.

Torreggiani A, Esposti AD, Tamba M, Marconi G, Fini G (2006a) Experimental and theroretical Raman investigation on interactions of Cu(II) with histamine. *J. Raman Spectros.* 37:291-298.

Torreggiani A, Tamba M, Manco I, Faraone-Mennella MR, Ferreri C, Chatgilialoglu C (2006b) Investigation of radical-based damage of RNase A in aqueous solution and lipid vesicles. *Biopolymers* 81:39-50.

Torreggiani A, Reggiani M, Manco I, Tinti A (2007) Vibrational investigation on the copper(II) binding mode of carcinine and its pH dependence. *J. Mol. Struct.* 834-836:294-301.

Tu A (1982) *Raman Spectroscopy in Biology.* Principles and Applications. New York: Wiley-Interscience.

Tucker SL, Thornton CR, Tasker K, Jacob C, Giles G, Egan M, Talbot NJ (2004) A fungal metallothionein is required for pathogenicity of *Magnaporthe grisea. Plant Cell* 16:1575-1588.

Tuma R (2005) Raman spectroscopy of proteins: from peptides to large assemblies. *J. Raman Spectrosc* 36:307-319.

Tuma R, Tsuruta H, Benevides JM, Prevelige PE, Jr., Thomas GJ, Jr. (2001) Characterization of subunit structural changes accompanying assembly of the bacteriophage P22 procapsid. *Biochemistry* 40:665-674.

Tuma R, Thomas GJ (2002) In: *Handbook of Vibrational Spectroscopy* (Chalmers JM, Griffiths PR, eds), vol 23, p 215. Chichester: Wiley.

Vallee BL, Falchuk KH (1993) The biochemical basis of zinc physiology. *Physiol. Rev.* 73:79-118.

Valls M, Bofill R, Romero-Isart N, Gonzalez-Duarte R, Abian J, Carrascal M, Gonzalez-Duarte P, Capdevila M, Atrian S (2000) Drosophila MTN: a metazoan copper-thionein related to fungal forms. *FEBS Lett* 467:189-194.

Vasak M (1991) Metal removal and substitution in vertebrate and invertebrate metallothioneins. *Methods Enzymol* 205:452-458.

Vergani L, Grattarola M, Borghi C, Dondero F, Viarengo A (2005) Fish and molluscan metallothioneins. *Febs J.* 272:6014-6023.

Viarengo A, Burlando B, Ceratto N, Panfoli I (2000) Antioxidant role of metallothioneins: a comparative overview. *Cell Mol. Biol.* (Noisy-le-grand) 46:407-417.

Villarreal L, Tio L, Atrian S, Capdevila M (2005) Influence of chloride ligands on the structure of Zn- and Cd-metallothionein species. *Arch. Biochem. Biophys.* 435:331-335.

Villarreal L, Tio L, Capdevila M, Atrian S (2006) Comparative metal binding and genomic analysis of the avian (chicken) and mammalian metallothionein. *Febs J.* 273:523-535.

Wagner G, Frey MH, Neuhaus D, Worgotter E, Braun W, Vasak M, Kagi JH, Wuthrich K (1987) Spatial structure of rabbit liver metallothionein-2 in solution by NMR. *Experientia Suppl.* 52:149-157.

Wagner G, Neuhaus D, Worgotter E, Vasak M, Kagi JH, Wuthrich K (1986) Nuclear magnetic resonance identification of "half-turn" and 3(10)-helix secondary structure in rabbit liver metallothionein-2. *J. Mol. Biol.* 187:131-135.

Wallar BJ, Lipscomb JD. (1996) Dioxygen Activation by Enzymes Containing Binuclear Non-Heme Iron Clusters, *Chem. Rev.*, 96, 2625-2658.

Wang H, Zhang Q, Cai B, Li H, Sze KH, Huang ZX, Wu HM, Sun H (2006) Solution structure and dynamics of human metallothionein-3 (MT-3). *FEBS Lett.* 580:795-800.

Wang Y, Hess D, Hunziker PE, Kagi JH (1996) Separation and characterization of the metal-thiolate-cluster domains of recombinant sea urchin metallothionein. *Eur. J. Biochem.* 241:835-839.

Wi S, Pancoska P, Keiderling TA (1998) Predictions of protein secondary structures using factor analysis on Fourier transform infrared spectra: effect of Fourier self-deconvolution of the amide I and amide II bands. *Biospectroscopy* 4:93-106.

Williams RW (1983) Estimation of protein secondary structure from the laser Raman amide I spectrum. *J. Mol. Biol.* 166:581-603.

Williams RW (1986) Protein secondary structure analysis using Raman amide I and amide III spectra. *Methods Enzymol.* 130:311-331.

Williams RW, Dunker AK (1981) Determination of the secondary structure of proteins from the amide I band of the laser Raman spectrum. *J. Mol. Biol.* 152:783-813.

Winge D, Dameron CT, Mehra RK (1992) Metal:Sulfide Quantum Crystallites in Yeast. In: *Metallothioneins* (Stillman MJ, Shaw CF, 3rd, KT S, eds), pp 257-270. New York: VCH.

Winge DR, Miklossy KA (1982) Differences in the polymorphic forms of metallothionein. *Arch. Biochem. Biophys.* 214:80-88.

Winge DR, Nielson KB, Gray WR, Hamer DH (1985) Yeast metallothionein. Sequence and metal-binding properties. *J. Biol. Chem.* 260:14464-14470.

You C, Mackay EA, Gehrig PM, Hunziker PE, Kagi JH (1999) Purification and characterization of recombinant *Caenorhabditis elegans* metallothionein. *Arch. Biochem. Biophys.* 372:44-52.

Zangger K, Oz G, Otvos JD, Armitage IM (1999) Three-dimensional solution structure of mouse [Cd^7]-metallothionein-1 by homonuclear and heteronuclear NMR spectroscopy. *Protein Sci.* 8:2630-2638.

Zhu C, Lü T, Zhang R, Zhao N, Liu J (2000) Modeling of kiwifruit metallothionein Kiwi503. *Chinese Science Bull.* 45:1413-1417.

In: Biopolymer Research Trends
Editor: Tamas S. Nemeth, pp. 49-85

ISBN: 978-1-60021-983-2

Chapter 2

A New Method of Internal Structural Analysis of Keratin Fibers Using Raman Spectroscopy

Akio Kuzuhara*

Central Research Laboratories, Mandom Corp., 5-12,
Juniken-cho, Chuo-ku, Osaka 540-8530, Japan

Abstract

In order to investigate in detail the influence of chemical modification on the internal structure of keratin fibers, which have a hierarchical structure, we have developed a new method for directly analyzing the structure of cross-sections at various depths of keratin fibers using Raman spectroscopy. This method involves embedding keratin fiber samples in an epoxy resin and microtoming the cured blocks on a microtome to 1-μm (white human hair) and 1.5-μm (black human hair) thickness, and then mounting the samples on a slide glass. The cross-sectional samples are then analyzed with a Raman microscope. Using this analytical technique, the Raman spectra of virgin black human hair, which had been impossible due to it's high melanin granule content, can be recorded. Also, the heterogeneous reaction between reducing agents (thioglycolic acid and L-cysteine), or a protein crosslinking agent (2-iminothiolane hydrochloride) and keratin fibers at the molecular level can be analyzed. Moreover, the secondary structure [the β-sheet and/or random coil (β/R) and the α-helix (α) contents] of cross-sections at various depths of keratin fibers changed by the chemical treatments (bleaching and permanent waving treatments), or chemical modification using 2-iminothiolane hydrochloride (2-IT) can be analyzed by amide I band analysis. Thus, the characterization of the cortex region, which consists of crystalline fibrous protein and the amorphous matrix is an effective method, since information about crystalline and amorphous protein structure can be obtained. Furthermore, the changes of the disulfide (-SS-) content, cysteic acid content, and random coil content show the level of damage on keratin fibers. It can be supposed that this method is a beneficial analytical tool to investigate more detailed internal structural changes due to the influence of not only external factors such as heating, permanent

* Correspondence to: A. Kuzuhara; email address: kuzuhara-a@mandom.co.jp or kuzuhara@mb.infoweb.ne.jp

waving, bleaching treatments, and exposure to sunlight, but also internal factors such as aging and nutritional deficiencies on human hair, since the Raman spectra of virgin black human hair keratin fibers can be analyzed. Structural analysis of keratin fibers can be done to a much higher level of detail than previously as Raman spectroscopy can be used.

INTRODUCTION

Keratin fibers, like wool and hair, have a hierarchical structure. The hierarchical structure of keratin fibers consists of two components, the cortex and the cuticle. The hierarchical structure of a fine wool fiber is shown in Figure 1. The cortex consists of spindle-shaped macrofibrils that have two main structures, the microfibril and the matrix, which are distinguished by their structures and amino acid compositions [1-6]. The microfibril is a crystalline fibrous protein which is mainly composed of α-helical proteins with a low cystine content. These structures are aligned along the fiber axis and embedded in an amorphous matrix with a high cystine content consisting of disulfide (-SS-) groups. And, -SS- groups form the tertiary cross-linkage in keratin fibers, contributing physical and mechanical properties as well as structural stability. Therefore, it is important to obtain information about -SS- groups, when investigating the influence of chemical modification on the structure of keratin fibers.

The direct characterization of keratin fibers has been performed using X-ray diffraction [1,7-12], solid state NMR [8,13], Raman [14-34] and Infrared spectroscopy [14,20,26]. Information about the amorphous region (the cuticle and matrix) of keratin fibers can not be obtained from X-ray diffraction, since the information obtained from X-ray diffraction only reflects the state of the high crystalline structure in keratin fibers. Solid state NMR has a low sensitivity, uses a lot of sample volumes, and it is not possible to obtain separate information about the cuticle and cortex structure of keratin fibers, which have a hierarchical structure.

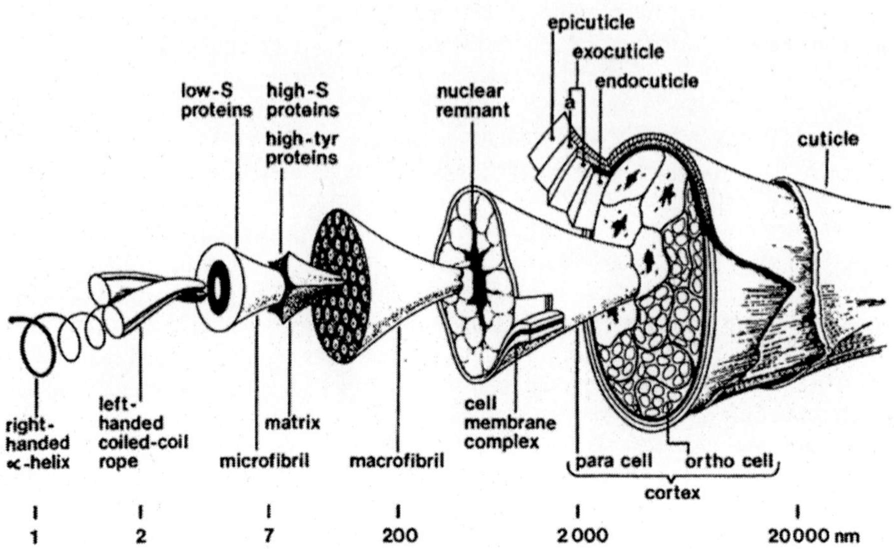

Figure 1. The hierarchical structure of a fine wool fiber [4].

On the other hand, when directly characterizing the cuticle and cortex structure of a single keratin fiber, the analytical technique using a Raman microscope is effective since it can be measured at a spot diameter of 1 μm.

The advantage of Raman spectroscopy for studying keratin fibers is that it is nondestructive, requires no sample extraction or purification, and provides information about -SS- groups through reduction and oxidation, which is impossible to measure using Infrared spectroscopy, since bands can be assigned to S-S and C-S vibrations of cystine. Also, structural information is provided by amide I and amide III vibrations, and the skeletal C-C stretch, which is only weakly active in the infrared absorption spectrum of keratin fibers.

The work of Frushour and Koenig has provided assignments for the side and main chain vibrations in wool keratin [14]. Raman spectroscopy has been used in previous work for fiber identification and characterization [15] and in structural studies of keratin fibers [16,17]. More recent work describes chlorination [18], investigation of human hair and other keratotic biopolymers [19], the analysis of merino wool cuticle and cortical cells [20], keratin orientation in wool and feathers [21], the reduction of wrinkle formation in wool with 2-IT [22,23], and chemical modification of keratin fibers using 2-IT [24]. Other recent work using Raman spectroscopy has investigated the structural changes in keratin fibers due to chemical treatments such as wool fabrics subjected to hydrogen peroxide bleaching [25], the photo-oxidation of wool [26], the influence of bleaching treatments [27], permanent waving treatments [28], and photo-oxidation on human hair [29].

However, up to now, Raman spectroscopic analysis had been limited to white human hair [27,28,30-33] which has no melanin granules, and blond human hair [29,34] which has few melanin granules. The characterization of virgin black human hair which has a high melanin granule content had been impossible, because of sample destruction due to laser exposure, and an increasing baseline resulting from fluorescence. Also, Raman spectroscopic analysis had only provided information about the surface (the cuticle) of keratin fibers. In order to improve these two weak points, near-infrared Raman spectroscopy and confocal Raman spectroscopy were proposed as the analytical methods for hair samples which contain melanin granules, however the Raman spectra of virgin black human hair had still been impossible due to it's high melanin granule content.

It is well known that human hair is changed and damaged by cumulative mechanical factors such as heating with a hair drier and brushing. Furthermore, there are chemical factors such as the permanent waving, bleaching and coloring treatments, as well as environmental factors such as exposure to sunlight and salt water. Compared with these external factors, studies on internal factors such as aging and nutritional deficiencies are still lacking comprehensiveness Therefore, if the Raman spectra of virgin black human hair keratin fibers can be recorded, Raman spectroscopy becomes a beneficial analytical tool to investigate the influence of these internal and external factors on virgin black human hair.

In this chapter, the author has described a new method for not only directly analyzing the structure of cross-sections at various depths of keratin fibers, without isolating the cuticle and cortex, but also measuring virgin black human hair keratin fibers which contain a lot of melanin granules using Raman spectroscopy, in order to investigate in detail the influence of chemical modification on the internal structure of keratin fibers. Moreover, the author has introduced evidence research on keratin fibers using this new method. Finally, the author has suggested other possible applications of this new analytical method.

METHODS

Materials

White Chinese hair (average fiber diameter: 78 μm) used as keratin fibers were purchased from Beaulax Co. (Tokyo, Japan). Also, virgin black human hair samples (sections of new growth hair: 2 mm from the scalp) from a group of Japanese females in their twenties and another group of Japanese females in their fifties were collected from the top of the head. The virgin black and white single human hair fibers were immersed in a solution of 0.5 wt % sodium laurylsulfate at a ratio of hair to solution of 1: 60. The hair bundles were soaked for 60 min at 40°C. Next, the hair bundles were washed in distilled water and then dried in air.

Raman Spectra

All Raman spectra were recorded on a Ramanor T-64000 Raman microscope system (Jobin Yvon, Longjumeau, France), which is comprised of an optical microscope adapted to a single grating spectrograph and a charge coupled device (CCD) array detector. The laser excitation was provided by an argon ion laser operating at a cross slit of 100 - 200 μm, laser power of 30 - 50 mW and a wavelength of 514.5 nm. The laser beam on the sample was focused to a spot diameter of 1 μm using a 100× microscope objective. Spectra were recorded by scanning the 200 - 2000 cm^{-1} region with a total acquisition time of 500 - 1000 seconds. A spectra resolution of 2.3 cm^{-1} was used. In order to prevent fluorescence, points at various depths of the cortex with the fewest possible melanin granules were selected for Raman spectra analysis. In addition laser power, cross slit width, and total acquisition time were optimized for each point to achieve a good signal/noise (S/N) ratio. By collecting three spectra from the samples, and taking an average of these, it was possible to ensure no sample degradation occurred, and that the spectrum obtained were quite reproducible. Furthermore, the cosmic ray was removed.

Normalization of Raman spectra of keratin fibers is often carried out based on the C-H band at 1450 cm^{-1} [27,28,30-32], amide I band at 1657 cm^{-1} [18], and the phenylalanine (Phe) peak at 1003 cm^{-1} [22-24,33]. In particular, normalization of Raman spectra of the keratin samples affected by chemical modification is effective when carried out based on the Phe peak, because Phe is not influenced by chemical modification. However, normalization based on the C-H band is better than Phe peak when trying to obtain more accurate data, because the Phe peak area is small compared to the C-H band peak area. In fact, the ratio of the peak area of the S-S band divided by the peak area of the C-H band was more accurate compared to that of the peak area of the S-S band divided by the peak area of Phe (not shown). Therefore, the C-H band or Phe peak should be chosen for normalization purposes.

The disulfide (-SS-) content of the hair samples was estimated from the ratio of the peak area of the S-S band (calculated from the peak to a baseline which was drawn between 470 and 560 cm^{-1}) divided by the peak area of the C-H band (calculated from the peak to a baseline which was drawn between 1375 and 1500 cm^{-1}) or Phe peak (calculated from the peak to a baseline which was drawn between 986 and 1020 cm^{-1}). The cysteic acid content of the hair samples was estimated from the ratio of the peak area of the S-O band (calculated from the peak to a baseline which was drawn between 1013 and 1095 cm^{-1}) divided by the peak area of the C-H band or Phe peak. The random coil content of the hair samples estimated from the ratio of the peak area of the Amide III (unordered) band (calculated from the peak to a baseline which was drawn between 1200 and 1288 cm^{-1}) divided by the peak area of the C-H band or Phe peak.

Moreover, the proportion of the eight band components of the hair samples was evaluated by spectral simulation of the amide I band region, assuming Gaussian line shapes and appropriate line width. (Amide I band analysis). The band frequency and line width of the eight components in the amide I band region are shown in Table I. According to Church et al.'s method [20], the band frequency of the eight components was selected. Here, the band frequency and line width of the eight components of all hair samples were fixed, while the band intensity of all hair samples was changed. Also, the component content (β/R) at 1671 cm^{-1} assigned to the β-sheet and/or random coil forms, and the component content (α) at 1652 cm^{-1} assigned to the α-helix form of the hair samples was compared by estimating the ratio of the peak area of each component divided by the peak area of the C-H band or Phe peak.

In the case of choosing the Phe peak for normalization purposes, the following estimated contents: -SS- content, cystic acid content, β/R content, α content and random coil content in the cuticle region only (depth of 1 μm from the fiber surface) were multiplied by the ratio of Phe content in the cuticle and cortex regions (Phe content of cuticle/ Phe content of cortex = 0.70), because the Phe content of the cuticle (11.5 mol/ 1000 mol total amino acids) is 30 % less than the Phe content of the cortex (16.5 mol/ 1000 mol total amino acids) [35].

The mean and standard deviation of -SS- content, random coil content, β/R content and α content in the cortex region of the hair samples were calculated from the respective contents measured at the five analysis points (depths of 5, 10, 15, 20, 25 and 30 μm) in the cortex region by assuming that the respective contents in the cortex region were constant.

Finally, the hair samples were embedded in an epoxy resin (Refine Tec Ltd., Yokohama, Japan), and the cured blocks were microtomed on a HM360 microtome (Microm international GmbH, Walldorf, Germany) to 1.00 (white hair samples) and 1.50 μm thickness (black hair samples), and mounted on a slide glass.

Table I. Band Frequency and Line Width of the Eight Components in the Amide I Band Region

Components	Band Frequency (cm^{-1})	Band Line Width (cm^{-1})
1	1725	25
2	1695	20
3	1671	35
4	1652	30
5	1630	20
6	1616	20
7	1605	20
8	1585	10

Here, the best suitable thickness (1.50 μm) for the black human hair samples was selected by measuring the Raman spectra of cross-sectional hair samples at varying thicknesses (0.75, 1.00, 1.25, and 1.50 μm) beforehand and choosing the thickness that produced the best Raman spectra for the samples. On the other hand, following the same procedure, the best suitable thickness (1.00 μm) for white human hair samples was selected. The cross-sectional samples were produced using black and white human hair, and sections of the hair at varying depths from the surface (spot diameter: 1 μm) were measured with a Raman microscope.

In the case of investigating the influence of chemical modification on the internal structure of keratin fibers, two adjoining cross-sections (one: untreated sample, the other one: chemically treated sample) of a single hair fiber should be compared, since not only the chemical and physical properties, but also the morphology of human hair fibers are different from fiber to fiber.

RESULTS AND DISCUSSION

Raman Spectra of Keratin Fibers

Measurement by Raman spectroscopy becomes a beneficial means of investigating the structural changes of cross-sections at various depths of human hair due to it being able to obtain information on the secondary structure of proteins and disulfide (-SS-) groups in keratin fibers.

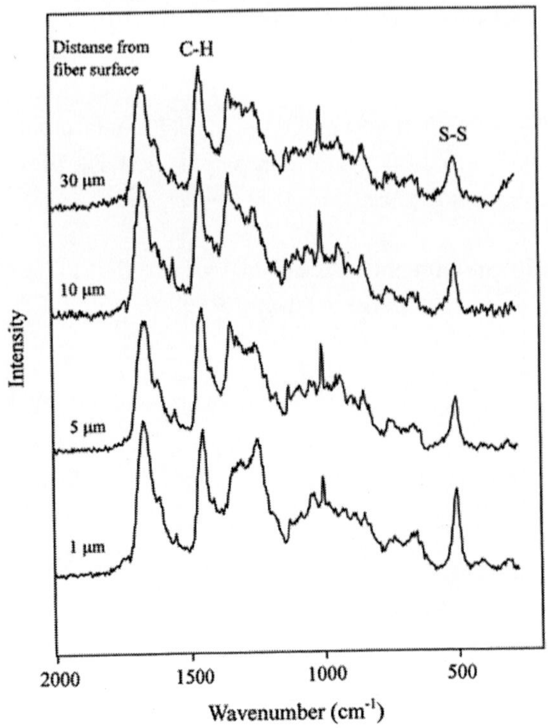

Figure 2. Raman spectra of the white human hair fiber at depths of 1, 5, 10 and 30 μm.

The bands of particular interest lie in the wave number range of 500-1800 cm^{-1}. These are the vibrations assigned to the S-S and C-S bonds of cystine, amino acids (tryptophan, tyrosine, and phenylalanine), the amide I and amide III vibrations, and the C-C skeletal stretching vibration of the α-helix. The Raman spectra of the white human hair fiber (Control) at depths of 1, 5, 10 and 30 μm are shown in Figure 2. The depth of 1 μm from fiber surface corresponds to the cuticle region and the depth of 5 ~ 30 μm from fiber surface corresponds to the cortex region. It is shown that the band shapes, as well as peak maximum frequencies, of the cuticle region were significantly different from those of the cortex region of the human hair fibers. The frequencies and tentative assignments of the virgin white human hair fibers at depths of 1 μm and 5 μm (corresponding to the cuticle and the cortex) from the fiber surface compared with those of wool are shown in Table II.

Table II. Frequencies and Tentative Assignments of Untreated Human Hair (Cuticle and Cortex Region) Compared with those of Wool

Human Hair		Wool		Assignment
Cuticle [a] (cm^{-1})	Cortex [b] (cm^{-1})	Ref. [18] (cm^{-1})	Ref. [14] (cm^{-1})	
1671	1666	1658	1653	Amide I
1613	1614	1615	1614	Tyr and Trp
1553	1552	1558	1553	Trp
1448	1446	1450	1448	CH_2 bending mode
ND [c]	1336	1340	1338	CH_2 bend, Trp
ND	1315	1316	1318	Cα-H bend
1245	1243	1245	1244	Amide III (unordered)
ND	1210	1209	1207	Tyr and Phe
ND	1174	1180	1176	Tyr
1123	1123	1126	1126	C-N stretch
1040	1040	-	-	Sulfonate S-O stretch
ND	1030	1034	1031	Phe
1001	1001	1006	1002	Phe
959	ND	959	952	CH_2 rock
ND	935	935	934	Skeletal C-C stretch (α)
884	880	883	881	Trp
851	851	852	851	Tyr
ND	750	752	752	Trp
664	664	665	661	Cys C-S stretch
642	642	644	642	Tyr
505	507	512	512	Cys S-S stretch g-g

[a] Depth of 1 μm from hair surface.

[b] Depth of 5 μm from hair surface.

[c] Non-detect.

The amide I peak maxima for the cortex region (depth of 5 μm from the fiber surface) was found to shift at 1665 cm^{-1}. Also, the skeletal C-C stretch (α) band at 938 cm^{-1}, assigned to the α-helical backbone, was observed in the cortex region only. This is in agreement with the findings by Fraser et al., in which the microfibril that exists in the cortex region is mainly composed of α-helical protein [4,7]. On the other hand, the amide I peak maxima for the cuticle region (depth of 1 μm from the fiber surface) was found to be at 1672 cm^{-1}, assigned to the β-sheet and/or random coil forms, but the skeletal C-C stretch (α) band, assigned to the α-helical backbone did not appear. Also, the amide III (unordered) band intensity, assigned to the random coil form, at 1250 cm^{-1} for the cuticle region was higher than that for the cortex region. These results indicate that the α-helix form does not exist in the hair cuticle, and that the cuticle has a more amorphous structure.

Curve-Fitting of Amide I Band Region

Structural information is provided by the amide I and amide III bands and the skeletal C-C stretch (α) band [14]. In particular, the β-sheet and/or random coil content (β/R) and the α-helix content (α) in keratin fibers can be estimated by amide I band analysis.

In this section, the proportion of the eight band components of the hair samples was evaluated by spectral simulation of the amide I band region, assuming Gaussian line shapes and appropriate line width (Amide I band analysis). According to Church et al.'s method [20], the band frequency of the eight components was selected. Here, the band frequency and line width of the eight components of all hair samples were fixed, while the band intensities of all hair samples were changed. The band component (β/R) observed at 1671 cm^{-1} has been assigned to the β-sheet and/or random coil forms, and the band component (α) at 1652 cm^{-1} has been assigned to the α-helix form [14,20,28,33]. The band component observed at 1695 cm^{-1} has been assigned to the amide groups of the asparagines and glutamine side chains [20,28,33,36,37], and the very weak band component at 1725 cm^{-1} has been assigned to the C=O stretching vibration of the protonated carboxylic acid groups of aspartic and glutamic acid side chains [20,28,33,38,39]. Moreover, four additional band components on the low wavenumber side of the amide I band complex, including the 1616 cm^{-1} band assigned to tyrosine and triptophan, were included in the fit [20,28,33].

The curve-fitting of the amide I band region of the cuticle Raman spectrum (depth of 1 μm from the fiber surface) of the white human hair based on these components is shown in Figure 3. The curve-fitting of this special region of the cortex Raman spectrum (depth of 20 μm from the fiber surface) of the white human hair based on these components is shown in Figure 4. The proportion of the band component (α) at 1652 cm^{-1}, assigned to the α-helix form, in the cortex region remarkably increased compared with that of the cuticle region, though the band component (β/R) observed at 1671 cm^{-1}, assigned to the β-sheet and/or random coil forms was rich in both the cuticle and cortex region.

Raman Spectra of Virgin Black Human Hair Keratin Fibers

Measurement by Raman spectroscopy becomes a beneficial means of investigating the structural changes of cross-sections at various depths of white human hair due to it being able to obtain information on the secondary structure of proteins and disulfide (-SS-) groups in keratin fibers. However, the characterization of virgin black human hair, which contains a high melanin granule content was impossible because of sample destruction due to laser exposure, and an increasing baseline resulting from fluorescence.

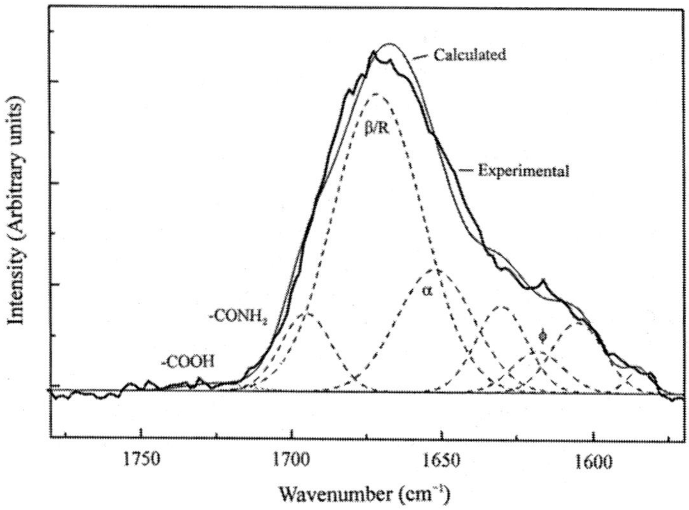

Figure 3. Curve-fitting of the amide I band region of the cuticle Raman spectrum (depth of 1 μm from the fiber surface) of the white human hair. Experimental: experimental Raman spectrum; Calculated: calculated Raman spectrum; ϕ: Tyr/Trp side chains; α: α-helix form: β/R: β-sheet and/or random coil forms; $CONH_2$: amide groups of asparagines and glutamine residues; and COOH: protonated carboxylic acid groups of aspartic and glutamic acid residues.

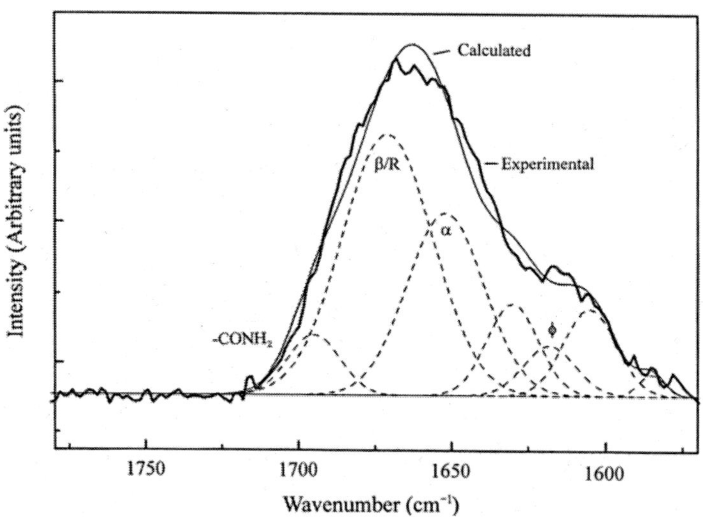

Figure 4. Curve-fitting of the amide I band region of the cortex Raman spectrum (depth of 20 μm from the fiber surface) of the white human hair. Experimental: experimental Raman spectrum; Calculated: calculated Raman spectrum; ϕ: Tyr/Trp side chains; α: α-helix form: β/R: β-sheet and/or random coil forms; and $CONH_2$: amide groups of asparagines and glutamine residues.

In this section, the cross-sectional virgin black human hair samples were analyzed at various depths using a Raman microscope. The cuticle (depth of 1 μm from the fiber surface) and cortex Raman spectra (depth of 5 μm from the fiber surface) of virgin black human hair from a Japanese female in her twenties (23 years old) are shown in Figure 5. The cuticle (depth of 1 μm from the fiber surface) and cortex Raman spectra (depth of 5 μm from the fiber surface) of virgin black human hair from a Japanese female in her fifties (56 years old) are shown in Figure 6. Similarly as in the case of the white human hair, the band shapes, as well as peak maximum frequencies, of the cuticle region were significantly different from those of the cortex region of the virgin black human hair fibers. As is shown, Raman spectra of the virgin black human hair were similar to that of white human hair. The amide I peak maxima for the cortex region (depth of 5 μm from the fiber surface) was found to shift at 1665 cm^{-1}. Also, the skeletal C-C stretch (α) band at 938 cm^{-1}, assigned to the α-helical backbone, was observed in the cortex region only. On the other hand, the amide I peak maxima for the cuticle region (depth of 1 μm from the fiber surface) was found to be at 1672 cm^{-1}, assigned to the β-sheet and/or random coil forms, but the skeletal C-C stretch (α) band, assigned to the α-helical backbone did not appear.

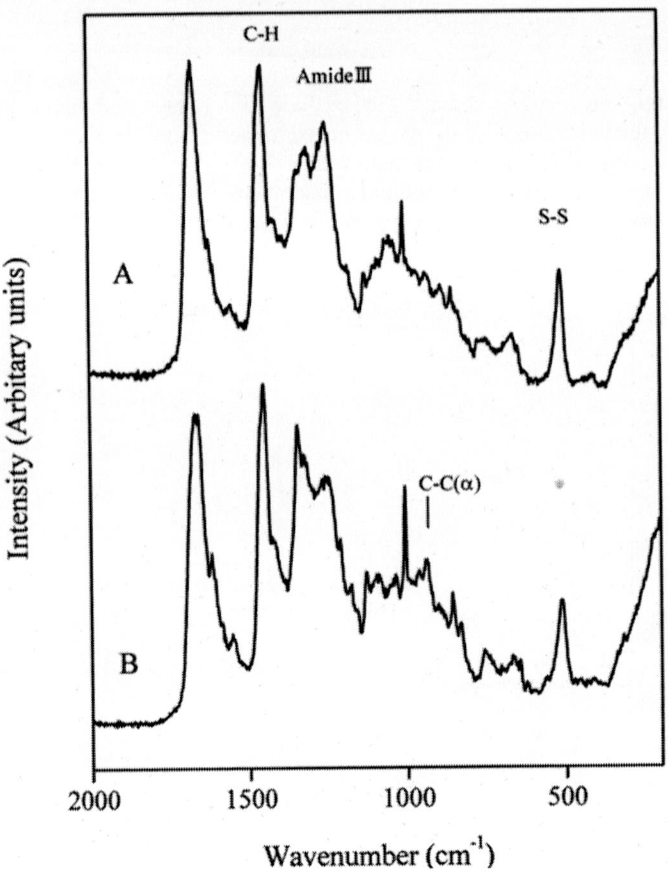

Figure 5. Cuticle and representative cortex Raman spectra of virgin black human hair from a Japanese female in her twenties (Sample 1: 23 years old): (A) cuticle region (depth of 1 μm from the fiber surface); (B) cortex region (depth of 5 μm from the fiber surface).

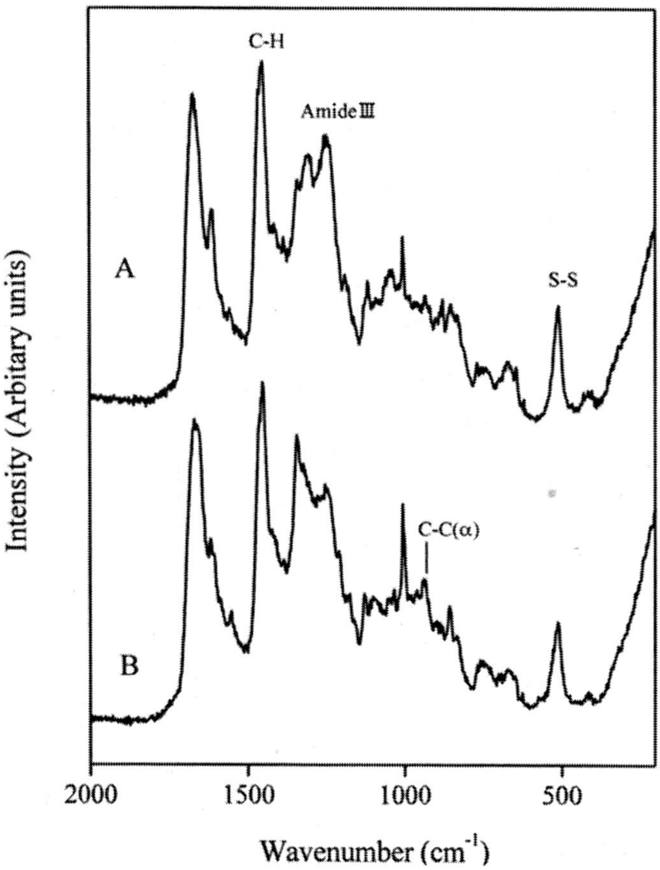

Figure 6. Cuticle and representative cortex Raman spectra of virgin black human hair from a Japanese female in her fifties (Sample 16: 56 years old): (A) cuticle region (depth of 1 μm from the fiber surface); (B) cortex region (depth of 5 μm from the fiber surface).

Also, the amide III (unordered) band intensity, assigned to the random coil form, at 1250 cm^{-1} for the cuticle region was higher than that for the cortex region. Moreover, the S-O band intensity, assigned to cysteic acid, at 1047 cm^{-1} for the cuticle was low, whereas cysteic acid did not exist in the cortex.

From this experiment, it has been shown that the Raman spectra of virgin black human hair which contains a high number of melanin granules can be recorded by cross-sectioning hair samples at a thickness of 1.5 μm, selecting points at various depths of the cortex with the fewest possible melanin granules, and optimizing laser power, cross slit, as well as total acquisition time.

Reproducibility of the Raman Bands of Two Adjoining Cross-sections of a Single Black Hair Keratin Fiber

In order to confirm whether it is possible to obtain information about the internal structure of human hair keratin fibers, the Raman bands of two adjoining cross-sections (distance: about 100 μm apart) of a single virgin black hair keratin fiber were compared by Raman spectroscopic analysis. Here, the normalization of Raman spectra of the keratin fibers was carried out based on the C-H band at 1450 cm^{-1}. The α content, the β/R content, the -SS-content, and the random coil content of the cortex region of the two adjoining cross-sections of the single keratin fiber at depths of 5 and 20 μm from the fiber surface are shown in Table III. The level of significance of each Raman band of the two adjoining cross-sections of the hair keratin fiber sample were P > 0.10. As a result, the reproducibility of the Raman bands of the two adjoining cross-sections of the single hair keratin fiber was clearly good.

Table III. α Content, β/R Content, -SS- Content, and Random Coil Content of Cortex Region of Two Adjoining Cross-sections of the Single Keratin Fiber at Depths of 5 and 20 μm from the Fiber Surface

Cross-Section	Distance from Fiber Surface-Time	Cortex Region			
		α	β/R	-SS-	Random Coil
1	5μm - 1	0.344	0.706	0.20	0.19
	5μm - 2	0.339	0.73	0.21	0.20
	5μm - 3	0.369	0.721	0.22	0.20
	20μm - 1	0.312	0.768	0.21	0.23
	20μm - 2	0.299	0.725	0.19	0.18
	20μm - 3	0.333	0.735	0.22	0.21
Mean ± SD		0.33 ± 0.023	0.73 ± 0.019	0.21 ± 0.011	0.20 ± 0.016
2	5μm - 1	0.369	0.727	0.22	0.19
	5μm - 2	0.356	0.768	0.25	0.24
	5μm - 3	0.312	0.731	0.22	0.22
	20μm - 1	0.362	0.709	0.21	0.22
	20μm - 2	0.353	0.714	0.21	0.20
	20μm - 3	0.362	0.713	0.17	0.23
Mean ± SD		0.35 ± 0.019	0.73 ± 0.020	0.21 ± 0.024	0.21 ± 0.017
P		0.17	0.63	0.67	0.17
Significance		NS	NS	NS	NS

Therefore, the influence of chemical modification using reducing and crosslinking agents on the internal structure of keratin fibers can be investigated by comparing two adjoining cross-sections (one: untreated sample, the other one: chemically treated sample) of a single hair fiber.

Analysis of Heterogeneous Reaction

Heterogeneous Reaction between Reducing Agents and Keratin Fibers

The setting treatment for wool fibers, and the permanent waving treatment for human hair fibers consists of two different processes, disconnection (the reduction process) and reconnection of disulfide (-SS-) groups (the oxidation process), and is widely used in the textile and hair styling industry. Also, the chemistry of the setting process and the changes in the chemical and physical properties of keratin fibers with reduction treatment have been widely studied [40-43]. However, studies on the mechanism connecting the chemical reaction between a reducing agent and the keratin fiber material, occurring on a molecular level, are still lacking comprehensiveness. Specifically, thioglycolic acid (TG) and L-cysteine (CYS) were used as a reducing agent in the first process (disconnection of -SS- groups). The penetration of reducing agents into human hair becomes the trigger of the waving process. Although TG performs well in the waving process of human hair, it is well known that hair treated with TG is damaged. On the other hand, it is known that hair treated with CYS is damaged less than hair treated with TG, although CYS does not perform well with regard to the wave formation of human hair. However, studies on the reason for this are still lacking comprehensiveness. Therefore, it is important to investigate how TG and CYS diffuse into keratin fibers and how the chemical reaction between the reducing agent and keratin fibers occurs.

First, we prepared cross-sectional samples of human hair treated with 6.00 wt % TG solution or 7.87 wt % CYS solution (at pH 9.0 and 25°C for 5 minutes, at a ratio of hair to solution of 1: 15). Next, the penetration of TG or CYS for the cross-sectional samples dyed with methylene blue was observed by optical microscopy. The photomicrograph of the white human hair cross-sectioned and then dyed with methylene blue is shown in Figure 7. The cuticle and the cortex of the untreated white human hair sample did not adsorb the methylene blue. However, the medulla, which exists in the center of the hair, adsorbed the methylene blue, since the medulla is rich in glutamic acid and consists of porous proteins. The photomicrographs of the white human hair treated with TG and CYS at 25°C and pH 9.0 for 5 minutes, then cross-sectioned and finally dyed with methylene blue, are shown in Figures 8 and 9, respectively. The white human hair treated with TG at 25°C and pH 9.0 for 5 minutes, adsorbed the methylene blue through the cuticle and partially into the cortex (Figure 8). On the other hand, the white human hair treated with CYS at 25°C and pH 9.0 for 5 minutes, adsorbed the methylene blue into the cuticle, but hardly into the cortex at all producing a definite boundary line of absorption (Figure 9). Which is to say that TG speedily diffuses into the human hair, whereas CYS remains in the cuticle.

Next, the heterogeneous reaction (the disconnection of -SS- groups) between the reducing agents (TG and CYS) and the keratin fibers was analyzed at the molecular level using Raman spectroscopy. The Raman spectra of the human hair fiber treated with TG (at 25°C and pH 9.0 for 5 minutes) at depths of 1, 5, 7, 10, 15 and 30 μm are shown in Figure 10. The peak intensity at 510 cm^{-1} assigned to the -SS- groups (the stretching vibration of S-S bond), decreased when progressing from center to fiber surface. On the other hand, the band at 932 cm^{-1}, assigned to C-C skeletal stretching of the α-helical backbone, does not disappear when progressing from the fiber center to the cortex surface which is at a depth of 5 ~ 30 μm from fiber surface.

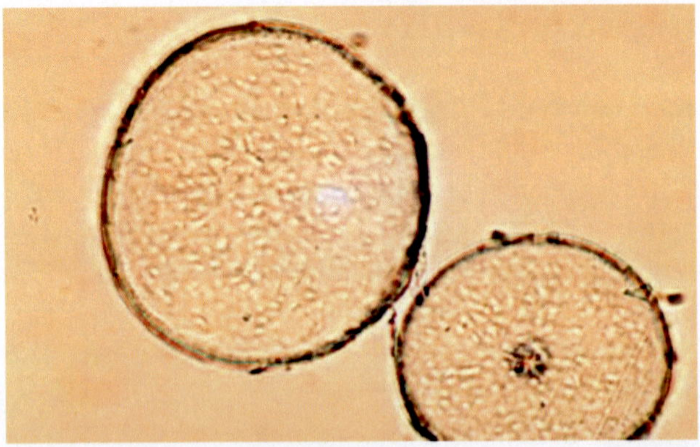

Figure 7. Photomicrograph of a white human hair cross-sectioned and then dyed with methylene blue.

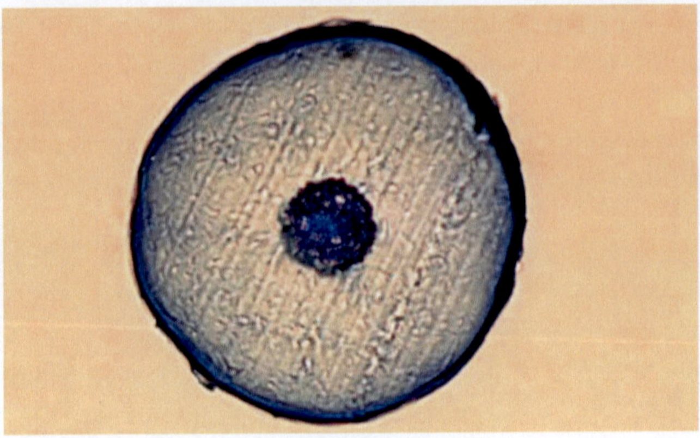

Figure 8. Photomicrograph of a white human hair treated with TG at 25°C and pH 9.0 for 5 minutes, then cross-sectioned and finally dyed with methylene blue.

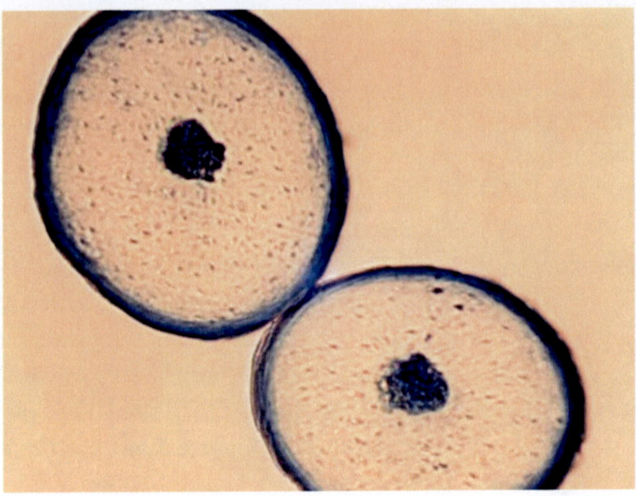

Figure 9. Photomicrograph of a white human hair treated with CYS at 25°C and pH 9.0 for 5 minutes, then cross-sectioned and finally dyed with methylene blue.

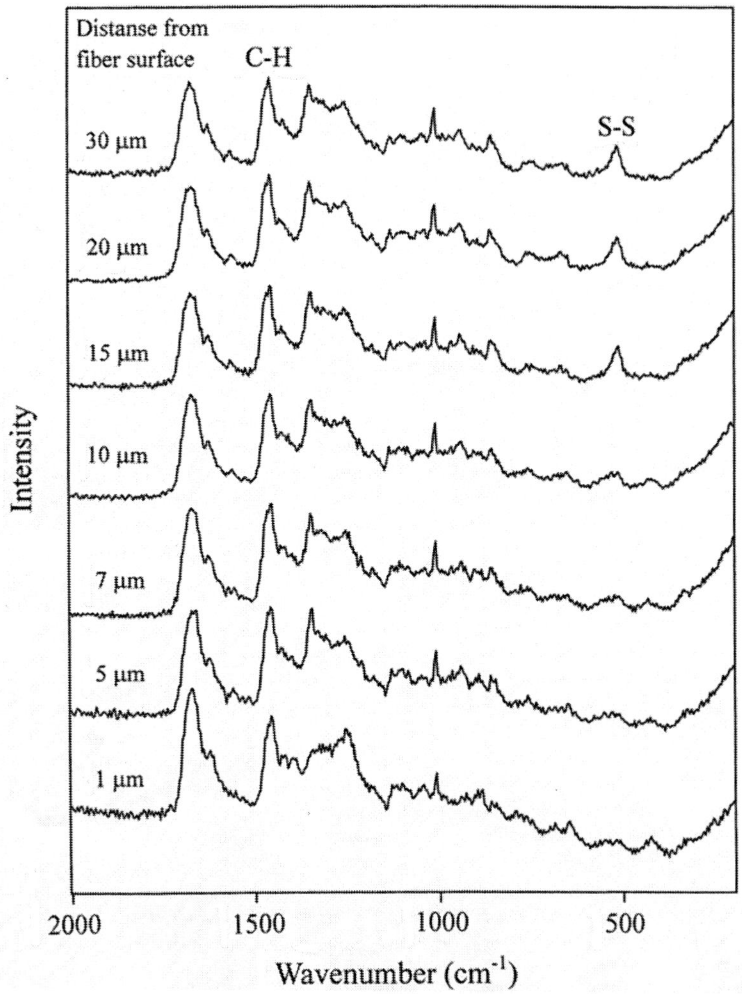

Figure 10. Raman spectra of the human hair fiber treated with TG (at 25°C and pH 9.0 for 5 minutes) at depths of 1, 5, 7, 10, 15 and 30 μm.

This suggests that the α-helical conformation is not influenced by the disconnection of -SS- groups. This result is in agreement with the opinion of Freser et al., in which the intramoleculer -SS- groups does not form in the α-helical backbone [44].

Depth profile, which shows the function between the disconnected relative concentration of -SS- groups of human hair treated with TG and CYS at 25°C and pH 9.0 for 5 minutes, and the distance from the fiber surface is shown in Figure 11. Here, it was assumed that the -SS- content is equally distributed in the cortex. The disconnection of -SS- groups of human hair treated with TG increased compared with that of human hair treated with CYS. Also, the -SS- groups of the human hair treated with TG were disconnected at a deeper hair depth than that of human hair treated with CYS. Considering the fact that the cuticle region ranged from fiber surface to 3 μm below fiber surface, CYS remained in the cuticle region of the virgin human hair, but CYS for the most part did not penetrate into the cortex region.

It was found that the hair treated with CYS was clearly less damaged as compared with the hair treated with TG, since CYS hardly penetrated into the cortex region of the human hair and -SS- groups into the cortex region were not disconnected for the most part. From these experiments, it can be concluded that the disconnection of -SS- groups existing in the cortex region caused by reducing agents largely influences hair damage.

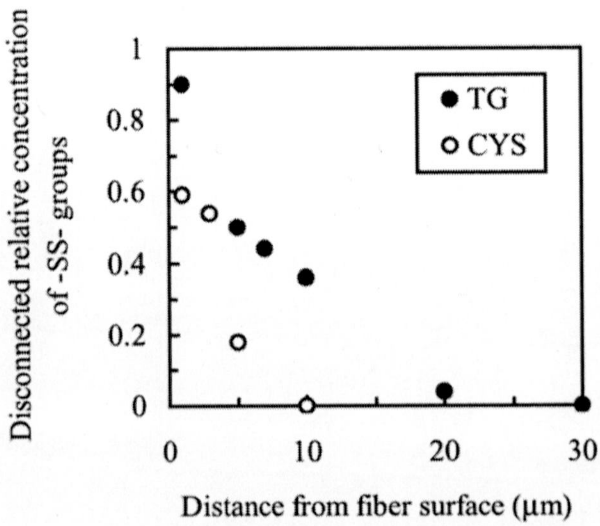

Figure 11. Depth profile, shows the function between the disconnected relative concentration of -SS- groups of human hair treated with TG and CYS at 25°C and pH 9.0 for 5 minutes, and the distance from fiber surface, of the hair samples.

Heterogeneous Reaction between Crosslinking Agent and Keratin Fibers

2-Iminothiolane hydrochloride (2-IT) is used as a protein crosslinking agent, and can introduce disulfide (-SS-) groups into proteins [45]. In particular, it shows good performance introducing new -SS- groups into keratin fibers. In previous studies [22,23], we reported a new creaseproof finish for wool using 2-IT, by introducing new -SS- groups, thus providing good wrinkle recovery and setting ability for wool fabrics. Also, we reported the chemical modification of keratin fibers using 2-IT was effective for a permanent hair-setting process [24]. In the case of using this method, human hair can be set in a permanent wave without damaging hair. Moreover, we analyzed the bond structure of the product obtained from the reaction of 2-IT and L-phenylalanine (Phe) used as the model compound of wool by ^{1}H-NMR and SIMS (secondary ion mass spectrometry). As a result, we confirmed that 2-IT reacts with the amino group of Phe, and the -SH group replaces the amino group [46]. The reaction mechanism of 2-IT is shown in Figure 12. The -SH groups introduced into proteins finally form -SS- groups through mild oxidation. However, studies on the reason for this are still lacking comprehensiveness. Therefore, it is important to investigate in detail the difference in the reaction mechanism of 2-IT with proteins existing in the cuticle and cortex regions.

In this section, in order to investigate in detail the influence of chemical modification using 2-IT on keratin fibers, the structure of cross-sections at various depths of white human hair, treated with 0.2 wt % 2-IT (at pH 8.0 and 50°C for 60 minutes, at a ratio of hair to solution of 1: 222) and then oxidized (6.0 wt % sodium bromate, at 25°C for 15 minutes, at a ratio of hair to solution of 1: 250), was directly analyzed using a Raman microscope.

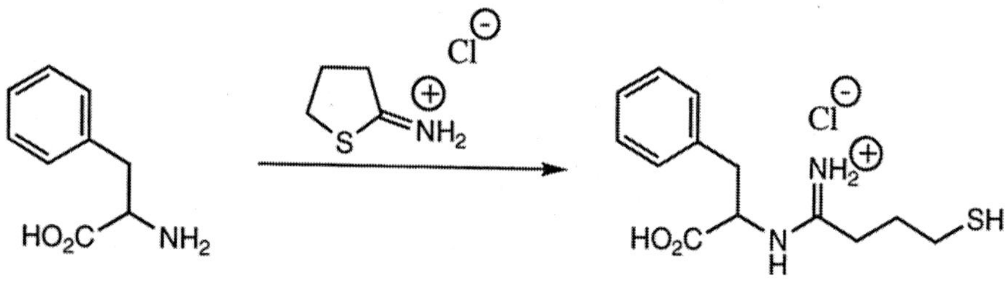

Figure 12. The reaction mechanism of 2-IT.

Here, normalization of Raman spectra of the keratin fibers was carried out based on the Phe peak, which was not influenced by chemical modification using 2-IT. Also, the mean and standard deviation of -SS- content, random coil content, β/R content and α content in the cortex region of hair samples were calculated from the respective contents measured at the five analysis points in the cortex region by assuming that the respective contents in the cortex region were constant.

First, the -SS- content at various depths of the hair fibers chemically modified using 2-IT was compared by Raman spectroscopy. The depth profile, which shows the function between the -SS- content (the ratio of the peak area: S-S band/ Phe peak) and the distance from fiber surface, of the hair samples [the untreated white human hair (Sample 1: Control), and the white human hair treated with 2-IT (Sample 2: 2-IT)] due to chemical modification is shown in Figure 13. The -SS- content of Sample 1 in the cuticle region was clearly higher than that of the cortex region. Also, the -SS- content of the cortex region at depths of between 3 and 30 μm from fiber surface of Sample 1 was constant (Mean ± standard deviation = 3.82 ± 0.52). This suggests that the -SS- content of Sample 1 is equally distributed in the cortex region. For the cuticle region, the -SS- content of Sample 2 remarkably increased compared with that of the Sample 1.

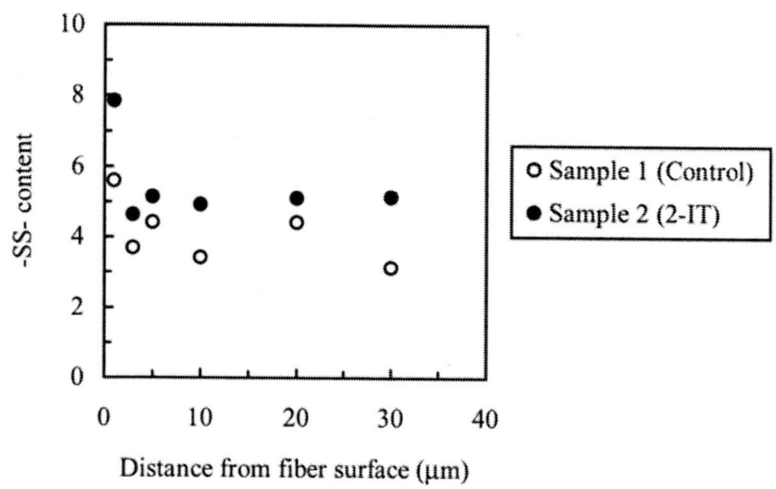

Figure 13. Depth profile, which shows the function between the -SS- content (the ratio of the peak area: S-S band/ Phe peak) and the distance from fiber surface, of the hair samples (Samples 1 and 2) due to chemical modification.

Also, the -SS- content (Mean ± standard deviation = 4.99 ± 0.19) existing throughout the cortex region of Sample 2 increased remarkably compared with that of Sample 1 (the level of significance calculated by statistical test: P = 0.003). This result indicates that 2-IT diffuses beyond the cuticle region, into the cortex region, and equally reacts with the free amino groups of proteins existing throughout the cortex region.

Next, the secondary structure at various depths of the hair fibers chemically modified using 2-IT was estimated by amide I band analysis. The depth profile, which shows the function between the β-sheet and/or random coil content (the ratio of the peak area: β/R band/ Phe peak) and the distance from fiber surface, of the hair samples due to chemical modification is shown in Figure 14. The β/R content in the cortex region of all hair samples was found to be almost constant. The mean and standard deviation (n=5) of the β/R content and α content in the cortex region of hair samples (Samples 1 and 2) is shown in Table IV. The β/R content of Sample 2 remarkably increased compared with that of Sample 1 (the level of significance statistically calculated from the five measured points in the cortex region: P = 0.007).

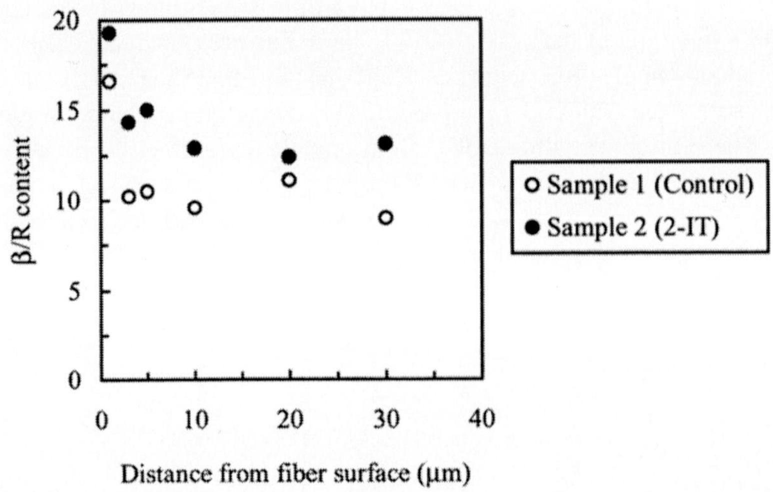

Figure 14. Depth profile, which shows the function between the β/R content (the ratio of the peak area: β/R component/ Phe peak) and the distance from fiber surface, of the hair samples (Samples 1 and 2) due to chemical modification.

Table IV. Mean and Standard Deviation of β/R Content and α Content in the Cortex Region (n=5: Depths of 3, 5, 10, 20, and 30 μm from the Fiber Surface) of Hair Samples (Samples 1 and 2)

Sample	β/R Content	α Content
1 (Control)	10.1 ± 0.74[a]	5.91 ± 0.46
2 (2-IT)	13.6 ± 0.95	5.74 ± 0.56

[a] Mean ± standard deviation.

On the other hand, the α-helix (α) content in the cortex region of Sample 2 did not increase compared with that of the Sample 1. This suggests that the formation of new -SS- groups resulting from chemical modification using 2-IT attributed to the increase in the β/R content in the cortex region. Thus, this indicates that new -SS- groups were introduced into the matrix which is contained in the cortex region.

Moreover, the random coil content at various depths of the hair fibers chemically modified using 2-IT was compared by Raman spectroscopy. The depth profile, which shows the function between the random coil content (the ratio of the peak area: amide III band/ Phe peak) and the distance from fiber surface, of the hair samples due to chemical modification is shown in Figure 15. The random coil content of Sample 1 in the cuticle region was clearly higher than that of the cortex region. Also, the random coil content of the cortex region at depths of between 3 and 30 μm from fiber surface of Sample 1 was constant (Mean ± standard deviation = 3.12 ± 0.39). This suggests that the random coil content of Sample 1 is equally distributed in the cortex region. For the cuticle region, the random coil content of Sample 2 remarkably increased compared with that of the Sample 1. Also, the random coil content in the cortex region of Sample 2 (Mean ± standard deviation = 4.53 ± 0.31) increased compared with that of the Sample 1 (the level of significance calculated by statistical test: P = 0.0004). These results were in an excellent agreement with the results (β/R content) of the previous amide I band analysis. The results from amide I band analysis and amide III band analysis indicate that the random coil content of some of the proteins existing throughout the cortex region of the white human hair, rather than the β-sheet content increased by performing chemical modification using 2-IT.

From these experiments, we concluded that the formation of new -SS- groups resulting from chemical modification using 2-IT induced the secondary structural changes of proteins existing throughout the cortex region.

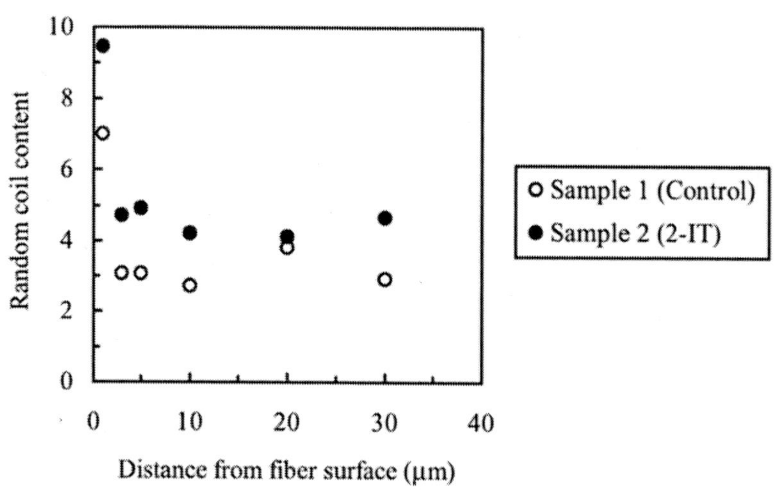

Figure 15. Depth profile, which shows the function between the random coil content (the ratio of the peak area: amide III band/ Phe peak) and the distance from fiber surface, of the hair samples (Samples 1 and 2) due to chemical modification.

Also, considering that the disconnection and reconnection of -SS- groups existing in the matrix is the basis of the permanent waving process, new -SS- groups were introduced into the matrix existing in the cortex region which in turn improved the permanent waving ability of the human hair. Following the same mechanism, it was concluded that wrinkle recovery and setting ability for wool fabrics improved by performing the chemical modification using 2-IT.

Analysis of Damaged Keratin Fibers

Influence of Bleaching Treatments on Keratin Fibers

Bleaching treatments for hair keratin fibers are widely used in the cosmetic industry to lighten the color of human hair, but they cause significant damage. The changes in the morphology of human hair resulting from these treatments have also been studied. It has been found that there is hole formation and abrasion effects of the cuticle surface [47,48], an increase in the porosity of the cortex [49], and decomposition of melanin granules [50]. TEM micrographs (50,000×) of a cross-section of the cuticle region of an untreated black human hair fiber and an excessively bleached black human hair fiber are shown in Figures 16 and 17. The cuticle region of the excessively bleached black human hair was significantly changed.

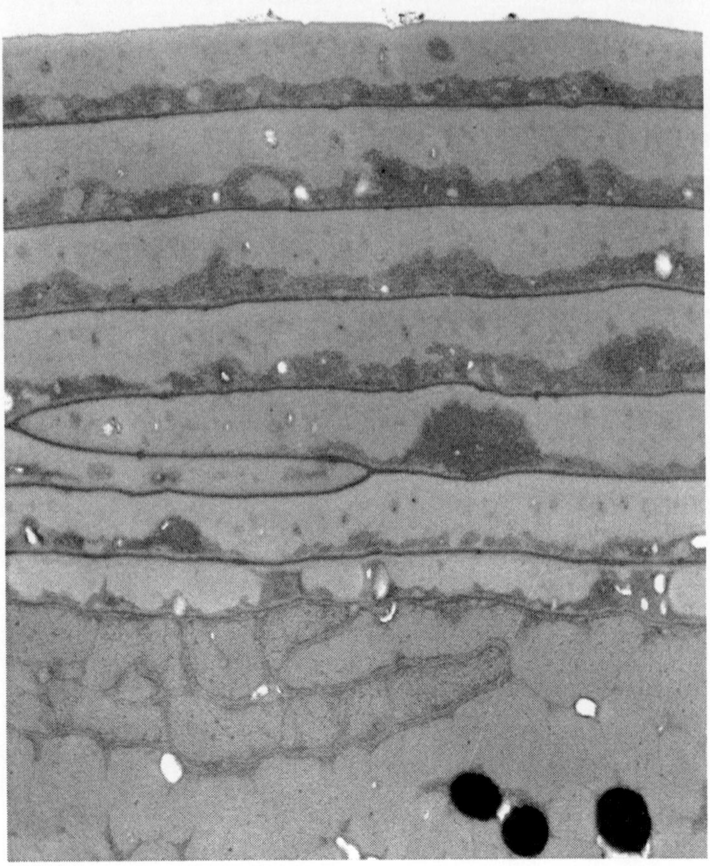

Figure 16. TEM micrograph (50,000×) of the cross-section of the cuticle region of the untreated black human hair fiber.

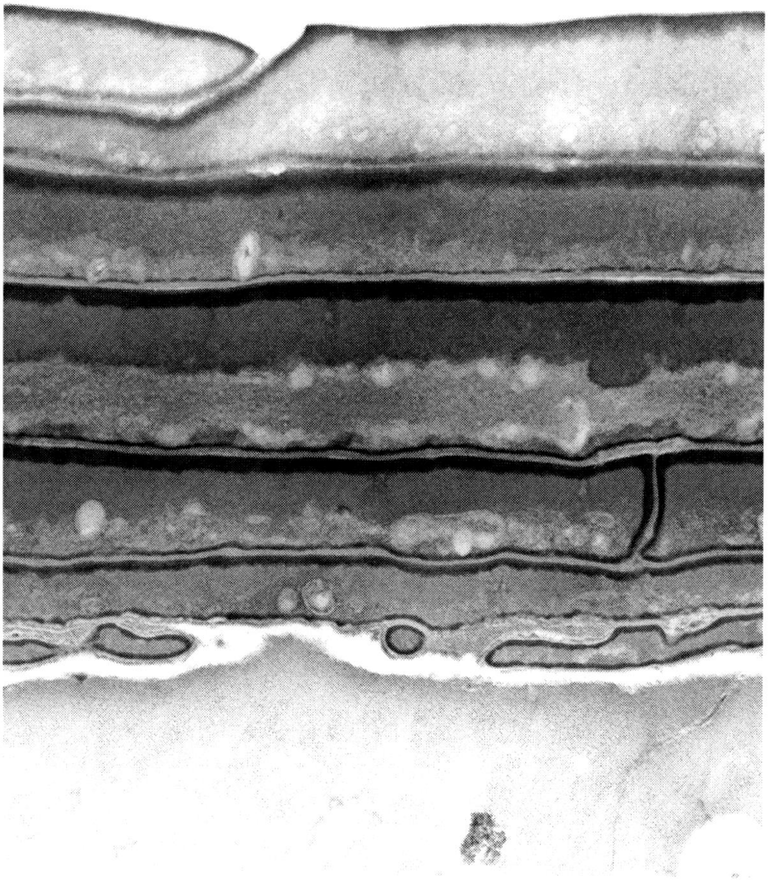

Figure 17. TEM micrograph (50,000×) of the cross-section of the cuticle region of the excessively bleached black human hair fiber.

Specifically, the reduction in electron density of the endocuticle (the dark areas) and the increase in electron density of the exocuticle (the light areas) accompanied with movement and elution of the protein in the cuticle and cortex were observed. Also, swelling of the cuticle and an increase in a cuticle length were observed. TEM micrographs (250,000×) of the cross-section of the cortex region of the untreated black human hair fiber and the bleached white human hair are shown in Figures 18 and 19. In the case of the untreated black human hair fiber, the microfibril (the black spots) and the matrix (the white spots) existing in the macrofibril were density packed. On the other hand, in the case of Sample 2, the disorder of the microfibril and matrix, and notable swelling of the matrix were observed. Also, the changes in the physical and mechanical properties of human hair resulting from these treatments have been studied. It has been found that there is a reduction in tensile strength [1,41,51,52], an increase in the rate of dye diffusion [48,49,53,54], an increase in coloring ability [49], and an increase in the wettability of the hair [47].

The changes in the chemical properties of human hair by performing bleaching treatments have been extensively studied. It has been found that there is a decrease in the 1/2-cystine content [1,35,41,54-56], an increase in the cysteic acid content [1,41,54-56], a decrease in the methionine and tyrosine [1,41,54], and an elution of proteins [57], when performing bleaching treatments.

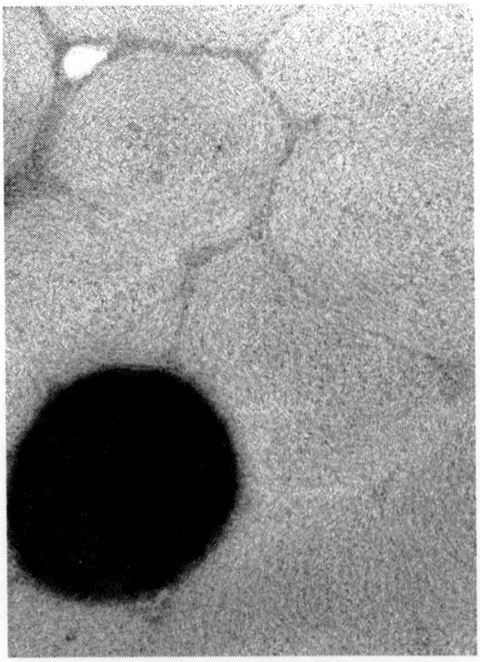

Figure 18. TEM micrograph (250,000×) of the cross-section of the cortex region of the untreated black human hair fiber.

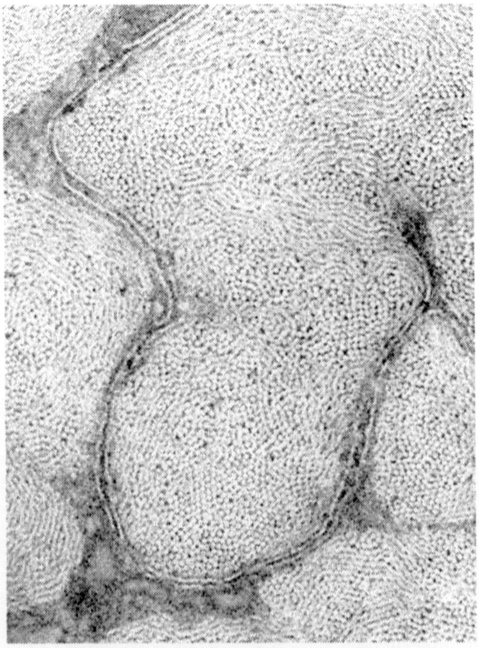

Figure 19. TEM micrograph (250,000×) of the cross-section of the cortex region of the excessively bleached black human hair fiber.

Especially, the oxidative cleavage of the -SS- groups that occurs during the chemical bleaching of human hair by current bleaching products is predominately an S-S fission process where the -SS- groups are finally converted to cysteic acid [1,41]. However, the secondary structural changes by performing bleaching treatment are still lacking comprehensiveness.

In this section, in order to investigate in detail the difference in the reaction mechanism of the bleaching treatment on proteins existing in the cuticle and cortex of human hair, the structure of cross-sections at various depths of bleached human hair was directly analyzed using a Raman microscope. So, cross-sectional samples of human hair treated with a bleaching cream (Gatsby Ex Hi-Bleach, Mandom Corp., Osaka, Japan) at 25°C for 30 minutes at a ratio of hair to solution of 1: 2 and then washed in distilled water for 1 minute were prepared. The same procedure was repeated 5 times (bleaching treatment). Finally, the hair sample treated with the bleaching cream was washed in distilled water for 1 minute, and then dried at room temperature. Here, the bleaching cream consists of three components and becomes 5.9 wt % hydrogen peroxide concentration and pH 10.3 when the three components are mixed. Also, other active ingredients, in the bleaching cream, which aid in bleaching are potassium persulfate, ammonium persulfate and sodium persulfate.

The cuticle Raman spectra (depth of 1 μm from the fiber surface) of the untreated white human hair fiber (Sample 1: Control) and the bleached white human hair (Sample 2: Bleached) is shown in Figure 20. As is shown, the S-S and C-S band intensities existing in the cuticle region of the virgin white human hair decreased, while the S-O band intensity at 1040 cm^{-1}, assigned to cysteic acid, increased by performing the bleaching treatment. This suggests that the -SS- groups existing in the cuticle region were cleaved and finally converted to cysteic acid by this process. Hogg et al. reported that the amide I band and C-H band intensities increased due to the effects arising from backbone conformational changes by subjecting wool fibers to excessive hydrogen peroxide bleaching [25]. However, in our experiment, this phenomenon could not be confirmed.

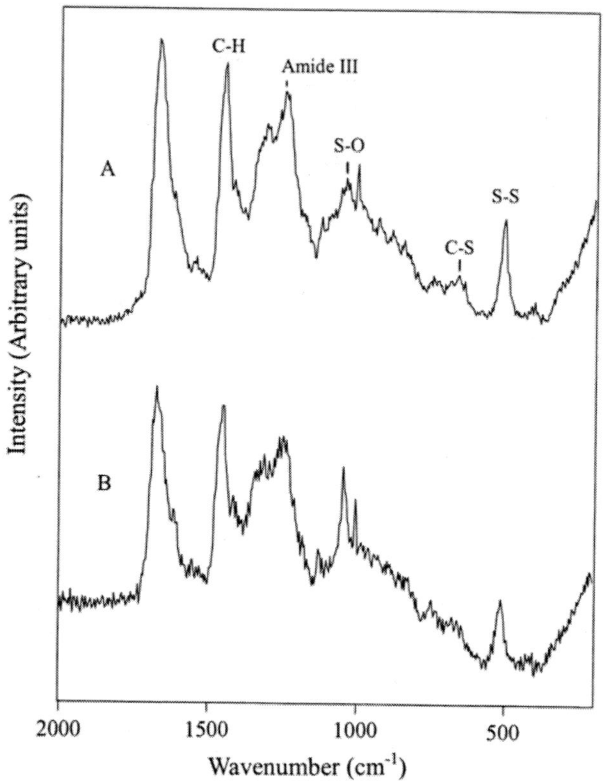

Figure 20. Cuticle Raman spectra (depth of 1 μm from the fiber surface) of the untreated white human hair fiber (Sample 1: Control) and the bleached white human hair (Sample 2: Bleached): (A) Sample 1, and (B) Sample 2.

The representative cortex Raman spectra (depth of 5 μm from the fiber surface) of the untreated white human hair fiber (Sample 1: Control) and the bleached white human hair (Sample 2: Bleached) is shown in Figure 21. Similarly as in the case of the cuticle region of Sample 1, the S-S band intensity decreased, while the S-O band intensity increased by performing the bleaching treatment. However, the decrease of the S-S band intensity existing in the cortex region of Sample 1 resulting from the bleaching treatment was low compared with that of the S-S band intensity existing in the cuticle region after bleaching. Therefore, the decrease of the C-S band intensity existing in the cortex region of the untreated white human hair (Sample 1) could not be confirmed. Also, the amide III (unordered) band intensity at 1243 cm^{-1}, assigned to random coil form, slightly increased, indicating that some of the proteins existing throughout the cortex region of the untreated white human hair changed to the random coil form by performing the bleaching treatment.

Next, the -SS- content at various depths of the hair fibers was compared by Raman spectroscopy. The depth profile, which shows the function between the -SS- content (the ratio of the peak area: S-S band/ C-H band) and the distance from fiber surface, of hair samples [the untreated white human hair (Sample 1: Control) and the bleached white human hair (Sample 2)] due to the bleaching treatment is shown in Figure 22. Here, normalization of Raman spectra of the keratin fibers was carried out based on the C-H band at 1450 cm^{-1}, in which the peak area is large. The -SS- contents of Samples 1, and 2 in the cuticle region (depth of 1 μm from the fiber surface) were higher than that of the cortex region (depths of 3 - 30 μm from the fiber surface).

Also, the -SS- content of the cortex region at depths of between 3 and 30 μm from the fiber surface of Sample 1 was constant (Mean ± standard deviation = 0.28 ± 0.026). Similarly as in the case of Sample 1, the -SS- contents of the cortex region at depths of between 3 and 30 μm from fiber surface of Samples 2 were constant. This suggests that the -SS- contents of Samples 1 and 2 are equally distributed in the cortex region. The -SS- content (Mean ± standard deviation = 0.23 ± 0.012) existing throughout the cortex region of Sample 2 decreased compared with that of the Sample 1 (the level of significance calculated by statistical test: P = 0.019).

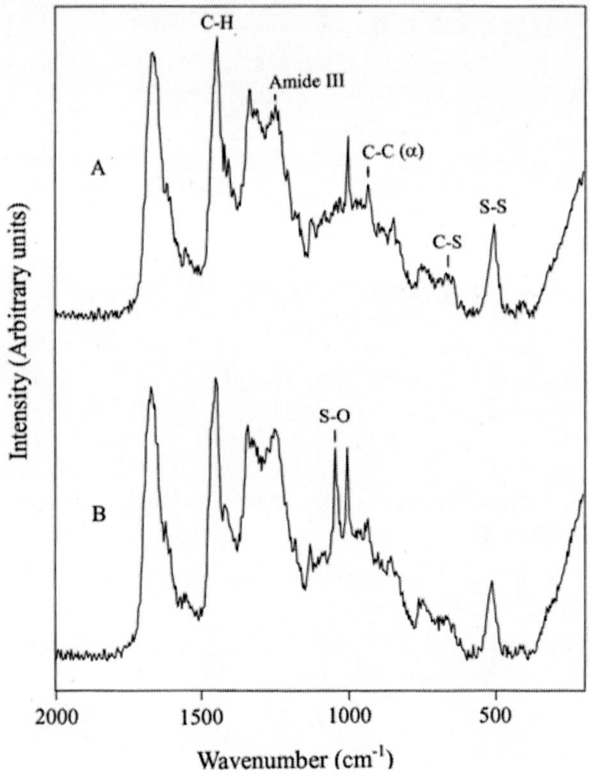

Figure 21. Representative cortex Raman spectra (depth of 5 μm from the fiber surface) of the untreated white human hair fiber (Sample 1: Control) and the bleached white human hair (Sample 2: Bleached): (A) Sample 1, and (B) Sample 2.

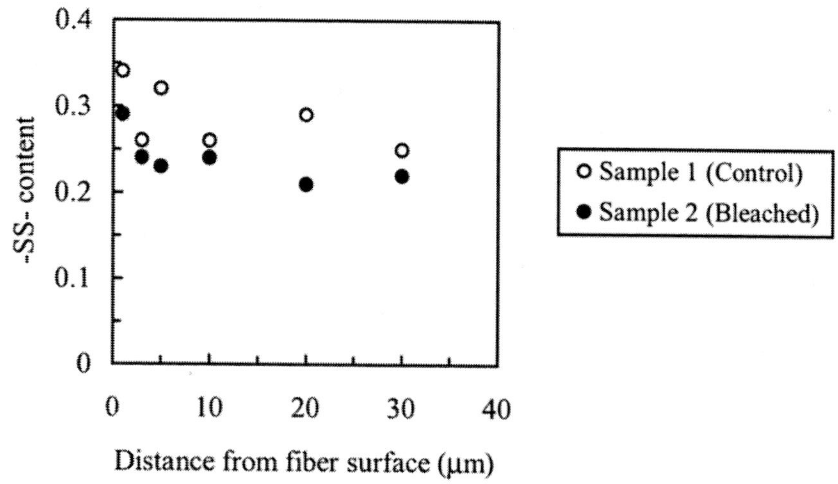

Figure 22. Depth profile, which shows the function between the -SS- content (the ratio of the peak area: S-S band/ C-H band) and the distance from fiber surface, of the hair samples due to the bleaching treatment.

Moreover, the cysteic acid content at various depths of the hair fibers was compared by Raman spectroscopy. The depth profile, which shows the function between the cysteic acid content (the ratio of the peak area: S-O band/ C-H band) and the distance from fiber surface, of hair samples due to the bleaching treatment is shown in Figure 23. The cysteic acid content of Sample 1 increased when progressing from center to fiber surface, which suggests that the partial -SS- groups existing in the surface of the untreated white human hair changed to cysteic acid through natural oxidation. The cysteic acid content existing in the cuticle region and throughout the cortex region of Sample 2 increased remarkably compared with that of the Sample 1. Also, similarly as in the case of Sample 1, the cysteic acid content of Sample 2 increased when progressing from center to fiber surface.

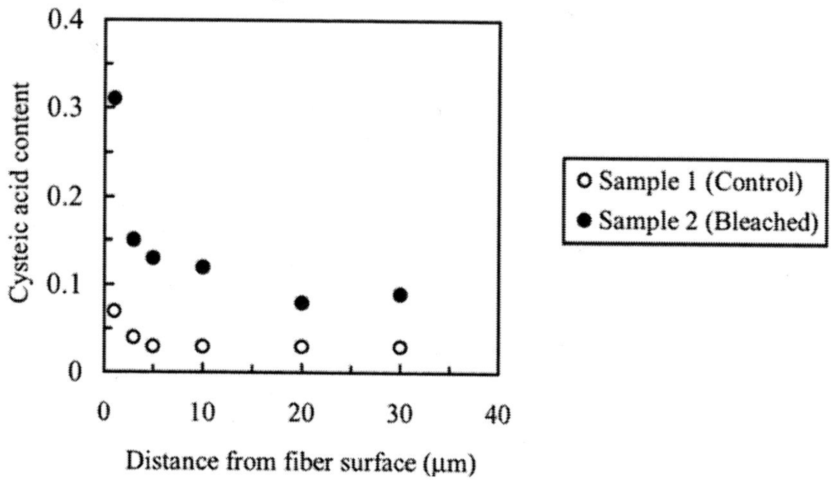

Figure 23. Depth profile, which shows the function between the cysteic acid content (the ratio of the peak area: S-O band/ C-H band) and the distance from fiber surface, of the hair samples due to the bleaching treatment.

Furthermore, the secondary structure at various depths of the hair fibers was estimated by amide I band analysis. The β-sheet and/or random coil content (β/R), and the α-helix (α) content in hair samples (Samples 1 and 2) at depths of 1, 3, 5, 10, 20, and 30 μm from the fiber surface are shown in Table V. The β/R and α contents in the cortex region of all hair samples were found to be almost constant. The mean and standard deviation (n=5) of the β/R content and α content in the cortex region of hair samples (Samples 1 and 2) are shown in Table V.

Table V. Mean and Standard Deviation (n=5) of the β/R Content and α Content in the Cortex Region (n=5: Depths of 3, 5, 10, 20, and 30 μm from the Fiber Surface) of Hair Samples (Samples 1 and 2)

Sample	β/R Content:	α Content:
1 (Control)	0.72 ± 0.018[a]	0.43 ± 0.012
2 (Bleached)	0.82 ± 0.028	0.32 ± 0.004

[a] Mean ± Standard deviation.

The β/R content in the cortex region of Sample 2 increased (the level of significance statistically calculated from the five measured points in the cortex region: $P = 0.005$), while the α content in the cortex region slightly decreased compared with that of the Sample 1 (the level of significance statistically calculated from the five measured points in the cortex region: $P < 0.001$). This suggests that the α-helix structure of some of the proteins existing throughout the cortex region of the untreated white human hair was changed to the β-sheet and/or random coil structures by performing the bleaching treatment.

Influence of Permanent Waving Treatments on Keratin Fibers

The permanent waving treatment for hair keratin fibers consists of two different processes, disconnection (the reduction process) and reconnection of -SS- groups (the oxidation process), and is widely used in the cosmetic industry. The changes in the chemical properties of human hair by performing permanent waving treatments have been extensively studied. It has been found that there is a slight decrease in the 1/2- cystine content [1,35], a slight increase in the cysteic acid content [1,35,54], an elution of proteins [57], and a reduction in the α-helix content [8], when performing the permanent waving treatments. The changes in the physical and mechanical properties of human hair resulting from these treatments have also been studied. It has been found that there is a reduction in tensile strength [1,41,51,52,58], and an increase in the swelling of the hair [1,41]. This research notwithstanding, molecular level studies on the mechanism leading to the reduction in tensile strength of permanent waved human hair, are still lacking comprehensiveness. Therefore, it is important to obtain information about the structural changes of hair keratin fibers, such as the change in -SS- groups, before and after conducting permanent waving treatments.

In order to investigate the mechanism leading to the reduction in tensile strength of permanent waved human hair, the cross-sectional structure at various depths of permanent waved white human hair was directly analyzed using a Raman microscope. Here, we prepared cross-sectional samples of human hair treated with 6.0 wt % TG (at pH 9.0, at 25°C for 15 and 60 minutes, at a ratio of hair to solution of 1: 250) and then oxidized (6.0 wt % sodium bromate, at 25°C for 15 minutes, at a ratio of hair to solution of 1: 250).

The cuticle Raman spectra (depth of 1 μm from the fiber surface) of the untreated white human hair (Sample 1) and the permanent waved white human hair (Sample 3) is shown in Figure 24.

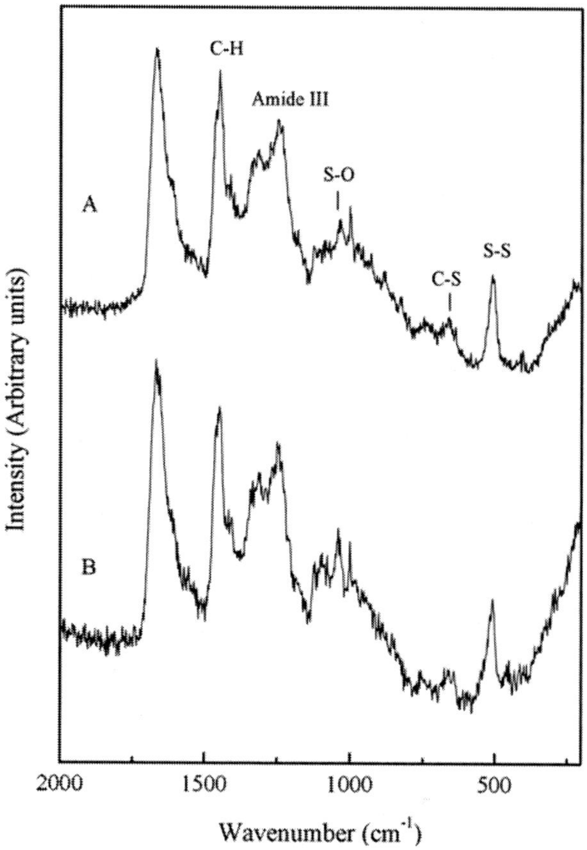

Figure 24. Cuticle Raman spectra (depth of 1 μm from the fiber surface) of the untreated white human hair (Sample 1) and the permanent waved white human hair (Sample 3): (A) Sample 1, and (B) Sample 3.

As is shown, the S-S and C-S band intensities existing in the cuticle region of the untreated white human hair were almost unchanged, while the S-O band intensity at 1040 cm^{-1}, assigned to cysteic acid, slightly increased by performing the permanent waving treatment. This suggests that as a result of the reduction process, most of the -SS- groups (producing -SH groups) existing in the cuticle region were disconnected, and that after oxidation most of the -SS- groups were reconnected while some of -SH groups were converted to cysteic acid.The representative cortex Raman spectra (depth of 5 μm from the fiber surface) of the untreated white human hair (Sample 1) and the permanent waved white human hair (Sample 3) is shown in Figure 25. It is shown that the band shape, as well as peak maximum frequency, of the cuticle region of the human hair fiber was significantly different from that of the cortex region of the human hair fiber. The S-S, S-C, and S-O band intensities existing in the cortex region of the virgin white human hair were almost unchanged by performing the permanent waving treatment. This suggests that the cleaved -SS- groups (-SH groups) existing in the cortex region as a result of the reduction process were reconnected by performing the oxidation process. Also, the amide III (unordered) band intensity at 1243 cm^{-1}, assigned to random coil form, slightly increased, indicating that some of the proteins existing throughout the cortex region of the untreated white human hair changed to the random coil form by performing the permanent waving treatment.

Next, to investigate the influence of the reduction treatment time on the cuticle and cortex of human hair, the disulfide (-SS-) and the cysteic acid contents at various depths of the hair fibers were compared by Raman spectroscopic analysis. Here, normalization of Raman spectra of the keratin fibers was carried out based on the C-H band at 1450 cm^{-1}, in which the peak area is large.

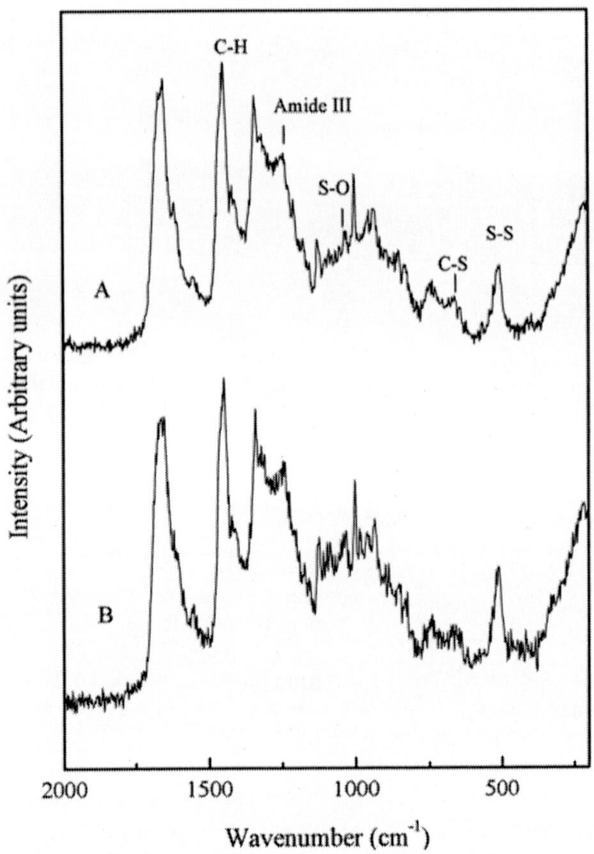

Figure 25. Representative cortex Raman spectra (depth of 5 μm from the fiber surface) of the untreated white human hair (Sample 1) and the permanent waved white human hair (Sample 3): (A) Sample 1, and (B) Sample 3.

The disulfide content in hair samples (Samples 1, 2, and 3) at depths of 1, 3, 5, 10, 20, and 30 μm from the fiber surface is shown in Table VI. The -SS- contents of Samples 2 and 3 decreased compared with that of Sample 1 for the cuticle region (depth of 1 μm from the fiber surface). On the other hand, the -SS- contents of Samples 2 (Mean ± standard deviation = 0.22 ± 0.017) and 3 (Mean ± standard deviation = 0.23 ± 0.021) were equal to that of Sample

1 (Mean ± standard deviation = 0.24 ± 0.023) for the cortex region (depths of 3-30 μm from the fiber surface). This result indicates that the -SS- content of the cortex region was almost unchanged, while -SS- content of the cuticle region was altered by performing the permanent waving treatment.

The cysteic acid content of hair samples (Samples 1, 2, and 3) at depths of 1, 3, 5, 10, 20, and 30 μm from the fiber surface is shown in Table VI. The cysteic acid content of Sample 1 increased when progressing from center to fiber surface. This suggests that the partial -SS- groups existing in the surface of the untreated white human hair changed to cysteic acid through natural oxidation. The cysteic acid content existing in the cuticle region and throughout the cortex region of Sample 2 increased compared with that of Sample 1. Also, the cysteic acid content existing in the cuticle region and throughout the cortex region of Sample 3 scarcely increased compared with that of Sample 2. This result indicates that the cysteic acid contents of the cuticle and cortex region increased by increasing the reduction treatment time.

Table VI. Disulfide Content, Cysteic Acid Content, β/R Content, α Content and Random Coil Content in Hair Samples (Samples 1, 2, and 3) at Depths of 1, 3, 5, 10, 20, and 30 μm from the Fiber Surface

Sample	Distance from Fiber Surface (μm)	Disulfide Content	Cysteic Acid Content	β/R Content	α Content	Random Coil Content
1 (Control)	1	0.35	0.07	0.93	0.38	0.43
	3	0.28	0.00	0.77	0.47	0.21
	5	0.24	0.00	0.67	0.52	0.17
	10	0.23	0.00	0.66	0.41	0.18
	20	0.21	0.00	0.65	0.41	0.16
	30	0.24	0.00	0.64	0.41	0.17
2 (R-15)	1	0.25	0.06	0.85	0.30	0.37
	3	0.22	0.03	0.83	0.41	0.25
	5	0.24	0.05	0.78	0.41	0.23
	10	0.23	0.04	0.81	0.39	0.24
	20	0.19	0.01	0.78	0.38	0.26
	30	0.23	0.02	0.80	0.39	0.24
3 (R-60)	1	0.32	0.10	0.93	0.44	0.45
	3	0.22	0.03	0.63	0.43	0.20
	5	0.27	0.07	0.66	0.44	0.16
	10	0.21	0.05	0.64	0.42	0.21
	20	0.22	0.03	0.62	0.39	0.20
	30	0.24	0.04	0.66	0.4	0.21

Table VII. Mean and Standard Deviation of β/R Content and α Content in the Cortex Region (n=5: Depths of 3, 5, 10, 20, and 30 μm from the Fiber Surface) of Hair Samples (Samples 1, 2, and 3)

Sample	β/R Content	α Content
1 (Control)	0.68 ± 0.046 [a]	0.44 ± 0.046
2 (R-15)	0.80 ± 0.019	0.40 ± 0.014
3 (R-60)	0.64 ± 0.017	0.41 ± 0.020

[a] Mean ± Standard deviation.

Moreover, the secondary structure at various depths of the hair fibers was estimated by amide I band analysis. The β-sheet and/or random coil content (β/R), and the α-helix (α) content in hair samples (Samples 1,2, and 3) at depths of 1, 3, 5, 10, 20, and 30 μm from the fiber surface are shown in Table VI. The β/R and α contents in the cortex region of all hair samples were found to be almost constant. The mean and standard deviation (n=5) of the β/R content and α content in the cortex region of hair samples (Samples 1, 2, and 3) are shown in Table VII.

The β/R content in the cortex region of Sample 2 increased (the level of significance statistically calculated from the five measured points in the cortex region: P = 0.001), while the α content in the cortex region slightly decreased compared with that of the Sample 1 (the level of significance statistically calculated from the five measured points in the cortex region: P = 0.08). This suggests that the α-helix structure of some of the proteins existing throughout the cortex region of the untreated white human hair were changed to the β-sheet and/or random coil structures by performing the oxidation treatment after the reduction treatment (15 minutes).

The β/R content of Sample 3 remarkably decreased compared with that of Sample 2 (the level of significance calculated by statistical test: P = 0.0000007), indicating that the β-sheet and/or random coil structures of some of the proteins existing throughout the cortex region of the untreated white human hair were eluted by performing the extended reduction treatment (60 minutes). Also, considering that -SS- content did not change by increasing the reduction time (from 15 minutes to 60 minutes), it was supposed that the β-sheet and/or random coil structures of some of the proteins not related to the matrix were eluted. On the other hand, the α content in the cortex region of Sample 3 did not change compared with that of the Sample 2, suggesting that the α-helix structure of some of the proteins existing throughout the cortex region of the untreated white human hair were not eluted by the extended reduction treatment (60 minutes).

Furthermore, the random coil content at various depths of the hair samples was compared by Raman spectroscopy. The random coil content in hair samples (Samples 1, 2, and 3) at depths of 1, 3, 5, 10, 20, and 30 μm from the fiber surface is shown in Table VI. The random coil content for Sample 1 in the cuticle region was clearly higher than that of the cortex region. Also, the random coil content of the cortex region at depths of between 3 and 30 μm from fiber surface of Sample 1 was constant (Mean ± standard deviation = 0.18 ± 0.017). This suggests that the random coil content of Sample 1 is equally distributed in the cortex region. Also, the random coil content in the cortex region of Sample 2 (Mean ± standard deviation =

0.24 ± 0.010) increased compared with that of the Sample 1 (the level of significance calculated by statistical test: P = 0.0002). Moreover, the random coil content in the cortex region of Sample 3 (Mean ± standard deviation = 0.20 ± 0.019) decreased compared with that of the Sample 2 (the level of significance calculated by statistical test: P = 0.0040). These results were in an excellent agreement with the results (β/R content) of the previous amide I band analysis.

The results from amide I band analysis and amide III band analysis suggests that the α-helix structure of some of the proteins existing throughout the cortex region of the untreated white human hair was changed to the random coil structure, rather than the β-sheet structure by performing the oxidation treatment after the reduction treatment (15 minutes).

TEM micrographs (50,000×) of the longitudinal section of the cortex region of the untreated white human hair fiber (Sample 1) and the permanent waved white human hair (Sample 3) are shown in Figures 26 and 27. In the case of Sample 1, the microfibril (black lines: 7 – 8 nm) and the matrix (white lines) existing in the macrofibril were regularly arranged along the fiber axis. On the other hand, in the case of Sample 3, disorder of the microfibril and matrix was observed. TEM micrographs (12,000×) of the longitudinal section of the cuticle region of the untreated white human hair fiber (Sample 1) and the permanent waved white human hair (Sample 3) are shown in Figures 28 and 29.

The cuticle region of Sample 3 was almost unchanged, except for the slight reduction in electron density of the endocuticle (the dark areas) and the slight increase in electron density of the exocuticle (the light areas). This result indicates that the structure of some of proteins existing in the cortex region (the microfibril, and matrix) of the untreated white human hair, rather than that of the cuticle region was changed by performing the permanent waving treatment.

Figure 26. TEM micrograph (50,000×) of the longitudinal section of the cortex region of the untreated white human hair fiber (Sample 1).

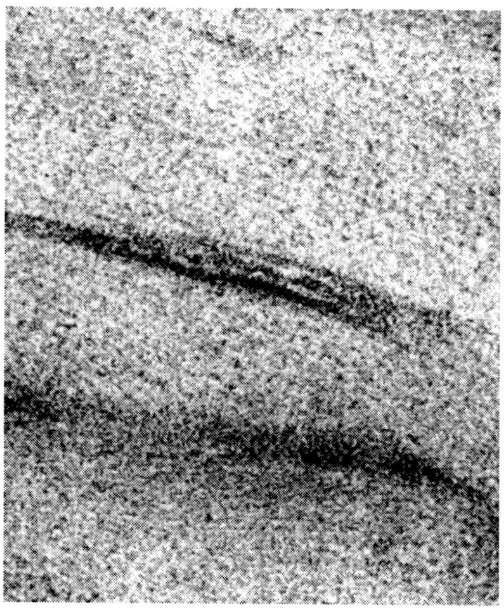

Figure 27. TEM micrograph (50,000×) of the longitudinal section of the cortex region of the permanent waved white human hair (Sample 3).

The damage degrees of the untreated and permanent waved human hair were compared. The tensile strength of the untreated and permanent waved white single fiber (Samples 1-3) measured at 25°C and 60 % RH is shown in Table VIII. The tensile strength of the untreated human hair (Sample 1: Control) decreased by increasing the reduction treatment time, indicating that the untreated human hair (Sample 1) was damaged by performing the permanent waving treatment.

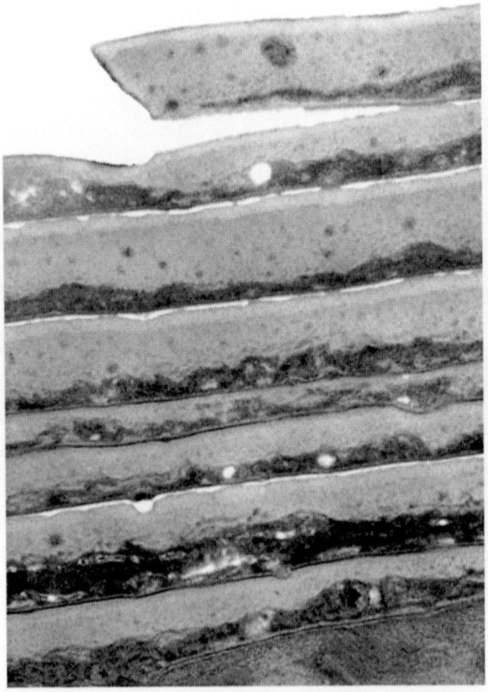

Figure 28. TEM micrograph (12,000×) of the longitudinal section of the cuticle region of the untreated white human hair fiber (Sample 1).

Figure 29. TEM micrograph (12,000×) of the longitudinal section of the cuticle region of the permanent waved white human hair (Sample 3).

Table VIII. Tensile Strength of the Untreated and Permanent Waved White Single Fiber at 25°C and 60 %RH (n=10)

Sample	Tensile strength[a]	
	(N/fiber × 10^{-2})	(N/m^2 × 10^8)
1 (Control)	127 ± 21	3.58 ± 0.49
2 (R-15)	95 ± 20	2.76 ± 0.53
3 (R-60)	85 ± 22	2.21 ± 0.29

[a] Mean ± Standard deviation.

Considering that the tensile strength of human hair is derived from the cortex region, and is not derived from the cuticle region [1,41,59], it can be assumed that the structure of some of the proteins existing in the cortex region (the microfibril, matrix, and -SS- conformation) of the untreated white human hair were changed, and some of the proteins (the random coil structure) were eluted from the cortex region, thereby leading to the remarkable reduction in the tensile strength of the white human hair after the permanent waving treatment.

As a result, the α-helix structure of some of the proteins existing in the microfibril of the cortex region was changed to the random coil structure by the performing the oxidation treatment after the reduction treatment (15 minutes). On the other hand, the -SS- content existing in the matrix of the cortex region, which forms the cross-linkages in the keratin hair fibers thereby contributing to physical and mechanical properties as well as structural stability, was almost unchanged, despite the remarkable reduction in the tensile strength of the white human hair following the permanent waving treatment. This result suggests that the disconnected -SS- groups existing in the matrix of the cortex region as a result of the reduction process did not return to their original -SS- conformation (the intermolecular -SS- linkage) before the permanent waving treatment despite conducting the oxidation process. Moreover, the proteins not related to the matrix, which have the random coil structure, were eluted from the cortex region of the untreated white human hair by performing the oxidation treatment after an extended reduction treatment (60 minutes). Also, transmission electron microscope observation has found that the macrofibril (the microfibril and matrix) existing in the cortex region of the untreated white human hair was remarkably disturbed by performing the permanent waving treatment.

From these experiments, we concluded that some of the proteins existing in the cortex region (the microfibril, matrix, and -SS- conformation) of the untreated white human hair were changed, and some of the proteins not related to the matrix, which have the random coil structure, were eluted from the cortex region, thereby leading to the remarkable reduction in the tensile strength of the white human hair after the permanent waving treatment.

CONCLUSION

It was revealed that the protein structural change of keratin fibers resulting from chemical treatments (reduction, introduction of -SS- groups using 2-IT, bleach, permanent wave etc.) at various depths of the cross-sectional hair samples could be directly characterized without isolating the cuticle and cortex, by Raman spectroscopy. In particular, we revealed that the Raman spectra of virgin black human hair, which had been impossible due to it's high melanin granule content could be recorded by our method. The key points of this method are to cross-section hair samples to a thickness of 1.50-μm, to select points at various depths of the cortex with the fewest possible melanin granules, and to optimize laser power, cross slit width, as well as total acquisition time. This chapter discusses that the -SS-, cysteic acid, random coil, β/R and α contents can be estimated, but does not statistically discuss the structural change of black human hair with aging. It has been shown that Raman spectroscopy becomes a beneficial analytical tool to investigate more detailed internal structural changes due to the influence of not only external factors such as heating, permanent waving, bleaching treatments, and exposure to sunlight, but also internal factors such as aging and nutritional deficiencies on black human hair, since the Raman spectra of virgin black human hair keratin fibers can be recorded. Moreover, it can be supposed that the measurement of the Raman spectra of soft keratin fibers which contain which contain melanin granules such as human skin tissue and petrified keratin fibers like horn, nail and animal hoof, as well as teeth can be applied using our method.

REFERENCES

[1] Robbins, CR. *Chemical and Physical Behavior of Human Hair*. 2rd ed. New York/Berlin/Heidelberg: Springer-Verlag; 1988.

[2] Lindley, H. The Chemical Composition and Structure of Wool. In: Asquith RS, editor. *Chemistry of Natural Protein Fibers*. New York: Plenum Press; 1977; pp 147-191.

[3] Arai, K. The Chemistry of Wool and Its Structure and Properties. *Sen-i Gakkaishi* 1989, 45 (12), 512-516.

[4] Fraser, RDB; Gillespie, JM; MacRae, TP; Marshall, RC. *Schematic diagram of a fine wool fiber* (CSIRO Division of Protein Chemistry); In the paper referenced by Marshall, RC; Gillespie, JM; McGuirk, GJ; Marler, JW; Reis, PJ; Rogen, IM; Whiteley, KJ. Proc 7th Int Wool Text Res Conf, Tokyo 1985, II, pp 36-44.

[5] Gillespie, JM. The proteins of Hair and Other Hard α-Keratins. In: Goldman RD, Steinert PM editors. *Cellular and Molecular Biology of Intermediate Filaments*. New York/London: Plenum Press; 1990; pp 95-128.

[6] Feughelman, M. Natural Protein Fibers. *J. Appl. Polym. Sci.* 2002, 83, 489-507.

[7] Fraser, RDB; MacRae, TP; Rogers, GE. Molecular Organization in Alpha-Keratin. *Nature* 1962, 193, 1052-1055.

[8] Nishikawa, N; Tanizawa, Y; Tanaka, S; Horiguchi, Y; Asakura, T. Structural Change of Keratin Protein in Human Hair by Permanent Waving Treatment. *Polymer* 1998, 39 (16), 3835-3840.

[9] Briki, F; Busson, B; Doucet, J. Organization of Microfibrils in Keratin Fibers Studied by X-Ray Scattering Modeling Using the Paracrystal Concept. *Biochim. Biophys. Acta*, 1998, 1429, 57-68.

[10] Kreplak, L; Doucet, J; Briki, F. Unraveling Double Stranded α-Helical Coiled Coils: An X-Ray Diffraction Study on Hard α-Keratin Fibers. *Biopolymers*, 2001, 58, 526-533.

[11] Kreplak, L; Doucet, J; Dumas, P; Briki, F. New Aspects of the α-Helix to β-Sheet Transition in Stretched Hard α-Keratin Fibers. *Biophys J.*, 2004, 87, 640-647.

[12] Kajiura, Y; Watanabe, S; Itou, T; Nakamura, K; Iida, A; Inoue, K; Yagi, N; Shinohara, Y; Amemiya, Y. Structural Analysis of Human Hair Single Fibers by Scanning Microbeam SAXS. *J. Struct. Biol.*, 2006, 155, 438-444.

[13] Yoshimizu, H; Ando, I. Conformational Characterization of Wool Keratin and S-(Carboxymethyl)kerateine in the Solid State by ^{13}C CP/MAS NMR Spectroscopy. *Macromolecules* 1990, 23, 2908-2912.

[14] Frushour, BG; Koenig, JL. In: Clark RJH, Hester RE editors. *Advances in Infrared and Raman spectroscopy: Vol. 1.* London: Heyden; 1975; pp 35-97.

[15] Lang, PL; Katon, JE; O'Keefe, JF; Schiering, DW. The Identification of Fibers by Infrared and Raman Microspectroscopy. *Microchem. J.* 1986, 34 (3), 319-331.

[16] Hsu, SL; Moore, WH; Krimm, S. Vibrational Spectrum of the Unordered Polypeptide Chain: A Raman Study of Feather Keratin. *Biopolymers* 1976, 15, 1513-1528.

[17] Shishoo, R; Lundell, M. Investigation of Structural Changes in Wool Fibers Due to Annealing. *J. Polym. Sci. Polym. Chem. Ed.* 1976, 14, 2535-2544.

[18] Carter, EA; Fredericks, PM; Church, JS; Denning, RJ. FT-Raman Spectroscopy of Wool – I. Preliminary Studies. *Spectrochem. Acta* 1994, 50A, 1927-1936.

[19] Williams, AC; Edwards, HGM; Barry, BW. Raman Spectra of Human Keratonic Biopolybers: Skin, Callus, Hair and Nail. *J. Raman. Spectros.* 1994, 25, 95-98.

[20] Church, JS; Corino, GL; Woodhead, AL. The Analisis of Merino Wool Cuticle and Cortical Cells by Fourier Transform Raman Spectroscopy. *Biopolymers*, 1997, 42, 7-17.

[21] Rintoul, L; Carter, EA; Stewart, SD; Fredericks, PM. Keratin Orientation in Wool and Feathers by Polarized Raman Spectroscopy. *Biopolymers* 2000, 57, 19-28.

[22] Kuzuhara, A; Hori, T. The Wrinkle Behavior of Wool Fabrics Introduced Thiol Groups and the Effects of Hg Ion Adsorption. *Sen-i Gakkaishi* 2000, 56, 69-75.

[23] Kuzuhara, A; Hori, T. Reducing Wrinkle Formation in Wool with 2-Iminothiolane Hydrochloride. *Textile Res. J.* 2002, 72, 285-289.

[24] Kuzuhara, A. Chemical Modification of Keratin Fibers Using 2-Iminothiolane Hydrochloride. *J. Appl. Polym. Sci.* 2003, 90, 3646-3651.

[25] Hogg, LJ; Edwards, HGM; Farwell, DW; Peters, AT. FT Raman Spectroscopic Studies of Wool. *J. Soc. Dyers Colour* 1994, 110, 196-199.

[26] Jones, DC; Carr, CM; Cooke, WD; Lewis, DM. Investigating the Photo-Oxidation of Wool Using FT-Raman and FT-IR Spectroscopies. *Textile Res. J.* 1998, 68 (10), 739-748.

[27] Kuzuhara, A. Analysis of Structural Changes in Bleached Keratin Fibers (Black and White Human Hair) Using Raman Spectroscopy. *Biopolymers*, 2006, 81, 506-514.

[28] Kuzuhara, A. Analysis of Structural Changes in Permanent Waved Human Hair Using Raman Spectroscopy. *Biopolymers*, 2007, 85, 274-283.

[29] Pande, CM. FT-Raman Spectroscopy – Applications in hair research. *J. Soc. Cosmetic. Chem.* 1994, 45, 257-268.

[30] Kuzuhara, A; Hori, T. Reduction Mechanism of Tioglycolic Acid on Keratin Fibers Using Microspectrophotometry and FT-Raman Spectroscopy. *Polymer*, 2003, 44, 7963-7980.

[31] Kuzuhara, A. Analysis of Structural Change in Keratin Fibers Resulting from Chemical Treatments Using Raman Spectroscopy. *Biopolymers*, 2005, 77, 335-344.

[32] Kuzuhara, A; Hori, T. Reduction Mechanism of L-Cysteine on Keratin Fibers Using Microspectrophotometry and Raman Spectroscopy. *Biopolymers*, 2005, 79, 324-334.

[33] Kuzuhara, A. Protein Structural Changes in Keratin Fibers Induced by Chemical Modification Using 2-Iminothiolane Hydrochloride: A Raman Spectroscopic Investigation. *Biopolymers*, 2005, 79, 173-184.

[34] Schlucker, S; Liang, C; Strehle, KR; DiGiovanna, JJ; Kraemer, KH; Levin, IW. Conformational Differences in Protein Disulfide Linkages between Normal Hair and Hair from Subjects with Trichothiodystrophy: A Quantitative Analysis by Raman Microspectroscopy. *Biopolymers,* 2006, 82, 615-622.

[35] Chao, J; Newson, AE; Wainwright, IM; Mathews, RA. Comparison of the Effects of Some Reactive Chemicals on the Proteins of Whole Hair, Cuticle and Cortex. *J. Soc. Cosmetic Chem.* 1979, 30, 401-413.

[36] Dhamelincourt, P; Ramiirez, FJ. Polarized Micro-Raman and FT-IR Spectra of L-Glutamine. *Appl. Spectrosc.* 1993, 47, 446-451.

[37] Thomas, G.J; Prescott, B; Day, LA. Structure Similarity, Diffrence and Variability in the Filamentaous Viruses Fd, If1, Ike, Pf1 and Xf: Investigation by Laser Raman Spectroscopy. *J. Mol.Biol.* 1983, 165, 321-365.

[38] Sengupta, PK; Krimm, S. Vibrational Analysis of Peptides, Polypeptides, and Proteins. XXXII. α-Poly(L-glutamic acid). *Biopolymers* 1985, 24, 1479-1491.

[39] Lopez Navarrete, JT; Hernandez, V; Ramirez, FJ. Vibrational Study of Aspartic Acid and Glutamic Acid Dipeptides. *J. Mol. Struct.* 1995, 348, 249-252.

[40] Gershon, SD; Goldberg, MA; Rieger, MM. Permanent Waving; In Balsam MS, Gershon SD, Rieger MM, Sagarin E, Strianse SJ editors. *Cosmetics Science and Technology: Vol. 2.* New York: Wiley; 1972; pp 167-250.

[41] Robbins, CR. *Chemical and Physical Behavior of Human Hair.* 4th ed. New York/Berlin/Heidelberg: Springer; 2001.

[42] Farnworth, AJ.; Lipson, M; McPhee, JR. The Development of Washable Non-Iron Effects in Pure Wool Fabrics. *Textile Res. J.* 1960, 30, 11-22.

[43] Feughelman, M. The Mechanical Properties of Permanently Set and Cystine Reduced Wool Fibers at Various Relative Humidities and the Structure of Wool. *Textile Res. J.* 1963, 33, 1013-1022.

[44] Fraser RDB., MacRae TP., Sparrow LG., Parry DAD. Disulphide Bonding in α-Keratin. *Int. J. Biol. Macromol.* 1988; 10: 106-112.

[45] Schramm, HJ; Dulffer, T. The Use of 2-Iminothiolane as a Protein Crosslinking Reagent. *Hoppe-Seyler's Z Physiol. Chem.* 1977, 358, 137-139.

[46] Kuzuhara, A; Hori, T. A New Creaseproof Finish for Wool Using 2-Iminothiolane Hydrochloride and Its Reaction Mechanism. *Textile Res. J.* 2002, 72, 526-530.

[47] Tate, ML; Kamath, YK; Ruetsch, SB; Weigmann, H–D. Quantification and Prevention of Hair Damage. *J. Soc. Cosmetic Chem.* 1993, 44, 347-371.

[48] Ruetsch, SB; Yang, B; Kamath, YK. Chemical and Photo-Oxidative Hair Damage Studied by Dye Diffusion and Electrophoresis. *J. Soc. Cosmetic Chem.* 2003, 54, 379-394.

[49] Kuzuhara, A. Influence of Urea on the Coloring Ability of a Low-Temperature Coloring Method of Keratin Fibers Using Polyethyleneimine. *J. Appl. Polym. Sci.* 2004, 91, 3827-3834.

[50] Kaplin, IJ; Schwan, A; Zahn, H. Effects of Cosmetic Treatments on the Ultrastructure of Hair. *Cosmet Toiletries* 1982, 97, 22-26.

[51] Wortman, F–J; Souren, I. Extensional Properties of Human Hair and Permanent Waving. *J. Soc. Cosmetic Chem.* 1987, 38, 125-140.

[52] Syed, AN; Ayoub, H. Correlating Porosity and Tensile Strength of Chemically Modified Hair. *Cosmet Toiletries* 2002, 117, 57-64.

[53] Kuzuhara, A; Hori, T. Influence of Urea on the Diffusion of Polyethylenimine in Human Hair. *Sen-i Gakkaishi* 2003, 59, 123-127.

[54] Robbins, CR; Kelly, C. Amino Acid Analysis of Cosmetically Altered Hair. *J. Soc. Cosmetic Chem.* 1969, 20, 555-564.

[55] Wolfram, LJ; Hall, K; Hui I. The Mechanism of Hair Bleaching. *J. Soc. Cosmetic Chem.* 1970, 21, 875-900.

[56] Erlemann, GA; Beyer, H. Influence of Hydrogen-Peroxide to the Chemical Structure of Human Hair. *J. Soc. Cosmetic Chem.* 1971, 22, 795-807.

[57] Sandhu, SS; Robbins, CR. A Simple and Sensitive Technique, Based on Protein Loss Measurement, to Assess Surface Damage to Human Hair. *J. Soc. Cosmetic Chem.* 1993, 44, 163-175.

[58] Beyak, R; Meyer, CF; Kass, GS. Elasticity and Tensile Properties of Human Hair. I. Single Fiber Test Method. *J. Soc. Cosmetic Chem.* 1969, 20, 615-625.

[59] Robbins, CR; Crawford, RJ. Cuticle Damage and the Tensile Properties of Human Hair. *J. Soc. Cosmetic Chem.* 1991, 42, 59-67.

In: Biopolymer Research Trends
Editor: Tamas S. Nemeth, pp. 87-118

ISBN: 978-1-60021-983-2

Chapter 3

COMPLEX NMR APPROACHES TO STUDYING CONFORMATIONAL DYNAMICS OF BIOPOLYMERS

Alexey G. Krushelnitsky
Kazan Institute of Biochemistry and Biophysics,
Russian Academy of Sciences
Kazan, Russia

ABSTRACT

The investigation of molecular motions in biological polymers has been one of the basic trends in molecular biophysics for a long time. Many physical methods have been applied to studying biomolecular mobility. However, in spite of the large amount of experimental data there are still some methodological problems that are not yet completely resolved. One of the most essential ones is the ambiguity of transition from the first-hand experimental parameters to the parameters characterizing molecular motions. The most poorly defined characteristic of a motion is its geometry. There are almost no experimental techniques, except computer simulation, that provide direct and unambiguous information on the motional geometry models. At the same time, this information in many cases can be of high importance for revealing molecular mechanisms of the protein biological function. In this contribution we describe the experimental approaches that may solve this problem. These approaches are based on the complex experimental NMR study. One of the main advantages of NMR in respect to other physical methods is that it allows using different magnetic nuclei and different magnetic interactions (dipole-dipole and quadrupole couplings, chemical shift anisotropy) for probing the same kind of molecular mobility. The comparative quantitative analysis of different types of NMR data obtained on the same sample may allow the discrimination of different motional models directly from the experimental data. This complex approach is demonstrated by a study of molecular dynamics of a model system, homopolypeptide poly-L-lysine, and backbone dynamics of a protein barstar in solid state. Limitations as well as perspectives of the development of this approach are discussed in detail.

INTRODUCTION

By now it has been widely recognized that the molecular dynamics of proteins as well as their three-dimensional structure is a key factor determining the molecular mechanism of the protein biological function. There are many examples demonstrating that were the proteins rigid they would not work (see reviews [1-7]). Thus, the molecular dynamics studies of proteins and other biopolymers have been a very popular trend of research in molecular biophysics for a long time and it is easy to predict that the number of works dealing with protein dynamics will significantly increase in the future. Many physical methods have been applied to the investigation of biopolymer dynamics, yet it is clear by now that the most powerful and informative experimental tool for this purpose is nuclear magnetic resonance (NMR). NMR provides the most representative and spatially selective dynamic information since magnetic nuclei are spread throughout a whole protein molecule and modern multi dimensional NMR techniques make possible site-specific line assignments in high resolution NMR spectra. Different types of NMR experiments enable covering an extremely wide frequency range of molecular dynamics – from picoseconds to seconds and even more. All this makes NMR a unique experimental method for studying molecular dynamics of biopolymers, although it is necessary to admit that this method is rather expensive and time-consuming in comparison with many other techniques.

NMR can be applied for studying biopolymers both in liquid and solid states. Both approaches have their own intrinsic methodical advantages and limitations. Liquid state methods enable easy achieving narrow lines in the spectra and high sensitivity, relatively simple sample preparation. It is also important that globular proteins are contained in the natural surrounding - water - and thus the problem of the influence of inter-protein interactions on the internal structural and dynamics properties can be neglected in the liquid state experiments. The advantages of the solid state NMR techniques applied to the molecular dynamics investigations is that they are applicable to non-solvable molecules (e.g. membrane proteins), they enable studying biomolecular dynamics as a function of hydration level (in solution it is impossible for the obvious reasons). However, the most essential advantage of the solid state experiments is that there is no Brownian tumbling of proteins in the solid state samples. The isotropic overall tumbling in solution averages the dipole-dipole, quadrupolar and chemical shift anisotropy (CSA) interactions out in the time scales longer than the correlation time of the Brownian tumbling. Thus, these magnetic interactions can not be used in the investigation of the slow internal dynamics which significantly limits capabilities of the liquid state NMR techniques. The correlation time of the Brownian tumbling for most proteins is around 10^{-8} s. Thus, the micro and millisecond time scales of internal dynamics of proteins in solution are practically inaccessible for most experimental methods, by the way, not only NMR. At the same time, this is a time scale of many biologically relevant events like catalysis, allosteric regulation, molecular recognition and binding, some stages of folding, etc. Thus, protein motions of this time scale are particularly interesting and important from the biological point of view.

There are some options to access slow protein conformational dynamics in solution by means of either chemical exchange methods [8] or the analysis of the residual dipolar couplings (RDC) measured in protein solutions with various aligning media [9,10]. However, the chemical exchange methods provide no information on the motional amplitude and the

RDC's on the contrary are not sensitive to the timescale of the internal motions. The RDC analysis seems to be an interesting and perspective technique for studying slow protein conformational dynamics in solution but it is still emerging and rapidly developing method which is far from being a routine yet. Although the solid state NMR experiments of course also have their own limitations, from the physical point of view they are best suited for studying molecular motions in all time scales.

We must note that there is still a debated issue on the influence of inter-protein interactions in the solid state on internal dynamic properties of proteins: the dynamic behavior of globular proteins in diluted solutions and solid (crystalline or powder) state in general can be different. Thus, it is unclear whether the data obtained in the solid state can be used for the interpretation of the properties of proteins in their natural surrounding - water. There is no an unambiguous convincing answer to this doubt yet since the direct comparison of the dynamic protein behavior in the liquid and solid state is practically impossible due to the overall tumbling in solution. However, there are experimental indications that even if the difference between the conformational dynamics in two states exists, it is not significant [11-13]. Also note that the protein spatial structures in crystals and solutions is in general very similar [14]. Thus, one can hardly claim that the solid state studies of proteins are biologically irrelevant.

Large amount of experimental NMR data on biopolymer dynamics have been accumulated by now. However, there are still some methodological problems of the analysis of these data that are not completely resolved yet. One of the most important problems is the ambiguity of the transition from the first-hand experimental parameters to the parameters characterizing molecular dynamics. The experimental NMR parameters (relaxation times, line shapes, etc.) are sensitive to molecular dynamics but the relation between these parameters and molecular dynamics characteristics (motional amplitudes, correlation times, activation energies, etc.) is ambiguous. That is why very often the experimental NMR parameters are being used as indirect indicators of molecular dynamics without detailed interpretation of the nature and properties of conformational motions. Such an approach may provide interesting information on dynamics when it is used on a comparative basis, i.e. if the keystone to the data analysis is a comparison of the same experimental parameters measured on a sample at different conditions (hydration, temperature, ligand binding, different locations within a protein, etc.). However, it is evidently clear that obtaining the detailed information on the amplitude, geometry, correlation time(s), form of the correlation function and physical nature of conformational motion would be of much higher significance since these data may reveal the molecular mechanisms of the protein biological function.

The most detailed and comprehensive quantitative information on molecular dynamics that can be obtained from a physical experiment is a correlation function of a motion. More detailed data can be obtained only from the computer simulations of molecular dynamics but this technique has obvious limitations which will not be discussed here. From the correlation function one may estimate the correlation time of a dynamic process and a portion of a certain physical value – in the case of NMR this is a magnetic interaction - that is being averaged by the molecular motion (Figure 1). This interaction is usually dipole-dipole internuclear interaction, quadrupolar interaction, if the quadrupolar nuclei are concerned, or CSA. The portion of the unaveraged interaction is often called order parameter S^2 introduced by Lipari and Szabo within the frames of model-free approach [15,16]. Order parameter is a measure of amplitude of a motion – to a first approximation, the less the order parameter, the more the motional amplitude. However, the order parameter provides no hint about the physical model of the motion. This is demonstrated by Figure 2: different motional models may equally explain the same value of the order parameter.

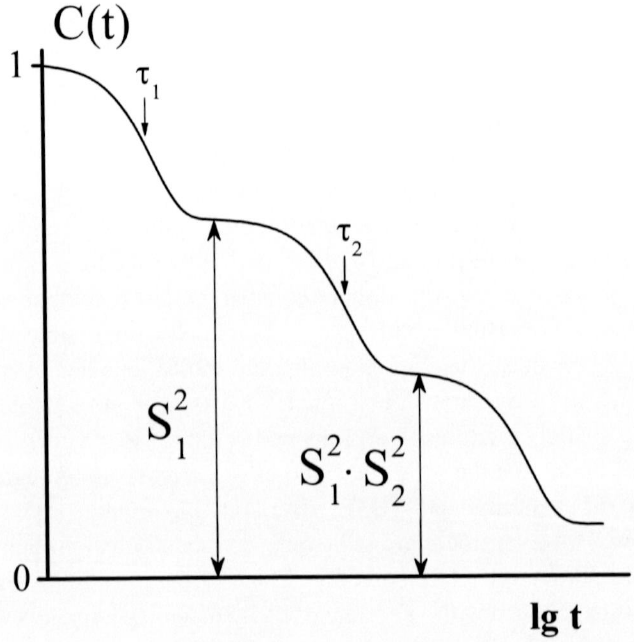

Figure 1. Schematic presentation of the normalized correlation function with stepwise averaging by several molecular motions having different correlation times.

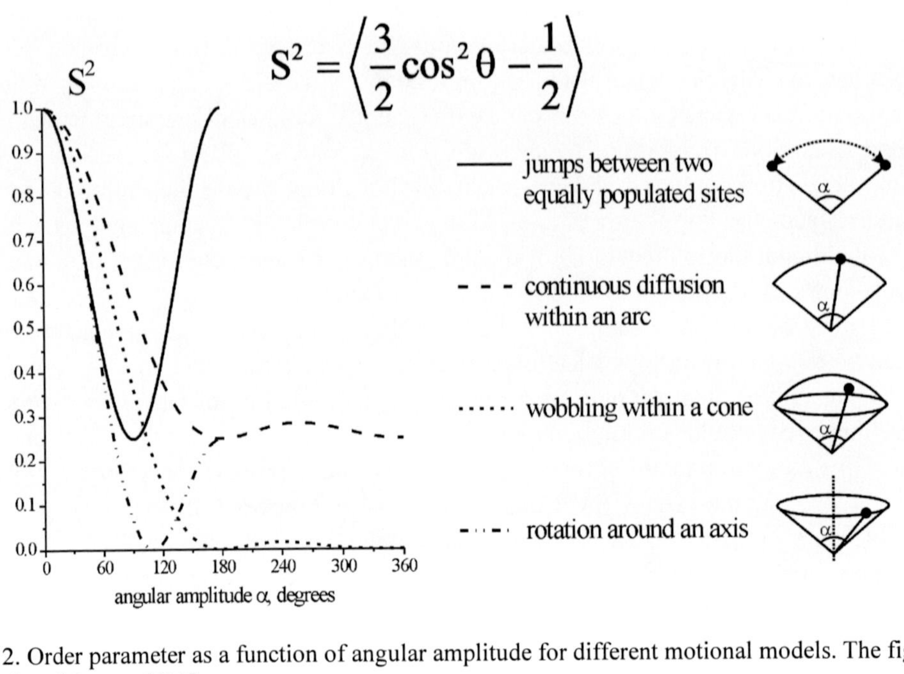

Figure 2. Order parameter as a function of angular amplitude for different motional models. The figure is reproduced from ref [32].

This is an important difference between structural and dynamic NMR experiments. In structural experiments one obtains experimental parameters (e.g. cross-peak intensities in NOESY experiments) that can unambiguously be recast to internuclear distances and various angles. This is not the case for dynamic studies - the transition from the experimental to dynamic parameters is ambiguous. This is a reason why many researches stop the data interpretation on the order parameter and do not go further. The main aim of the present contribution is to outline some of the methodical NMR approaches that may provide such detailed information. Briefly, these approaches can be formulated as follows: combined quantitative analysis of different order parameters characterizing the same motion but obtained from different NMR experiments. As we will demonstrate by two different examples below, such a comparative analysis may provide more detailed and definite information on physical nature of conformational dynamics as compared to applied up to now routine methods.

Solid Poly-L-Lysine: ^{13}C and ^{1}H NMR Relaxometry

Synthetic homopolypeptide poly-L-lysine was chosen as a convenient model system for studying biopolymer dynamics and its hydration response. Standard one-dimensional cross-polarization magic angle spinning (CPMAS) spectrum on natural abundance ^{13}C nuclei permits resolving all carbons in the polylysine (Figure 3) at all hydration levels studied (from 0 to 0.2 g water / 1 g polypeptide) which makes isotopic enrichment and multi-dimensional NMR techniques unnecessary. The secondary structure of polylysine and its dependence on hydration is known for a long time [17], at low hydration levels it has predominantly β-sheet structure.

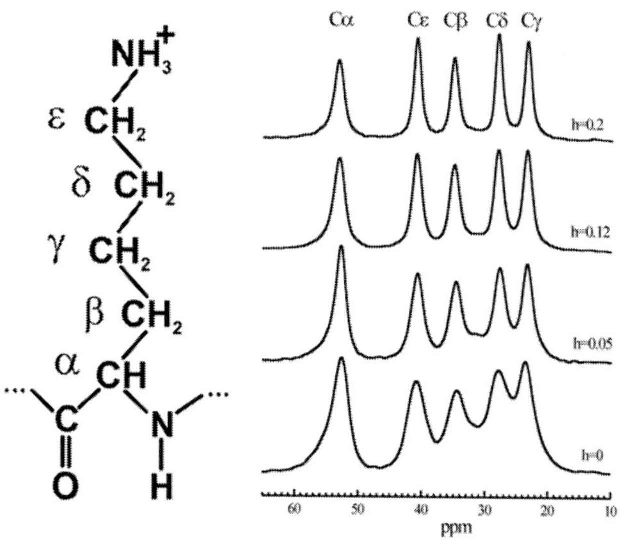

Figure 3. Chemical structure of polylysine (left) and aliphatic domain of natural abundance ^{13}C CPMAS spectra of polylysine at various hydration levels (right). The hydration level (h) is expressed in g H_2O per 1 g dry polypeptide. The figure is reproduced from ref [18].

This polypeptide is also convenient for the proton relaxation experiments: eight out of nine nonexchangeable protons are located on a side chain; and polylysine has no methyl groups, and thus, high-frequency internal motions are not hidden by high-amplitude fast methyl proton rotation. This is important because fast spin diffusion between protons in solid organic substances does not allow us to obtain selective dynamic information as in the case of natural abundance ^{13}C experiments. The results presented below were published in refs [18,19].

Primary Data Analysis

Obtaining the correlation function of motion from the relaxation experiments in complex systems like biopolymers is in most cases complicated and ambiguous procedure. Single relaxation experiment cannot provide the form of the correlation function: obtaining a function from a single value (or limited number of values) is in general case ill-defined problem. Thus, one may analyze only the relaxation times themselves or use these or those *a priori* assumptions to get some dynamic parameters. To reduce the number of *a priori* assumptions we have performed a wide set of relaxation experiments and analyzed simultaneously the whole set of the relaxation times measured on the same sample. Specifically, in carbon experiments we measured T_1 relaxation times at carbon resonance frequencies 50.3 and 100.5 MHz, $T_{1\rho}$ relaxation times with the resonance offset of the spin-lock field at spin-lock frequencies 115-125 and 205-225 kHz and $T_{1\rho}$ relaxation times with the low value of the spin-lock field (5.5 kHz) and proton decoupling during the carbon spin-lock pulse. The application of the resonance offset of the spin-lock pulse and proton decoupling during the carbon spin-lock pulse allow neglecting the interfering spin-spin contribution to the relaxation rate [20] and expanding the frequency range of the $T_{1\rho}$ measurements. These modifications of the $T_{1\rho}$ experiment were described in detail in [21] and will not be discussed here. As for the proton experiments, we have measured T_1's at the resonance frequency 200 MHz and off-resonance $T_{1\rho}$'s at two spin-lock fields, 105±3 and 194±4 kHz. The spin-lock field resonance offset for homonuclear $T_{1\rho}$ relaxation experiment was introduced by Jones [22]. The pulse sequence of the proton off-resonance $T_{1\rho}$ relaxation experiment was described in [23]. In the case of the proton experiments the hydration of the polypeptide was performed using D_2O instead of H_2O. The relaxation times were measured for the same sample at several different temperatures from 0° C to 55° C. Carbon relaxation times could be obtained for each carbon separately, whereas fast spin-diffusion between protons made possible obtaining only the averaged over all protons in the polypeptide value of the proton relaxation times. The polylysine sample at each hydration level was characterized by a set of 15-20 carbon relaxation times and approximately the same number of proton relaxation times. Such a relatively large number of the relaxation experiments performed on the same sample and combined quantitative treatment of the whole set of data enabled reducing the number of assumptions in the analysis (although some assumptions of course remained, see below) and obtaining more definite and reliable information on the form of the correlation function of motion on a wide timescale. The typical examples of the frequency-temperature dependencies of the relaxation times are shown in Figure 4.

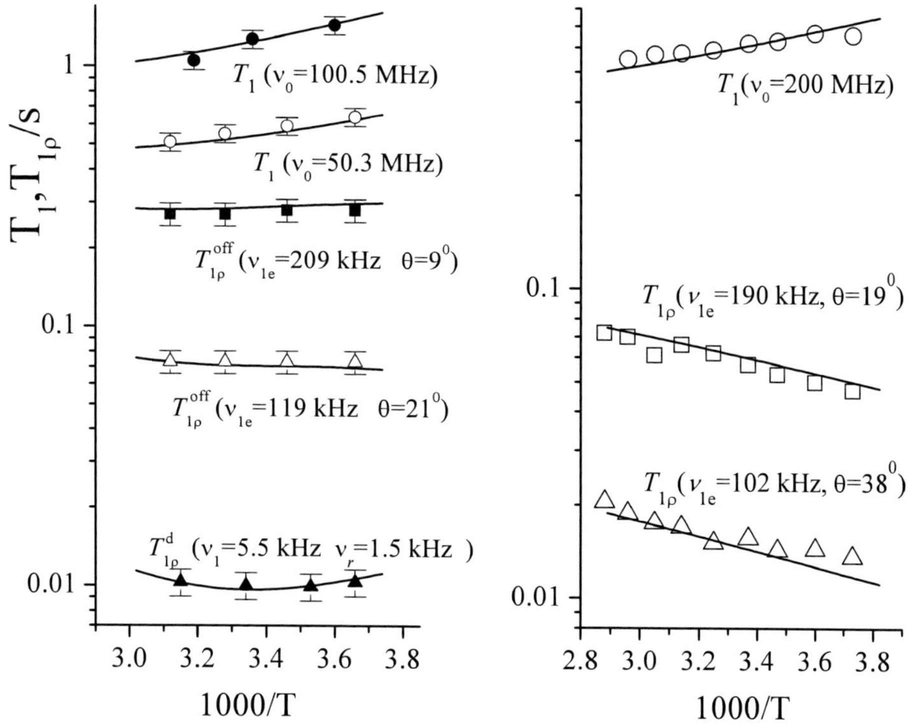

Figure 4. Typical examples of the frequency-temperature dependencies of the relaxation times. Left: carbon relaxation times measured for β-carbon in the dry sample. $T_{1\rho}^{off}$ and $T_{1\rho}^{d}$ denote the relaxation times $T_{1\rho}$ measured with the resonance offset of the spin-lock pulse and proton decoupling during the on-resonance carbon spin-lock pulse, respectively. Right: proton relaxation times measured for the dry sample. Size of the symbols on the right plot correspond to the experimental error. Conditions of the experiments are denoted on the plots. Solid lines are the fitting curves calculated according to the formalism described in the text. The figure is reproduced from refs [18,19].

The mathematical formalism of the analysis of the relaxation times is based on the correlation function formalism and model-free approach. The dominating mechanism of both carbon and proton spin-lattice relaxation is dipole-dipole inter-nuclear interaction. The carbon relaxation times T_1 and $T_{1\rho}$ are determined by the hetero-nuclear ^{13}C-^{1}H interaction and can be expressed as

$$\frac{1}{T_1}=\frac{K_d}{10}\big(J(\omega_H-\omega_C)+3J(\omega_C)+6J(\omega_H+\omega_C)\big) \tag{1}$$

$$\frac{1}{T_{1\rho}^{off}}=\frac{1}{T_1}+\sin^2\theta\left[\frac{1}{T_{1\rho}^{\Delta}}-\frac{1}{2T_1}\right] \tag{2}$$

$$\frac{1}{T_{1\rho}^{\Delta}}=\frac{K_d}{10}\big(2J(\omega_{1e})+3J(\omega_H)\big) \tag{3}$$

where K_d is the dipole-dipole interaction constant ($2.12 \cdot 10^{10}$ s^{-2} per one ^1H-^{13}C bond), $J(\omega)$ is a motional spectral density function reflecting the ^{13}C-^1H internuclear vector reorientation, $\omega_H/2\pi$ and $\omega_C/2\pi$ are the proton and carbon resonance frequencies in the laboratory frame, respectively, $\omega_{1e}/2\pi$ is the carbon resonance frequency in the tilted rotating frame. θ is the angle between B_0 and effective B_{1e} magnetic fields:

$$tg\theta = \frac{\gamma_C B_1}{2\pi\Delta\nu} \tag{4}$$

where $\Delta\nu$ is the resonance offset of the spin-lock field. In the limiting cases, $\theta=0^0$ and $\theta=90^0$, Eq. (2) yields the standard expressions for the spin-lattice relaxation time T_1 (Eq. 1) and for the on-resonance $T_{1\rho}$, respectively.

As for the proton-decoupled $T_{1\rho}$ carbon relaxation, there has been no quantitative approach to describe this experiment yet. Thus, for the analysis of these data we used a semi-empirical expression

$$\frac{1}{T_{1\rho}^d} = \frac{4X_a K_d}{15}\left(\frac{1}{6}J(\omega_1+2\omega_r)+\frac{1}{3}J(\omega_1+\omega_r)+\frac{1}{3}J(\omega_1-\omega_r)+\frac{1}{6}J(\omega_1-2\omega_r)\right) \tag{5}$$

where ω_1 and ω_r are the ^{13}C spin-lock and MAS rate circular frequencies and X_a is a phenomenological coefficient taking into account the partial averaging of the ^{13}C-^1H dipolar interaction by proton decoupling. Since the value of X_a cannot be quantitatively interpreted, the analysis of the absolute values of the relaxation times $T_{1\rho}^d$ is senseless; however, the slope of the temperature dependencies of these relaxation times may provide quantitative information on the correlation time of the slow internal motions. The Eqs. (1-5) were derived and discussed in [21].

The proton relaxation times are determined by the homo-nuclear ^1H-^1H dipolar couplings and can be expressed as:

$$\frac{1}{T_1} = \frac{2}{3}K_{HH}\left(J(\omega_0)+4J(2\omega_0)\right) \tag{6}$$

$$\frac{1}{T_{1\rho}^{off}} = \frac{1}{T_1} + \sin^2\theta\left[\frac{1}{T_{1\rho}^{\Delta}} - \frac{3}{4T_1}\right] \tag{7}$$

$$\frac{1}{T_{1\rho}^{\Delta}} = K_{HH}\left(\cos^2\theta\cdot J(\omega_{1e})+\sin^2\theta\cdot J(2\omega_{1e})+\frac{3}{2}J(\omega_0)\right) \tag{8}$$

where K_{HH} is the rigid lattice proton second moment, all other parameters have the same meaning as in Eqs. (1-3). The expression for the relaxation time $T_{1\rho}^{off}$ determined by the homo-nuclear dipolar interaction was derived in [22].

As for the spectral density functions, following the model-free approach we assume it to be in the form

$$J(\omega) = (1-S^2)\cdot\frac{\beta}{\omega}\cdot\frac{(\omega\tau_0)^{\beta}}{1+(\omega\tau_0)^{2\beta}} \tag{9}$$

where τ_0 is the correlation time of motion and β is a phenomenological parameter characterizing a width of the correlation time distribution which varies between 0 (infinitely wide distribution) to 1 (delta function). This form of the spectral density function corresponds to the well-known Fuoss-Kirkwood distribution function [24]. It is important to note that there might be two types of distribution: inhomogeneous distribution (local inhomogenity over a sample that leads to a distribution of dynamic parameters over different internuclear vectors) and homogeneous distribution (complex shape of the correlation function of a single internuclear vector reorientation due to a complex nature of motion). This problem was analyzed in detail in [18]. The Eq. (9) is valid if the inter-nuclear vector undergoes only one type of motion. However, if there are two independent motions with different time ranges such that $\tau_1 >> \tau_2$ where τ_1 and τ_2 are the mean correlation times of the motions, the spectral density function becomes

$$J(\omega) = (1-S_2^2)\cdot\frac{\beta_2}{\omega}\cdot\frac{(\omega\tau_2)^{\beta_2}}{1+(\omega\tau_2)^{2\beta_2}} + S_2^2(1-S_1^2)\cdot\frac{\beta_1}{\omega}\cdot\frac{(\omega\tau_1)^{\beta_1}}{1+(\omega\tau_1)^{2\beta_1}} \tag{10}$$

In our analysis we assume that the order parameter S^2 and the distribution width parameter β are temperature independent and that the temperature dependence of correlation times is governed by the Arrhenius law

$$\tau_0 = \tau_{293K}\, e^{\left(\frac{E_a}{R}\left(\frac{1}{T}-\frac{1}{293K}\right)\right)} \tag{11}$$

where τ_{293K} is the correlation time at 293 K, E_a is an activation energy, R is the universal gas constant and T is the absolute temperature. Following the formalism described above, each molecular motion can be characterized by a set of four dynamic parameters: order parameter, correlation time, distribution width parameter and activation energy. These dynamic parameters were determined from the relaxation times by a computer minimization (fitting) of the following expression:

$$RMSD = \sqrt{\frac{1}{N}\sum_{i=1}^{N}\left(\frac{T^i_{\exp} - T^i_{sim}}{T^i_{\exp}}\right)^2} \tag{12}$$

where N is the number of experimental points in the data set, $T_{\exp}$ and T_{sim} are the experimental and simulated T_1, $T_{1\rho}^{off}$ and $T_{1\rho}^{d}$ relaxation times measured at different resonance frequencies, spin-lock fields and temperatures. This procedure of the relaxation times treatment is very similar to that very often performed in the liquid-state NMR relaxation studies of the protein dynamics – simultaneous fitting of the T_1, T_2 and heteronuclear NOE measured at two, three or sometimes four resonance frequencies according to the Lipari-Szabo formalism. The main difference between the liquid and solid state experiments is that there is no Brownian tumbling in the solid samples and due to $T_{1\rho}$ relaxation times one may sample the internal motions in the microsecond timescale. The analysis of the ^{13}C data has shown that in most cases the data set can be reasonably described assuming that the polypeptide undergoes two independent motions in the nanosecond and micro-millisecond timescales: the description of the data assuming only one type of motion leads to a worse quality of the fitting and unreasonable values of some dynamic parameters. The results of the analysis are presented in detail below.

The most essential assumption of the formalism described above is the temperature independence of the order parameters. Although the temperature range of the relaxation experiments is relatively narrow, the variation of S^2 of the fast motion within the temperature range from 0° C to 50° C could be of the order of 10-20% (see [25] and references therein). To check, whether such a temperature dependence of the fast motion order parameter may affect the fitting results, we have fitted the same set of the data assuming the simplest linear temperature dependence of S^2:

$$S^2(T) = S^2(273^0 K)\cdot(1-k_t\cdot(T-273^0 K)) \tag{13}$$

where T is the absolute temperature, k_t is the slope of the temperature dependence. We assumed k_t to be 0.002 and 0.004 K^{-1}, which correspond to the 10% and 20% difference of the order parameters at 0° C and 50° C, respectively. The results of the fitting for γ-carbon in the dry sample at different k_t's are presented in Table 1. It is seen that the absolute values of some dynamic parameters (e.g. the order parameter and the correlation time of the fast motion) are indeed dependent on the k_t. However, it is also seen that this dependence is not dramatic and the fitting quality decreases (RMSD increases) with increasing k_t. Thus, in spite of the certain ambiguity of the determination of the absolute values of some dynamic parameters we think that assumption of the temperature independent order parameters is methodologically more correct. The reasons for making such an assumption are as follows: the real temperature dependence of the order parameters is not known and at present theoretically unpredictable; such dependence does not lead to dramatic changes of the fitting results; the analysis presented below relies mainly on the comparison of the order parameters, the precise absolute values are less important.

The results of the ^{13}C relaxation times analysis (Figure 5) show that in dry polylysine there are two low amplitude motions in dry polypeptide - fast and slow – with the correlation times of the order of 10^{-9} s and 10^{-4} s, respectively. As the hydration level of the polypeptide increases, the polypeptide internal mobility obviously increases as well.

Table 1. Dynamic parameters obtained from the fitting the relaxation times of γ-carbon in the dry sample. Columns A, B and C correspond to the coefficient k_t in the Eq. (13) 0, 0.002 and 0.004 K^{-1}, respectively. The order parameter of the fast motion in columns B and C corresponds to the temperature 20°C

	Dynamic parameters	A	B	C
Fast motion	S^2	0.86±0.01	0.80±0.02	0.75±0.02
	τ, ns	3.3±0.2	1.5±0.2	0.4±0.05
	β	0.63±0.04	0.4±0.02	0.34±0.02
	E_a, kJ/mol	24±2	21±3	22±2
Slow motion	S^2	0.84±0.04	0.86±0.09	0.75±0.08
	τ, μs	43±20	220±200	80±75
	β	0.12±0.03	0.12±0.07	0.05±0.02
	E_a, kJ/mol	90±23	75±22	110±15
RMSD		0.036	0.039	0.061

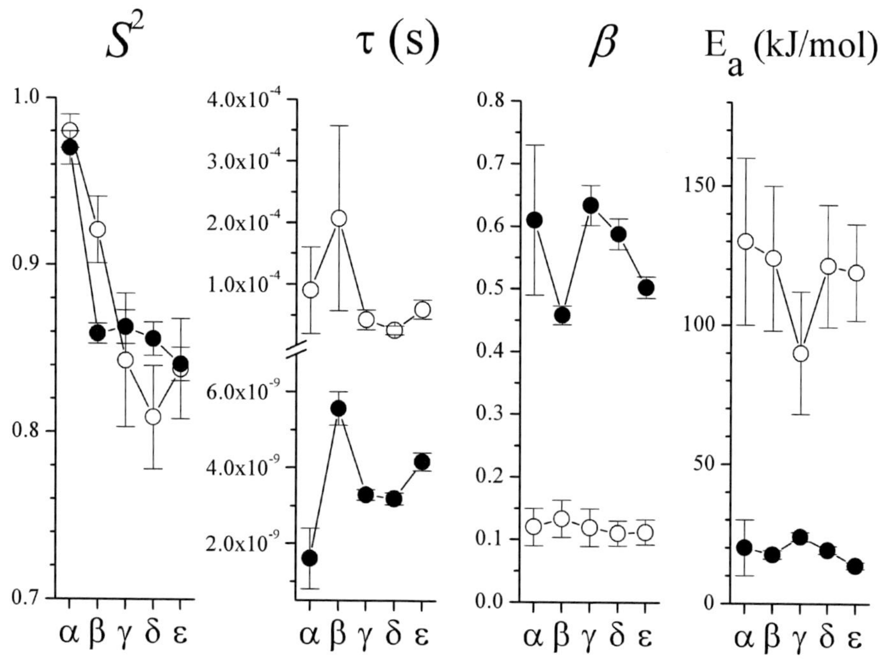

Figure 5. Dynamic parameters obtained from the fitting the ^{13}C relaxation times of all aliphatic carbons in dry polypeptide. Solid and open circles correspond to the parameters characterizing fast and slow motions, respectively. The figure is reproduced from ref [18].

However, the hydration response of the backbone and side chain motions is different: the backbone dynamics reveal only a modest increase of the order parameters, whereas the side chain dynamics show about five orders of magnitude decrease of the correlation time of the slow motion, so that at hydration levels higher than 10 % two motions become experimentally indistinguishable and the relaxation data for the side chain carbons could be well described by only one type of motion. The hydration dependence of the order parameters and correlation times of α and δ carbons (all side chain carbons reveal similar dynamic properties) are shown in Figure 6. It is interesting to note that the hydration response of the internal dynamics of a native globular protein studied by the same methodical approach is very different [23]. The comparative analysis of the hydration dependence of the dynamic parameters in various biological polymers may provide valuable information on the interaction between water and biopolymers and the influence of hydration shell on the internal structural and dynamic properties of biopolymers [23]. This is an interesting topic, however, it is outside the scope of the present contribution.

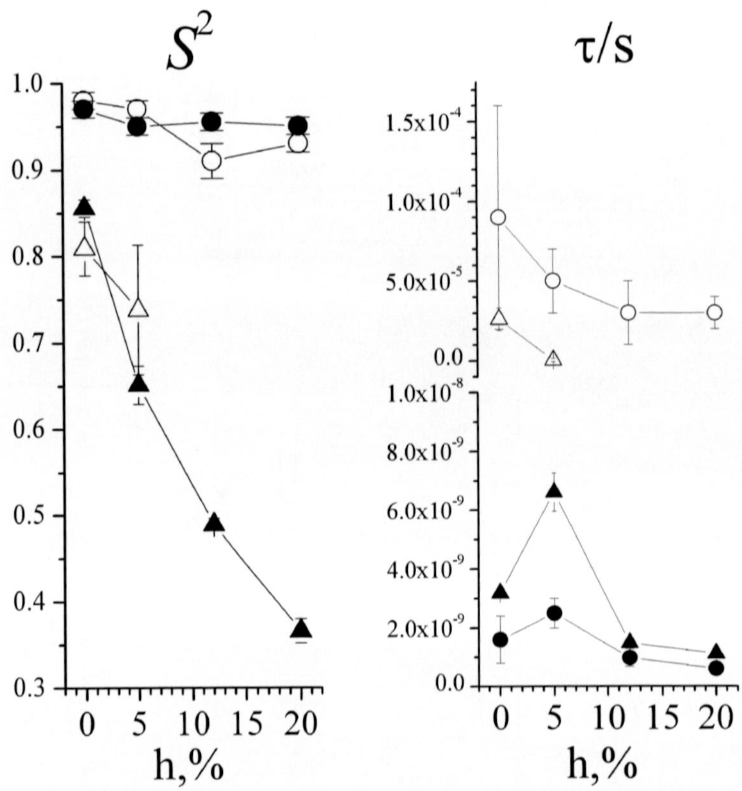

Figure 6. The hydration dependence of the order parameters and correlation times for α (circles) and δ (triangles) carbons. Solid and open symbols correspond to the fast and slow motions, respectively. At h=12% and 20% the data for δ-carbon (as well as for all other side chain carbons) could be described by only one (fast) motion.

Considering the data presented in Figure 5, we would like to make a following comment on the absolute values of the activation energy of the slow motion. The absolute values of E_a around 100-150 kJ/mole are abnormally high, they are close to the energy of the splitting a covalent bond. Nevertheless, such values of the activation energy are often observed in synthetic polymers as well [26]. This apparent contradiction can be explained by the fact that the obtained E_a values characterize actually the slope of the temperature dependence of the correlation time of the motion within a relatively narrow temperature range. This slope can be attributed to the height of the activation barrier only if the temperature dependence of the correlation time has an Arrhenius form (i.e. a straight line in the lg(τ) – 1/T coordinates) over all the temperature range. In the case of the slow conformational motions in densely packed polymers this is not the fact. The activation barrier for the slow motions is determined by sterical hindrances of the neighboring structural elements of the polymer. Due to the thermal motion the shape and the height of the potential barrier around each kinetic unit is constantly changing and obviously the amplitude of this change is temperature dependent. This qualitatively explains the non-Arrhenius behavior of the temperature dependence of the correlation time of the slow motion and the abnormally high values of the activation energy. This problem was considered in detail by Slutsker and co-workers [27,28]. It has been shown that the real activation energy of the slow conformational motion could be several times lower than the value determined by the slope of the temperature dependence of the correlation time. Thus, this contradiction can be reasonably explained on the qualitative level; however, we are not aware of any microscopic models explaining the non-Arrhenius behavior of the slow conformational motions in densely packed polymers quantitatively.

Comparative Analysis of the Carbon and Proton Order Parameters

As mentioned above, the main drawback of the description of the molecular dynamics within the frames of the model-free approach is an ambiguous interpretation of the order parameters and practically no information on the physical nature of internal mobility. To approach the more detailed comprehension of the molecular mobility in polylysine let us now consider the comparative analysis of the carbon and proton relaxation data. The carbon and proton relaxation times are determined by a motion of different inter-nuclear vectors, ^{13}C-^{1}H and ^{1}H-^{1}H, respectively. Thus, even for the same molecular motion the order parameters determined from the carbon and proton relaxation experiments in general case can be different. The main idea of the approach described here is that the comparative analysis of the ^{13}C-^{1}H and ^{1}H-^{1}H order parameters for the same motion may limit the uncertainty of the data interpretation and lead to well-grounded conclusions on the physical models of the motion.

We must note that for organic substances, the difference between carbon and proton relaxation has one more important feature. In CH_2 and CH_3 groups the C-H and H-H distances are 1.08 Å and 1.78 Å, respectively. At the same time, the closest distance between atoms belonging to different chemical groups is approximately 2.3-2.5 Å. Since the dipole-dipole interaction strength is inversely proportional to internuclear distance to the power six, it is easy to estimate that the carbon relaxation is almost completely determined by covalently bound protons within the same chemical group - CH, CH_2 or CH_3. For proton relaxation, however, a substantial share of the dipole-dipole interaction comes from the interaction between protons of different chemical groups that may belong to different parts of a biopolymer chain. For instance, in globular proteins about 30-40% of the proton second moment comes from the interaction between protons belonging to different chemical groups [29]. One may thus conclude that the carbon relaxation reflects the reorientational motion of the backbone and side chains only whereas the proton relaxation provides, in addition, information on the relative movements of different parts of a polypeptide chain that are not associated with large amplitude reorientations of these parts.

The proton relaxation data (Figure 4, right) could be well described by only one (fast) type of motion with the correlation time in the nanosecond timescale. The slow (micro-millisecond timescale) motion was detected in carbon relaxation experiments primarily due to the measurements of the proton-decoupled $T_{1\rho}^{d}$ relaxation time. This type of the relaxation experiments is most sensitive to the slow molecular motions in the micro-millisecond timescale. The proton off-resonance $T_{1\rho}$ relaxation times poorly sample such a low frequency range of molecular dynamics and thus, the parameters of the slow motion could not be directly determined from the proton relaxation times.

Another problem of the analysis of the proton data is unknown value of the proton second moment. As mentioned above, an appreciable part of the proton second moment may originate from the interaction between protons belonging to different chemical groups and even different chains. Thus, if the packing pattern of the polylysine chains in the solid sample is not known, the value of the protein second moment cannot be determined and hence, the motional order parameter cannot be determined as well. To solve this problem, we proposed the following model of the polylysine packing pattern. This model was used in the computer Monte-Carlo simulations of the polylysine dynamics. We consider two parallel subtending three-stranded β-sheets arranged in respect to each other in such a way that the side chains of each β-sheet are inserted in between side chains of the opposite β-sheet as shown in Figure 7. Thus, the whole structure consisted of six peptide chains. To speed up the simulations, we reduced the number of atoms of our model by using Lys-Gly-Lys-Gly-Lys-Gly peptide instead of polylysine chain. This does not influence the results since we are not interested in side chains facing outside. Such a structural model was proposed on the basis of the following data:

1) polylysine takes predominantly an anti-parallel β-sheet structure;
2) the ^{13}C fast motion order parameters are large and have similar values for all side chain carbons, which means that each side chain is equally confined by surrounding atoms along its length (see Figure 5);
3) the proton second moment must be as large as possible and thus the proton density must be as large as possible.

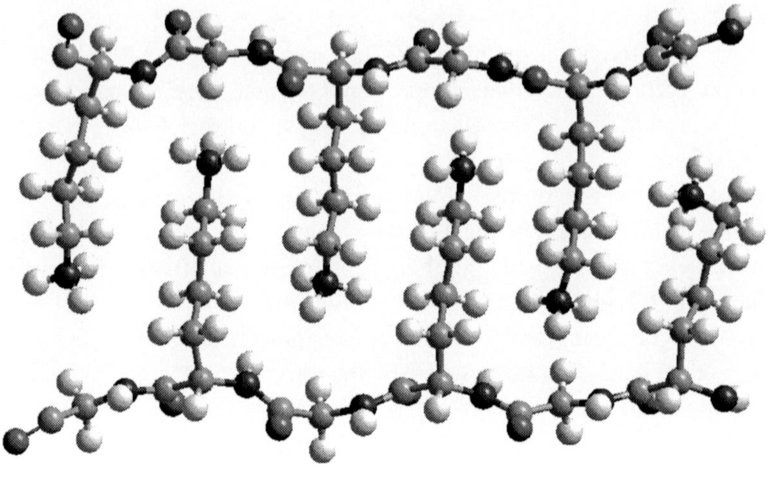

Figure 7. Part of the molecular structure used in the Monte-Carlo simulations: two $(Lys\text{-}Gly)_3$ chains belonging to subtending β-sheets. The whole structure consists of six such chains forming two three-stranded β-sheets. The figure is reproduced from ref [19].

The latter requirement follows from the estimation of the proton order parameter from the relaxation times: the less the second moment the more the difference between carbon and proton order parameters which is difficult to explain (see details in [19]. From the Monte-Carlo simulations we could estimate the proton second moment and both carbon and proton order parameters. The motion of the four lysine side chains located in the middle of the structure (Figure 7) was analyzed only. Each of these four side chains is surrounded by two side chains of the opposite β-sheet and by two side chains belonging to adjacent peptides of the same three-stranded β-sheet. Thus, the core of this six-chains polylysine structure resembles the densely packed dry polylysine sample. To assure a zero net charge of the whole structure, the electric charges of the ending NH_3 groups were set to zero.

The proton second moment was calculated according to the formula

$$K_{HH} = \frac{9}{20}\frac{1}{MN}\hbar^2\gamma^4\sum_{k=1}^{M}\left(\sum_{i\neq j}^{N}\frac{1}{\mathrm{r}_{ij}^6}\right)_k \tag{14}$$

where M is the number of steps in the Monte-Carlo trajectory, N is the number of protons, r_{ij} is the distance between i-th and j-th protons in the molecular structure, γ is the proton gyromagnetic ratio, k defines the number of the step in the Monte-Carlo trajectory. The proton order parameter was calculated according to the definition:

$$S_{\mathrm{HH}}^2 = \frac{\left(K_{HH}\right)_{\mathrm{averaged}}}{\left(K_{HH}\right)_{\mathrm{rigid\ lattice}}} \tag{15}$$

The denominator in the Eq. (15) is determined by the Eq. (14) and the averaged value of the proton second moment could be calculated as

$$\left(K_{HH}\right)_{\mathrm{averaged}} = \frac{2}{M(M+1)}\frac{1}{\mathrm{N}}\sum_{\mathrm{n=1}}^{M}\sum_{\mathrm{m=n}}^{M}\sum_{\mathrm{i\neq j}}^{N}\frac{1}{\left(\mathrm{r}_{\mathrm{ij}}^3\right)_{\mathrm{m}}}\frac{1}{\left(\mathrm{r}_{\mathrm{ij}}^3\right)_{\mathrm{n}}}\frac{1}{2}\left(3\cos^2(\theta_{\mathrm{ij}})_{\mathrm{mn}}-1\right) \tag{16}$$

where $\left(\mathrm{r}_{\mathrm{ij}}^3\right)_{\mathrm{n}}$ is the third power of the distance between i-th and j-th protons in the molecule in the n-th step of the Monte-Carlo trajectory and $(\theta_{ij})_{mn}$ is the angle between the internuclear vector connecting i-th and j-th protons in the trajectory step number n and the same vector in the step number m. The carbon order parameter for each carbon in the side chain was calculated according to the equation

$$S_{\mathrm{CH}}^2 = \frac{2}{M(M+1)}\frac{1}{N}\sum_{\mathrm{n=1}}^{M}\sum_{\mathrm{m=n}}^{M}\sum_{\mathrm{i=1}}^{N_{\mathrm{H}}}\frac{1}{2}\left(3\cos^2(\theta_{\mathrm{i}})_{\mathrm{mn}}-1\right) \tag{17}$$

where $(\theta_i)_{mn}$ is the angle between the internuclear vector connecting a carbon and the i-th covalently bound proton in the trajectory step number n and the same vector in the step number m. N_H is equal 1 for Cα carbon and 2 for Cβ, Cγ, Cδ and Cε carbons. The Monte-Carlo simulations were conducted using the Amber forcefield in vacuum at 300 K with the help of HyperChem software. Several simulations were performed using slightly different initial structures, see details in [19].

Figures 8 and 9 presents the proton and carbon order parameters for the side chain carbons determined from the NMR experiments and Monte-Carlo simulations for the fast and slow motions, respectively. Since the fast spin diffusion between protons averages the relaxation rates of various protons in the polypeptide one may analyze only the averaged values and thus, the averaged proton order parameters are presented in Figs. 8 and 9 by the shaded areas. As mentioned above, the proton data do not allow determination of the slow motion dynamic parameters. Yet, there is a possibility to determine the minimum possible value of the order parameter of the slow motion. For this, we introduce the slow motion in the fitting the proton relaxation times with the fixed correlation time, activation energy and the distribution width parameter determined from the carbon relaxation data (Figure 5). Then we determined the minimum value of the slow motion order parameter that does not change appreciably the fitting quality of the proton relaxation time.

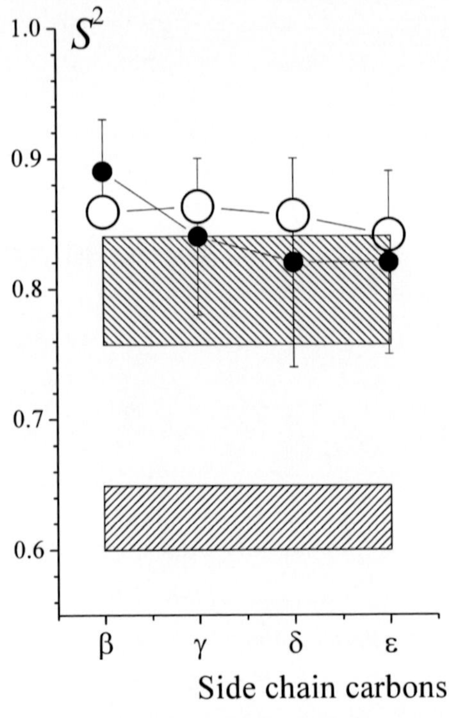

Figure 8. Carbon and proton fast motion order parameters for the dry polylysine sample obtained from NMR relaxation experiments and Monte-Carlo simulations. Open circles are the carbon experimental order parameters, see Fig 5. Filled circles are the carbon order parameters obtained from the Monte-Carlo simulations. Error bars for the solid circles define the spread of the values for different simulations. The upper shaded area defines the average proton order parameter determined from the Monte-Carlo simulations. The lower shaded area is the experimental proton order parameter obtained from the proton relaxation experiments. The figure is reproduced from ref [19].

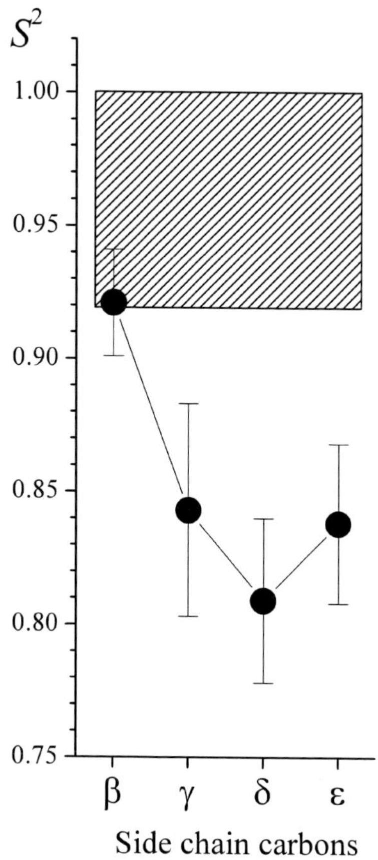

Figure 9. The order parameter of the slow motion for the dry polylysine sample obtained from the carbon (circles, see Figure 5) and proton (shaded area) NMR relaxation experiments. The figure is reproduced from ref [19].

This value corresponds to the lower border of the shaded area in Figure 9. If the order parameter of the slow motion is lower (amplitude is higher) then the fitting becomes appreciably worse and the variation of the fast motion parameters cannot improve it. The order parameters of the slow motion could not be determined from the computer simulations of the polylysine dynamics since the simulation trajectory was definitely too short to sample such a slow motion.

The combination of the ^{1}H-^{1}H and ^{1}H-^{13}C order parameters for the same motions presented in Figures 8 and 9 allows one not only postulating the existence of molecular motions but also suggesting certain physical models of these motions that would explain both carbon and proton data. Now let us consider such possible models.

It is seen that the carbon order parameters of the fast motion determined from the relaxation times and computer simulations are in a good agreement. This confirms that the polylysine structure (Figure 7) used in the computer simulations is a reasonable model of the packing pattern of the polypeptide. The proton order parameter determined from the simulations is also close to the carbon's ones (Figure 8). At the same time, the difference between the proton order parameters determined from the simulations and relaxation experiments is too large (Figure 8) and cannot be explained by the inconsistency of the analysis. The only reasonable way to explain such a difference is to take into account the features of the carbon and proton relaxation times described above: the carbon relaxation times reflect only the reorientational dynamics of the polymer chain whereas the proton data in addition to that provide the information on the relative translational movements of different parts of the chain without appreciable reorientations.

It is worth to mention that a very similar phenomenon of the polypeptide and protein dynamics was observed about 20 years ago by Nusser and co-workers [30]. In this work, the internal dynamics of several proteins and polypeptides in the solid state was studied by means of field cycling proton relaxation, line shape analysis and Jeener-Broekaert experiments performed on labile backbone deuterons. Like carbon relaxation times, deuteron ones are also sensitive only to the reorientational motions.

The deuteron experiments have shown that the protein backbone is practically immovable, whereas the proton relaxation data reveal appreciable amplitude of the internal motions. The deuteron and proton data characterize the dynamics of a backbone and mainly side chains, respectively, and thus, such a direct comparison of motional amplitudes determined from these experiments is not fully correct. Nevertheless, the difference between the deuteron and proton order parameters is too high to be explainable by this. To explain the obvious contradiction between deuteron and proton experiments the authors suggested the model of local dilations that cause almost no reorientation of chemical bonds but lead to significant changes of inter-proton distances. We believe that the difference between carbon and proton order parameters of the fast motion observed in our experiments (Figure 8) could be most reasonably explained by a very similar model.

We suppose that in the case of the polylysine structure such dilations could be local stretchings of the distance between neighboring β-sheet planes. While diffusing along the backbone, such a defect induces sliding back-and-forth motion of side chains as shown in Figure 10. Such kinds of motion would cause rather small reorientations of chemical bonds of side chains, but definitely lead to much more effective averaging of the dipole-dipole interaction between protons belonging to different side chains. Our calculations show that this interaction accounts for about 20-25% of the total value of proton second moment. Thus, the assumption of such a specific motion based on the results by Nusser et al. [30] may explain - at least qualitatively - the experimentally observed difference between proton and carbon order parameters. We could not detect this motion in computer simulation because the length of the chains (Figure 7) was obviously too short. This motion is not associated with overpassing high-energy barriers caused by any sterical hindrances and thus, the activation energy of this motion is relatively small and close to the activation energy of the rotation around dihedral angles of the polylysine side chains.

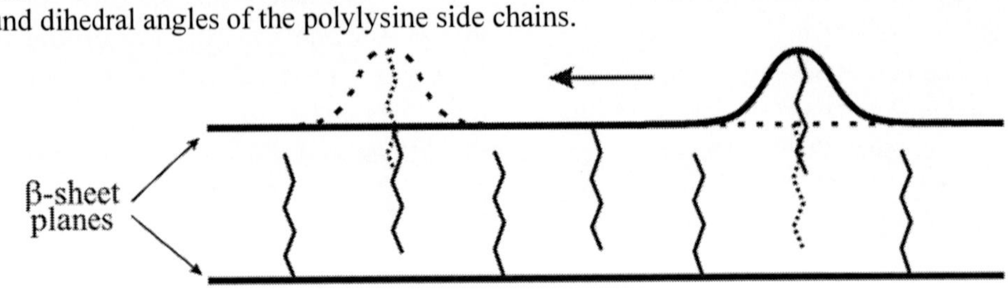

Figure 10. Schematic presentation of a diffusing defect in the polylysine structure consisting of foliated beta-sheets planes. The figure is reproduced from ref [19].

So, the comparative analysis of the proton and carbon order parameters of the fast motion leads to the conclusion on the two different motions in the same time scale (nanosecond range of correlation times).

The first motion is small amplitude librations of atoms around their average positions within the steric hindrances limits. This motion is reflected in the carbon relaxation experiments and computer simulations of the polylysine dynamics. This is a standard type of molecular motion present in all polymer systems without exception. The second type of motion could be revealed by the comparison of the carbon and proton order parameters: the proton order parameter is appreciably lower than the carbon one which indicates the existence of another motion which is not seen in the carbon experiments. Taking into account the different sensitivity of the carbon and proton relaxation experiments to the translational movements of the polymer chains (or different parts of the same chain) in respect to each other, we could explain the difference in the order parameters by introducing the "defect diffusion" type of motion shown in Figure 10.

As for the slow motion, we observe the inverse picture – the proton order parameter is appreciably higher than the carbon one (Figure 9). That means that the amplitude of angular reorientation of C-H vectors of side chain methylene groups is larger than that of the H-H vectors of the same groups. Theoretically, it is easy to suggest a model of a methylene group reorientation explaining such a difference of the proton and carbon order parameters. For this, an axis of the rotation of the group must be parallel (or close to that) to the H-H direction. In this case the H-H vector of the methylene group does not experience large amplitude reorientations, whereas the C-H vectors do. (An opposite situation, i.e. when $S^2_{CH} > S^2_{HH}$, cannot be interpreted in a similar way since there are two C-H directions in the methylene group and thus it is impossible to define a rotation axis which would explain a large amplitude rotation of the H-H vector and small amplitude rotation of both C-H vectors.) Since the H-H vectors of the polylysine side chains in the trans-conformations are parallel to the direction of the backbone, the small amplitude rotation of the β-sheet structures around the axis parallel to the direction of the backbone may explain the difference between the proton and carbon order parameters of the slow motion. However, such an explanation seems to us hardly probable since in this case the carbon order parameters would be the same for all side chain carbons which is not the case (Figure 5). The long correlation time of the slow motion is obviously connected with overpassing relatively high energy barriers and thus this motion should have most probably jump-like nature. We believe that the most probable model of the slow motion is jump-like trans-cis transitions of side chain conformation as shown in Figure 11. These transitions are the correlated simultaneous 180°-jumps of χ_2 and χ_4 dihedral angles (Figure 11, left) and the 180°-jump of χ_3 dihedral angle (Figure 11, right). In this case, the H-H vectors of the side chain methylene groups change their direction on 180° which does not affect the relaxation rates (because $\cos^2(0^0)=\cos^2(180^0)$, see the equation in Figure 2).

Such motion causes partial averaging of only that part of the proton second moment that comes from interaction between protons belonging to different methylene groups (of the same side chain and different side chains). However, it is clear that the overall proton order parameter in this case would be larger than that of carbons. Figure 11 can also explain why the slow motion order parameter of β-carbon is appreciably higher than that of other side chain carbons (Figure 5): the cis-trans transitions shown in this figure cause reorientation of all methylene groups except beta-group. Angular reorientations of C-H vectors in such transitions are large, and thus, large order parameters of slow motion can be explained only by a low population of cis-conformations. High activation energy and long correlation times of the slow motion are obviously the consequences of sterical hindrances caused by neighboring side chains that have to be overpassed during the conformational transitions shown in Figure 11. Upon hydration, the distance between β-sheet planes increases since water molecules are attracted to the hydrophilic NH_3 groups of polylysine and this obviously requires additional space between β-sheet planes. This leads to reducing sterical hindrances for the slow motion and, as a consequence, dramatic decreasing the correlation time of this type of motion (Figure 6).

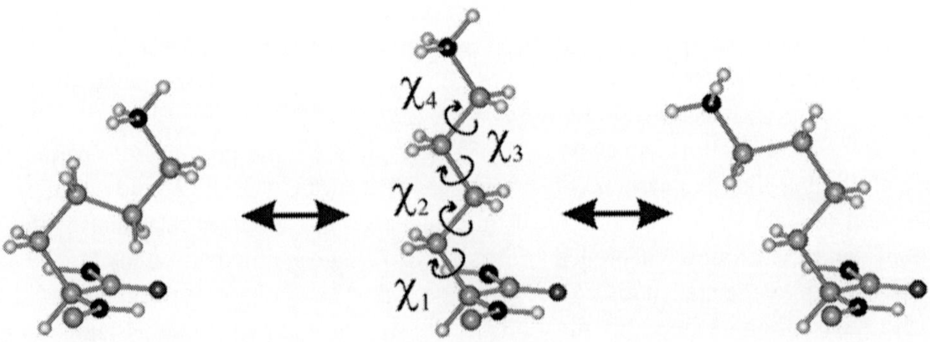

Figure 11. Possible trans-cis transitions of lysine side chain conformation that explain the experimentally observed inequation $S_{CH}^2 < S_{HH}^2$. The figure is reproduced from ref [19].

Thus, the comparative analysis of the carbon and proton order parameters in solid polylysine enabled to identify and to suggest physical models of three different types of molecular dynamics. It is clear that such a description of molecular dynamics is more physical and transparent than characterizing molecular motions by dimensionless order parameters only. We emphasize that this would not be possible with the analysis of the proton and carbon data separately.

Simultaneous Relaxation and Exchange Data Analysis

In addition to the comparative analysis of the motion of different internuclear vectors, there is a possibility of the comparative analysis of the motion of internuclear vector and CSA tensor which we will consider in this chapter. Methodically, this approach is very similar to the approach described above and it also enables determination of the motional models directly from the experimental data. This work was published few years ago [31].

The motion of the CSA tensor could be explored by two NMR techniques: lineshape analysis and the solid state exchange spectroscopy. These experiments are suitable for detecting motions with the correlation times in the microsecond range and faster (lineshape analysis) and millisecond range and slower (exchange spectroscopy) [32]. The significant advantage of the exchange experiments is that they allow obtaining a correlation function of the CSA tensor reorientation directly from the experiment. In the case of the lineshape analysis this is impossible. There are several 2D and 1D modifications of the solid state NMR exchange experiments [33,34]. However, all the exchange pulse sequences contain the same three functionally important time intervals: a preparation period for the initial frequency labeling of the signal; a variable mixing time during which the magnetization is stored along the z (the magnetic field) direction and the exchange may (or may not) occur; and the recording of the FID (secondary frequency labeling). The exchange as used here is a change of the CSA tensor orientation in respect to the B_0 field which changes the resonance frequency of a nucleus. In the 2D versions of the pulse sequences the exchange process can be detected by the appearance of the off-diagonal cross-peaks in the 2D spectrum. In the 1D versions the exchange process can be detected by decreasing the line intensity in the 1D spectrum with increasing mixing time. The mathematical formalism of the line intensity calculation is rather complex [34], however, it can be reduced to the following schematic expression:

$$\begin{matrix} Line \\ intensity \end{matrix} = Function \left(\begin{matrix} \begin{matrix} Experimental \\ parameters: \end{matrix} & \begin{matrix} Molecular \\ parameters: \end{matrix} \\ & \sigma - CSA\ tensor \\ \omega_L - resonance\ frequency & N_S - the\ number\ of\ the\ CSA\ tensor\ orientations \\ \omega_R - MAS\ rate & \\ t_1 - evolution\ time & \alpha_i, \beta_i, \gamma_i, i = 1, N_S - Euler\ angles\ for\ all\ orientations \\ \tau_m - mixing\ time & \\ & K_{ij} - exchange\ matrix \end{matrix} \right) \quad (18)$$

In this equation the molecular parameters actually define the model of motion – its geometry (number of the orientations and the Euler angles) and the time scale (exchange matrix K_{ij}, defining the transition probability per unit time from the orientation i to the orientation j). The main advantage of the 1D exchange experiments is that they work faster, save much machine time and the dynamic information they provide is essentially the same as can be obtained from the 2D experiments. The most powerful and universal 1D exchange technique up to now is a pulse sequence CODEX [35,36].

A very convenient system for a comparative relaxation and exchange NMR study are the ^{15}N nuclei located on a protein backbone. The NMR relaxation of these nuclei is determined by a dipole-dipole interaction with one covalently bound proton. At the same time, these nuclei possess nearly axially symmetric CSA tensor. The direction of the symmetry axis of the CSA tensor is only 20° from the N-H bond [37] that we will neglect in the analysis.

The basic idea of our analysis is to apply the model-free approach not only to the relaxation, but also to the exchange experiment as well. The interpretation of the experimentally obtained autocorrelation function of the CSA tensor reorientation (i.e. the mixing time dependence of the line intensity) is very similar to that of the internuclear vector: if the motional model is known (or assumed) then the correlation function can be unambiguously calculated using Eq. (18), however, if the model is not known the interpretation can be performed only within the frames of the model-free approach. The long mixing time limit of the correlation function, which by analogy with the relaxation experiments we call the exchange order parameter, can be equally explained assuming various motional models. As demonstrated below, the quantitative comparison of the relaxation and exchange order parameters for the same motion may provide criteria for the discrimination of the motional models directly from the experiment.

The relaxation and exchange order parameters are determined differently, however, they both are dependent on the distribution function $\rho(\theta,\varphi)$ which characterizes the orientational distribution of the N-H vector (or the main axis of the CSA tensor, which we assume to be the same). The function $\rho(\theta,\varphi)$ actually determines the geometry of motion. With the use of this function, the relaxation and exchange order parameters can be expressed in the following way:

$$S^2_{\rm relax} = \int_0^{\pi}\int_0^{2\pi}\int_0^{\pi}\int_0^{2\pi} \rho(\theta_1,\varphi_1)\rho(\theta_2,\varphi_2)\frac{3(\vec{n}(\theta_1,\varphi_1)\cdot\vec{n}(\theta_2,\varphi_2))^2-1}{2}d\varphi_1 d\theta_1 d\varphi_2 d\theta_2 \quad (19)$$

$$S^2_{\rm exch} = \int_0^{\pi}\int_0^{2\pi}\int_0^{\pi}\int_0^{2\pi} \rho(\theta_1,\varphi_1)\rho(\theta_2,\varphi_2)\frac{I(\theta_1,\varphi_1,\theta_2,\varphi_2)}{I(\theta_1,\varphi_1,\theta_1,\varphi_1)}d\varphi_1 d\theta_1 d\varphi_2 d\theta_2 \quad (20)$$

where $\vec{n}(\theta,\varphi)$ is a unit vector with the orientation defined by the polar angles θ and φ, $I(\theta_1,\varphi_1,\theta_2,\varphi_2)$ is a signal intensity in the exchange experiment on the stipulation that before and after the mixing time the symmetry axis of the CSA tensor had orientations defined by the polar angles θ_1,φ_1 and θ_2,φ_2, respectively. The value $I(\theta_1,\varphi_1,\theta_2,\varphi_2)$ can be calculated using the f-function formalism [34]. (We must admit that the Eq. (20) was published in [31] with an error although this did not lead to essentially wrong results.) We performed a set of model calculations of the exchange and relaxation order parameters for four different models of motion presented in Table 2 at different angle amplitudes. The results are presented in Figure 12. However, these results become more interesting and transparent if the relaxation and exchange order parameters are plotted as a function of each other, see Figure 13. It is clearly seen that the ratio of the relaxation and exchange order parameters is different for different motional models. This opens up a principal possibility to distinguish between motional geometries directly from the experiment. Figure 13 also clearly demonstrates the limitation of this method: the motion should have appreciably high amplitude (the relaxation order parameter must be lower than ~0.95), otherwise the curves for all motional models coincide.

The combined relaxation and exchange NMR investigation was applied for studying the molecular dynamics of the backbone of the protein barstar in a free and bound to another protein, binase, states [31]. Comparison of the dynamic parameters in two different biological states may indicate the biological relevance of the molecular motions.

Figures 14 and 15 present the experimental data – the temperature dependencies of the T_1 and $T_{1\rho}$ relaxation times and the mixing time dependencies. The experiments were performed on the rehydrated (6%) protein samples with 15% ^{15}N-enrichment of barstar. The 15% instead of 100% enrichment was chosen to dilute the ^{15}N spin system in the protein and thus to reduce the spin diffusion rate. Spin diffusion produces an additional component in the mixing time dependence in the exchange experiments and thus makes the analysis difficult and ambiguous [38]. The one-dimensional ^{15}N spectrum of powder protein sample reveals of course no site-specific resolution and thus we could obtain only the averaged dynamic information for the whole protein backbone. In spite of that, these experiments enabled (i) demonstrating the advantages of the complementarity of the relaxation and exchange methods, and (ii) comparing the parameters of molecular dynamics of barstar in two states that might become interesting for investigation of molecular details of the binding process.

Table 2. Motional models for the order parameters simulations (δ(x) is Dirac's delta-function). θ_a defines the angle amplitude of motion in all cases

Model	Orientational distribution function
a. Two equally populated sites	$\rho(\theta,\varphi)=\delta(\varphi)(0.5\delta(\theta)+0.5\,\delta(\theta-\theta_a))$
b. Two 0.2/0.8 populated sites	$\rho(\theta,\varphi)=\delta(\varphi)(0.2\delta(\theta)+0.8\,\delta(\theta-\theta_a))$
c. Diffusive reorientations within a planar angle	$\rho(\theta,\varphi)=\dfrac{\delta(\varphi)}{\theta_a}$ at $0<\theta<\theta_a$ $\rho(\theta,\varphi)=0$ at $\theta_a<\theta<\pi$
d. Wobbling in a cone	$\rho(\theta,\varphi)=\dfrac{\sin\theta}{2\pi(1-\cos\theta_a)}$ at $0<\theta<\theta_a$ $\rho(\theta,\varphi)=0$ at $\theta_a<\theta<\pi$

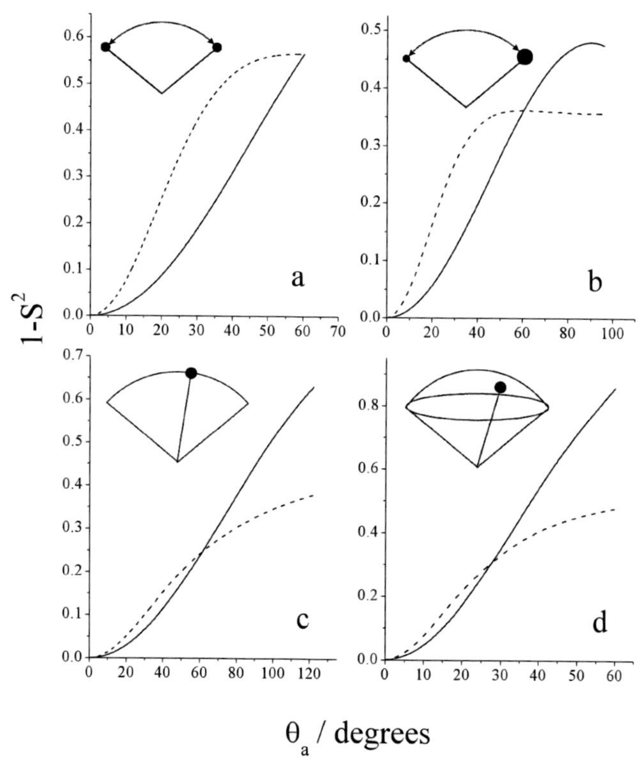

Figure 12. NMR relaxation (solid lines) and exchange (dashed lines) order parameters at various angle amplitudes for the motional models described in Table 2 calculated according to the Eqs. (19,20). Parameter S^2_{exch} was calculated at MAS rate 3 kHz, Δσ=160 ppm, one rotor period evolution delay in the CODEX sequence, resonance frequency for ^{15}N nuclei 40.5 MHz. The figure is reproduced from ref [31].

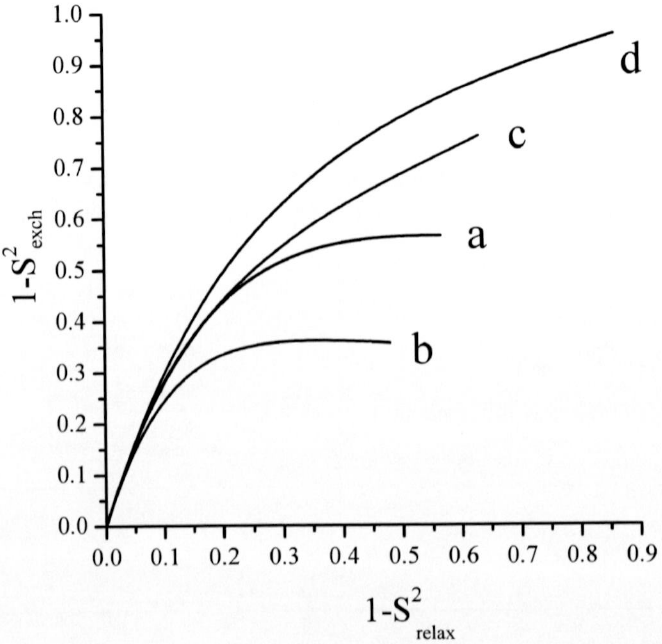

Figure 13. The same data as shown in Figure 12 plotted in ($1-S^2_{relax}$) vs. ($1-S^2_{exch}$) coordinates. The figure is reproduced from ref [31].

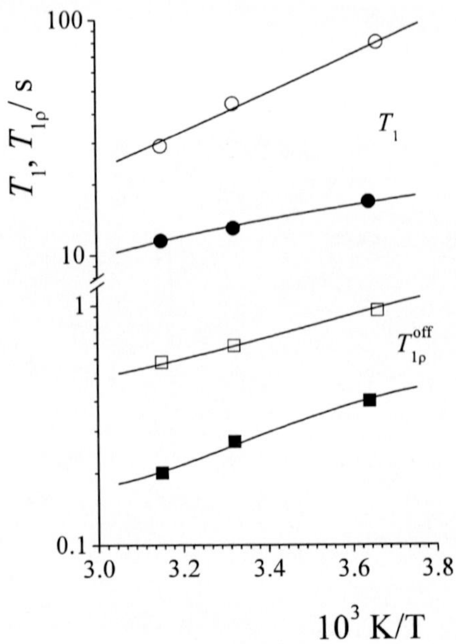

Figure 14. Temperature dependencies of the T_1 and off-resonance $T_{1\rho}$ relaxation times of backbone ^{15}N nuclei measured for free barstar (solid symbols) and barstar-binase complex (open symbols). The size of the symbols corresponds to the experimental error. Resonance frequency for ^{15}N nuclei 40.5 MHz, effective spin-lock frequency in the off-resonance $T_{1\rho}$ experiments was 60 kHz, the angle between B_0 and B_{1e} fields (see Eq. 4) was 17°. The figure is reproduced from ref [31].

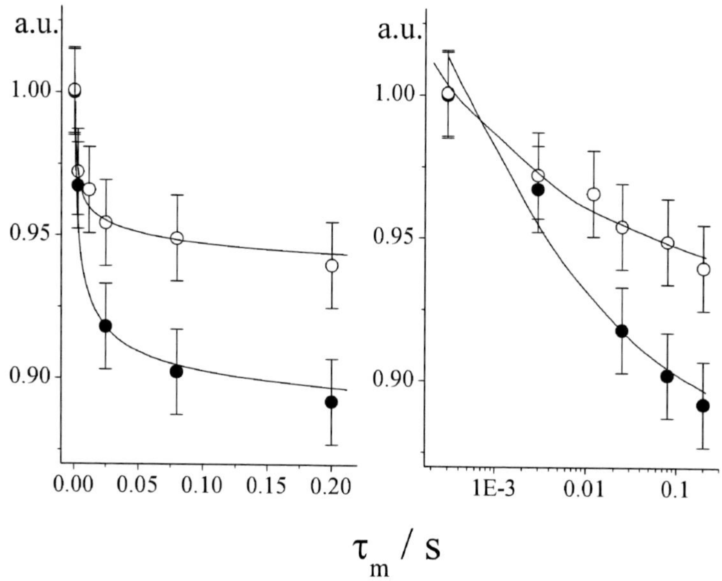

Figure 15. The CODEX mixing time dependences of the backbone ^{15}N nuclei signal for free barstar (solid symbols) and barstar-binase complex (open symbols) in linear (left) and logarithmic (right) time scale. Solid lines are the fitting curves. The fitting curves simulated for the two-site jumps, "fan" and wobbling in a cone models are very similar. The experiments were performed at 25^0 C, the experimental conditions were the same as used in the simulations (Figure 12). The figure is reproduced from ref [31].

The joint relaxation and exchange data analysis can be performed in two ways. First, one may define the exchange and relaxation order parameters independently from two sets of data and then define the motional model from the comparison of these order parameters. This is the most straightforward method of the analysis, however, it does not work well if the data sets are limited like in our case (see Figs. 14 and 15). High ambiguity of the order parameters obtained from such data does not allow reliable determination of the motional model. Thus, we used a simultaneous fitting of both exchange and relaxation data. While fitting, the following expression for the root mean square deviation was minimized:

$$\mathrm{RMSD}=\sqrt{\frac{1}{\mathrm{N_R}+\mathrm{N_C}}\left[\sum_{\mathrm{i=1}}^{\mathrm{N_R}}\left(\frac{T_{\mathrm{exp}}^{\mathrm{i}}-T_{\mathrm{sim}}^{\mathrm{i}}}{T_{\mathrm{exp}}^{\mathrm{i}}}\right)^2+\sum_{\mathrm{i=1}}^{\mathrm{N_C}}\left(\frac{I_{\mathrm{exp}}^{\mathrm{i}}-I_{\mathrm{sim}}^{\mathrm{i}}}{I_{\mathrm{exp}}^{\mathrm{max}}-I_{\mathrm{exp}}^{\mathrm{min}}}\right)^2\right]} \tag{21}$$

where N_R and N_C are the number of experimental points in the NMR relaxation and CODEX experiments, T_{exp} and T_{sim} are the experimental and simulated relaxation times, I_{exp} and I_{sim} are the experimental and simulated line intensities in the CODEX experiment, I_{exp}^{max} and I_{exp}^{min} are the maximal and minimal experimental line intensities in the CODEX experiment, respectively. The simulated relaxation times were calculated according to the formalism described in the previous section. The only difference was the value of K_d (Eqs. 1-3) which was taken as large as $0.52\cdot10^{10}$ s^{-2} for the ^{1}H-^{15}N interaction. The mixing time dependencies in the CODEX experiment were fitted according to the formula:

$$I(\tau_m) = K_N\left(S_{exch}^2 + (1 - S_{exch}^2)\int_0^\infty \rho\left(\frac{\tau}{\tau_0}\right) e^{\left(-\frac{\tau_m}{\tau}\right)} d\tau \right) \tag{22}$$

where K_N is the normalization coefficient, $\rho\left(\frac{\tau}{\tau_0}\right)$ is the Fuoss-Kirkwood distribution function, exactly the same function as was used in the relaxation times analysis. There is no analytical expression for the integral in Eq. (22), thus this integral was calculated numerically.

The data analysis has shown that one type of motion could not reasonably describe the data and thus we assumed existence of two types of backbone motions with different time scales, as we did while analyzing the polylysine data, see above. The number of the unknown fitting parameters in this case is nine – four parameters (order parameter, correlation time, distribution width parameter and activation energy) for the fast motion and five parameters for the slow motion. The slow motion is described by five parameters since there are two order parameters corresponding to the relaxation and exchange experiments, all other parameters (correlation time, distribution width parameter and activation energy) we assume for these experiments to be the same. Introducing two order parameters for the fast motion is senseless since the exchange experiment is insensitive to the fast (nanosecond time scale) motion. Yet, it is possible to reduce the number of fitting parameters by one taking into account the model calculations described above. Using the dependencies presented in Figure 15, S_{exch}^2 can be expressed as a function of S_{relax}^2 (these dependencies can be approximated by any suitable arbitrary function).

Thus, we fitted the data for each motional model separately. The fitting results are presented in Tab. 3. We do not present the results for the model *b* since it produces unreasonably high amplitude for the slow motion.

Table 3 demonstrates that all three motional models considered give very similar values of the dynamic parameters and practically the same data fitting quality. The protein backbone is a rigid structure, hence the motional amplitude is small and the order parameters fall in the region where different motional models give the same ratio of the relaxation and exchange order parameters.

Thus, one cannot determine the most suitable model of motion directly from the experimental data which was the main aim of this study. The model "wobbling in a cone" seems to be most reasonable for description of these data since it gives the least angular amplitude of the N-H vectors reorientation. However, this analysis evidently demonstrates that such an approach would be most effective for the study of high amplitude motions. In spite of the failure to determine the motional model, the determination of the dynamic parameters from the simultaneous analysis of the relaxation and exchange data is much more exact and reliable than using these two types of the experiments separately. It is also interesting to mention that the average angular amplitude of the barstar backbone slow reorientations decreases 1.5 times upon binding to binase independently on the model assumed in the analysis, see Table 3. This is an expected result, however the quantitative estimation of this change in dynamics was not done before. Such kind of the backbone mobility is likely functionally important: As hypothesized by Fersht and co-workers [39,40], a breathing motion of barstar expands the binding loop and the second helix in contact with the active site of barnase and thus, the internal dynamics modulates the conformation of the binding site for a better affinity to barnase (binase and barnase are similar proteins that both tightly bind to barstar).

Table 3. Dynamic parameters, obtained from the simultaneous fitting the relaxation times and the mixing time dependencies (Figures 14 and 15) for the motional models a, c and d (see Table 2). The correlation times correspond to the temperature 20° C. Angle amplitude (θ_a) was determined from the dependencies shown in Figure 12

	Two-site jumps (model a)		Diffusion within a planar angle (model c)		Wobbling in a one (model d)	
	free barstar	barstar-binase	free barstar	barstar-binase	free barstar	barstar-binase
S^2_{relaxS}	0.85±0.01	0.94±0.005	0.84±0.015	0.94±0.005	0.895±0.02	0.94±0.01
θ_a	26.5°±1°	16.5°±1°	48°±3°	28°±1°	15.5°±1.5°	11.5°±1°
τ_{OS}, ms	0.12±0.08	3.0±1.7	0.1±0.04	2.5±1.5	1.2±1.0	2.8±1.2
β_S	0.24±0.01	0.18±0.015	0.24±0.01	0.18±0.015	0.25±0.02	0.19±0.02
E_{aS}, kJ/mole	105±15	200±40	105±9	190±40	160±40	195±45
S^2_{relaxF}	0.86±0.045	0.96±0.015	0.84±0.05	0.965±0.015	0.89±0.03	0.96±0.02
τ_{OF}, ns	4±15	130±20	1.0±0.5	120±30	120±100	100±40
β_F	0.12±0.05	0.78±0.15	0.12±0.08	0.71±0.14	0.23±0.1	0.73±0.16
E_{aF}, kJ/mole	10±5	9.5±3	8.5±5	11.5±4	8±4	10±5
RMSD, %	6.2	3.0	6.3	2.9	5.7	2.9

Conclusions and Perspectives

Here we presented two examples of the combined analysis of different magnetic interactions aiming to describe the geometry of molecular motions in biopolymers in detail. The first example makes use of the comparison of the ^{13}C-^{1}H and ^{1}H-^{1}H internuclear vectors motion in polylysine and the second example concerns the study of protein backbone dynamics by simultaneous analysis of the ^{15}N-^{1}H internuclear vector and the ^{15}N CSA tensor reorientations. These methodical approaches are still associated with a number of *a priori* assumptions, nevertheless they in principle provide a possibility of discrimination of various motional models directly from the experimental data which is not possible using most of the routine techniques. This has a primary importance for biological macromolecules since their biological function has an intimate connection with their molecular dynamics. Thus, the knowledge of the physical nature of the molecular motion may reveal molecular details of the biological function with a possibility of subsequent intentional modification of biological properties. The idea of the joint quantitative analysis of different magnetic interactions for determination of motional models was utilized before in solutions, see [41] and references therein. As for the solid state experiments, only now the methodical development of the NMR techniques enables intensive practical applications of this idea.

We would like to outline few directions along which the complex NMR approaches to studying biopolymer dynamics, in our opinion, will develop in near future. First, very useful would be the incorporation into analysis the lineshape data of the static or slowly rotating samples. As mentioned above, the lineshape data provide information on the CSA tensor reorientations in the μs and faster time scales. Although this method does not allow direct determination of the correlation function of motion, like in the exchange experiments, the combined analysis of the relaxation, exchange and lineshape data may reduce the number of assumptions in the analysis and provide more reliable and definite information, especially using the temperature dependencies of the lineshape data. Second, we expect high potentialities of the complex NMR studies of molecular dynamics performed on ^{2}H nuclei. These nuclei are best suited for all three types of the NMR experiments mentioned above – relaxation, lineshape analysis and exchange. There is a large number of the ^{2}H relaxation and especially lineshape analysis experiments in proteins, however, we are not aware of any exchange experiment performed on proteins using deuterons. The main methodical advantage of deuteron experiments is a large and axially symmetric quadrupolar splitting and the quadrupolar mechanism of the spin-lattice relaxation of deuterons. Thus, the assumptions on the axial symmetry of the CSA tensor and the coincidence of the symmetry axis of the CSA tensor and the internuclear vector that have been made in the analysis of the ^{15}N data (see above) are not necessary in the case of ^{2}H nuclei since both spin-lattice relaxation and the qudrupolar splitting are determined by the reorientation of the electric field gradient in respect to B_0 field. And the third direction concerns the possibility of obtaining the dynamic information on a site-specific basis in proteins. This is the most important point since only the site-specific dynamic information on proteins may help revealing the molecular mechanisms of their biological function.

The site-specific resolution obviously requires high spectral resolution and individual line assignments. For proteins in the solid state this is not a simple task, although presently there are experimental techniques to achieve site-specific spectral resolution and assignment. The most simple way to obtain this is a preparation of selectively ^{15}N/^{13}C/^{2}H enriched proteins as was done e.g. in [42]. However, the obvious disadvantage of this approach is a poor sampling of a protein molecule. The alternative to this approach is a developing very rapidly over the last decade high resolution multidimensional NMR spectroscopy on totally ^{15}N/^{13}C enriched proteins in the microcrystalline form [43,44]. These techniques enable practically full sampling of a protein molecule and obtaining highly selective information. So far the multidimensional heteronuclear solid state NMR experiments in microcrystalline proteins were applied for obtaining mainly structural information [45-47] During the last few years these NMR approaches have been applied to measuring on a per residue basis such parameters as ^{15}N T_1's [48,49], ^{2}H T_1's and residual quadrupolar couplings [50] and ^{13}C-^{1}H residual dipolar coupling [51]. However, the limited number of the measured parameters in these works does not provide information on geometry of a motion; these data allow only the comparative characterization of internal dynamics in different protein locations, more detailed quantitative description of the dynamics requires either some assumptions or incorporation into analysis independent data. The complex NMR approaches described above combined with the modern multidimensional techniques would help to obtain more definite and spatially selective information on protein dynamics. Yet, there is a methodical problem in these investigations that must be mentioned. This problem is spin diffusion. Spin diffusion is a very useful phenomenon in structure determination but it is a severe obstacle in dynamics studies: in the relaxation experiments it equalizes the relaxation rates of different nuclei, and in the exchange experiments spin diffusion, as was mentioned above, causes appearance of an additional component in the exchange decay that overlaps the component corresponding to the molecular motion. Only the lineshape experiments are unaffected by spin diffusion since T_2 is shorter than the spin-diffusion rate. In totally ^{13}C-enriched proteins, spin diffusion is very fast and there is practically no way to obtain site-specific dynamic information using relaxation or exchange NMR experiments; the only solution here is the selective ^{13}C enrichment. As for the ^{15}N and ^{2}H nuclei, spin diffusion is not always too fast to disable obtaining site-specific dynamic information, but in many cases it makes the analysis of the data more complicated and ambiguous [13,38,52,53]. An effective and promising option to combine the total ^{15}N protein enrichment with the suppression of the spin diffusion effect appeared recently. This option is the proton-detected ^{15}N-^{1}H correlation spectroscopy of perdeuterated proteins with ~10% backsubstitution of exchangeable protons [54]. The low proton density in such sample combined with a relatively high MAS rate ensures the negligible spin diffusion rate between ^{15}N nuclei in the protein. At the same time, using the sensitive nuclei (protons) for detection of the signal allows obtaining spectra within the reasonable time limits, which is a crucial issue in the time consuming relaxation and exchange experiments. On the other hand, it is clear that this idea would work only for ^{15}N nuclei. In the NMR dynamic studies using ^{13}C or ^{2}H nuclei, which are more suitable for studying the side chain dynamics, the selective isotopic enrichment seems to be unavoidable, otherwise the spin diffusion would not be non-negligible and the quantitative analysis of the relaxation and exchange data would become ambiguous.

In conclusion, we would like to note that the main obstacle for implementing the complex NMR approaches to studying molecular dynamics of biopolymers and proteins is their high time consumption. Performing a wide set of various NMR experiments at different conditions (resonance frequency, temperature, etc.) may require several months of measuring time not to mention the data processing and analysis. Thus, one may hardly expect that these techniques will become routine procedures in near future. Yet, we are confident that only a more complicated and sophisticated methodical approaches can ensure a breakthrough in understanding the molecular mechanisms of involvement of the conformational dynamics in the biological function of biopolymers.

REFERENCES

[1] Gabel, F; Bicout, D; Lehnert, U; Tehei, M; Weik, M and Zaccai, G. Protein dynamics studied by neutron scattering. *Quart. Rev. Biophys.*, 2002 35, 327-367.

[2] Wand, AJ. Dynamic activation of protein function: A view emerging from NMR spectroscopy. *Nature Struct. Biol.*, 2001 8, 926-931.

[3] Sinha, N and Smith-Gill, SJ. Protein structure to function via dynamics. *Prot. Pept. Lett.*, 2002 9, 367-377.

[4] Daniel, RM; Dunn, RV; Finney, JL and Smith, JC. The role of dynamics in enzyme activity. *Annu. Rev. Biophys. Biomol. Struc.*, 2003 32, 69-92.

[5] Kern, D and Zuiderweg, ERP. The role of dynamics in allosteric regulation. *Curr. Opin. Struct. Biol.*, 2003 13, 748-757.

[6] Dodson, G and Verma, CS. Protein flexibility: its role in structure and mechanism revealed by molecular simulations. *Cell. Mol. Life Sci.*, 2006 63, 207-219.

[7] Karplus, M and Kuriyan, J. Molecular dynamics and protein function. *Proc. Nat. Acad. Sci. USA*, 2005 102, 6679-6685.

[8] Palmer, AG; Kroenke, CD and Loria, JP. Nuclear magnetic resonance methods for quantifying microsecond-to-millisecond motions in biological macromolecules. *Meth. Enzymol.*, 2001 339, 204-238.

[9] Blackledge, M. Recent progress in the study of biomolecular structure and dynamics in solution from residual dipolar couplings. *Prog. Nucl. Magn. Reson. Spectros.*, 2005 46, 23-61.

[10] Tolman, JR and Ruan, K. NMR residual dipolar couplings as probes of biomolecular dynamics. *Chem. Rev.*, 2006 106, 1720-1736.

[11] Tamura, A; Matsushita, M; Naito, A; Kojima, S; Miura, KI and Akasaka, K. Dynamics of the three methionyl side chains of Streptomyces subtilisin inhibitor. Deuterium NMR studies in solution and in the solid state. *Prot. Sci.*, 1996 5, 127-139.

[12] Krushelnitsky, A; Gogolev, Y; Golbik, R; Dahlquist, FW and Reichert, D. Comparison of the internal dynamics of globular proteins in the microcrystalline and rehydrated lyophilized states. *Biochim. Biophys. Acta*, 2006 1764, 1639-1645.

[13] Reif, B; Xue, Y; Agarwal, V; Pavlova, MS; Hologne, M; Diehl, A; Ryabov, YE and Skrynnikov, NR. Protein side-chain dynamics observed by solution- and solid-state NMR: comparative analysis of methyl ^{2}H relaxation data. *J. Am. Chem. Soc.*, 2006 128, 12354-12355.

[14] Wuthrich, K. NMR - this other method for protein and nucleic-acid structure determination. *Acta Crystallogr. D Biol. Cryst.*, 1995 51, 249-270.

[15] Lipari, G and Szabo, A. Model-free approach to the interpretation of nuclear magnetic resonance relaxation in macromolecules. 1. Theory and range of validity. *J. Am. Chem. Soc.*, 1982 104, 4546-4559.

[16] Lipari, G and Szabo, A. Model-free approach to the interpretation of nuclear magnetic resonance relaxation in macromolecules. 2. Analysis of experimental results. *J. Am. Chem. Soc.*, 1982 104, 4559-4570.

[17] Blout, ER and Lenormant, H. Reversible configurational changes in poly-L-lysine hydrochloride induced by water. *Nature*, 1957 179, 960-963.

[18] Krushelnitsky, A; Faizullin, D and Reichert, D. Hydration dependence of backbone and side chain polylysine dynamics: A C-13 solid-state NMR and IR spectroscopy study. *Biopolymers*, 2004 73, 1-15.

[19] Krushelnitsky, A and Reichert, D. Complex ^{1}H, ^{13}C-NMR relaxation and computer simulation study of side-chain dynamics in solid polylysine. *Biopolymers*, 2005 78, 129-139.

[20] Vanderhart, DL and Garroway, AN. ^{13}C NMR rotating frame relaxation in a solid with strongly coupled protons: Polyethylene. *J. Chem. Phys.*, 1979 71, 2773-2787.

[21] Krushelnitsky, A; Kurbanov, R; Reicher, D; Hempel, G; Schneider, H and Fedotov, V. Expanding the frequency range of the solid-state $T_{1\rho}$ experiment for heteronuclear dipolar relaxation. *Solid State Nucl. Magn. Reson.*, 2002 22, 423-438.

[22] Jones, GP. Spin-lattice relaxation in rotating frame - weak-collision case. *Phys. Rev.*, 1966 148, 332-335.

[23] Krushelnitsky, A and Reichert, D. Response of lysozyme internal dynamics to hydration probed by ^{13}C and ^{1}H solid-state NMR relaxation. *Appl. Magn. Reson.*, 2004 27, 501-518.

[24] Beckmann, PA. Spectral densities and nuclear spin relaxation in solids. *Phys. Rep.*, 1988 171, 85-128.

[25] Johnson, E; Palmer, AG and Rance, M. Temperature dependence of the NMR generalized order parameter. *Prot. Struc. Func. Bioinf.*, 2007 66, 796-803.

[26] Fedotov, VD and Schneider, H. Structure and dynamics of bulk polymers by NMR methods. Berlin Heidelberg: Springer-Verlag; 1989.

[27] Slutsker, AI; Polikarpov, YI and Vasil'eva, KV. Determination of the activation energy for complicated relaxation processes. *Phys. Sol. State (Fizika Tverdogo Tela)*, 2002 44, 1604-1610.

[28] Slutsker, AI; Polikarpov, YI and Vasil'eva, KV. On the determination of the energy of activation of relaxation transitions in polymers by differential scanning calorimetry. *Tech. Phys. (Zhurnal Tehnicheskoi Fiziki)*, 2002 47, 880-885.

[29] Krushelnitsky, AG; Fedotov, VD; Spevacek, J and Straka, J. Dynamic structure of proteins in solid state. ^{1}H and ^{13}C NMR relaxation study. *J. Biomol. Struct. Dyn.*, 1996 14, 211-224.

[30] Nusser, W; Kimmich, R and Winter, F. Solid-state NMR study of protein/polypeptide backbone fluctuations interpreted by multiple trapping diffusion of dilating defects. *J. Phys. Chem.*, 1988 92, 6808-6814.

[31] Krushelnitsky, AG; Hempel, G and Reichert, D. Simultaneous processing of solid-state NMR relaxation and 1D-MAS exchange data: the backbone dynamics of free vs. binase-bound barstar. *Biochim. Biophys. Acta*, 2003 1650, 117-127.

[32] Krushelnitsky, A and Reichert, D. Solid-state NMR and protein dynamics. *Prog. Nucl. Magn. Reson. Spectros.*, 2005 47, 1-25.

[33] Schmidt-Rohr, K and Spiess, HW. Multidimensional solid-state NMR and polymers. London: Academic Press; 1994.

[34] Luz, Z; Tekely, P and Reichert, D. Slow exchange involving equivalent sites in solids by one-dimensional MAS NMR techniques. *Prog. Nuc.l Magn. Reson. Spectros.*, 2002 41, 83-113.

[35] deAzevedo, ER; Hu, WG; Bonagamba, TJ and Schmidt-Rohr, K. Centerband-only detection of exchange: Efficient analysis of dynamics in solids by NMR. *J. Am. Chem. Soc.*, 1999 121, 8411-8412.

[36] deAzevedo, ER; Hu, WG; Bonagamba, TJ and Schmidt-Rohr, K. Principles of centerband-only detection of exchange in solid-state nuclear magnetic resonance, and extension to four-time centerband-only detection of exchange. *J. Chem. Phys.*, 2000 112, 8988-9001.

[37] Oas, TG; Hartzell, CJ; Dahlquist, FW and Drobny, GP. The amide N-15 chemical-shift tensors of 4 peptides determined from C-13 dipole-coupled chemical-shift powder patterns. *J. Am. Chem. Soc.*, 1987 109, 5962-5966.

[38] Krushelnitsky, A; Reichert, D; Hempel, G; Fedotov, V; Schneider, H; Yagodina, L and Schulga, A. Superslow backbone protein dynamics as studied by 1D solid-state MAS exchange NMR spectroscopy. *J. Magn. Reson.*, 1999 138, 244-255.

[39] Buckle, AM; Schreiber, G and Fersht, AR. Protein-protein recognition: crystal structural analysis of a barnase-barstar complex at 2.0-E resolution. *Biochemistry*, 1994 33, 8878-8889.

[40] Wong, KB; Fersht, AR and Freund, SMV. NMR N-15 relaxation and structural studies reveal slow conformational exchange in Barstar C40/82A. *J. Mol. Biol.*, 1997 268, 494-511.

[41] Idiyatullin, D; Daragan, VA and Mayo, KH. A simple method to measure (CH2)-C-13 heteronuclear dipolar cross-correlation spectral densities. *J. Magn. Reson.*, 2004 171, 4-9.

[42] Cole, HBR and Torchia, DA. An NMR study of the backbone dynamics of staphylococcal nuclease in the crystalline state. *Chem. Phys.*, 1991 158, 271-281.

[43] Hughes, CE and Baldus, M. Magic-angle-spinning solid-state NMR applied to polypeptides and proteins. *Annu. Rep. NMR Spectros.*, 2005 55, 121-158.

[44] McDermott, AE. Structural and dynamic studies of proteins by solid-state NMR spectroscopy: rapid movement forward. *Curr. Opin. Struct. Biol.*, 2004 14, 554-561.

[45] Castellani, F; van Rossum, B; Diehl, A; Schubert, M; Rehbein, K and Oschkinat, H. Structure of a protein determined by solid-state magic-angle-spinning NMR spectroscopy. *Nature*, 2002 420, 98-102.

[46] Zech, SG; Wand, AJ and McDermott, AE. Protein structure determination by high-resolution solid-state NMR spectroscopy: Application to microcrystalline ubiquitin. *J. Am. Chem. Soc.*, 2005 127, 8618-8626.

[47] Böckmann, A; Lange, A; Galinier, A; Luca, S; Giraud, N; Juy, M; Heise, H; Montserret, R; Penin, F and Baldus, M. Solid state NMR sequential resonance assignments and conformational analysis of the 2 x 10.4 kDa dimeric form of the Bacillus subtilis protein Crh. *J. Biomol. NMR*, 2003 27, 323-339.

[48] Giraud, N; Böckmann, A; Lesage, A; Penin, F; Blackledge, M and Emsley, L. Site-specific backbone dynamics from a crystalline protein by solid-state NMR spectroscopy. *J. Am. Chem. Soc.*, 2004 126, 11422-11423.

[49] Giraud, N; Blackledge, M; Goldman, M; Böckmann, A; Lesage, A; Penin, F and Emsley, L. Quantitative analysis of backbone dynamics in a crystalline protein from nitrogen-15 spin-lattice relaxation. *J. Am. Chem. Soc.*, 2005 127, 18190-18201.

[50] Hologne, M; Faelber, K; Diehl, A and Reif, B. Characterization of dynamics of perdeuterated proteins by MAS solid-state NMR. *J. Am. Chem. Soc.*, 2005 127, 11208-11209.

[51] Lorieau, JL and McDermott, AE. Conformational flexibility of a microcrystalline globular protein: order parameters by solid-state NMR spectroscopy. *J. Am. Chem. Soc.*, 2006 128, 11505-11512.

[52] Krushelnitsky, A; Bräuniger, T and Reichert, D. ^{15}N spin diffusion rate in solid-state NMR of totally enriched proteins: the magic angle spinning frequency effect. *J. Magn. Reson.*, 2006 182, 339-342.

[53] Giraud, N; Blackledge, M; Böckmann, A and Emsley, L. The influence of nitrogen-15 proton-driven spin diffusion on the measurement of nitrogen-15 longitudinal relaxation times. *J. Magn. Reson.*, 2007 184, 51-61.

[54] Chevelkov, V; Rehbein, K; Diehl, A and Reif, B. Ultrahigh resolution in proton solid-state NMR spectroscopy at high levels of deuteration. *Angew. Chem. Int. Ed.*, 2006 45, 3878-3881.

In: Biopolymer Research Trends
Editor: Tamas S. Nemeth, pp. 119-143

ISBN: 978-1-60021-983-2

Chapter 4

THE FECO UNIT VIBRATIONS AS A PROBE OF THE STRUCTURE AND DYNAMICS OF THE ACTIVE SITE OF HEME PROTEINS: COMBINED QUANTUM CHEMICAL, VIBRONIC AND SPECTROSCOPIC STUDY

Solomon S. Stavrov*

Sackler Institute of Molecular Medicine,
Department of Human Molecular Genetics and Biochemistry,
Sackler Faculty of Medicine,
Tel Aviv University, Ramat Aviv, P.O.B.39040,
Tel Aviv 69978, Israel

ABSTRACT

ZINDO quantum chemical calculations and vibronic theory of activation are used to study the effect of different distortions of the active center of carbonyl complexes of heme proteins and external electric fields on the magnitude of the C-O vibrational frequency and its relationship with the changes in the Fe-C frequency. It is shown that the experimentally observed negative linear correlation between these two frequencies stems from the variation of the electric field of the heme environment. Study of the effect of the electric field of the distal histidine on the C-O frequency allowed assigning a number of the CO infrared absorption sub bands of carboxymyoglobin to specific orientations and tautomeric states of the histidine. The results on the field dependence of the C-O vibrational frequency suggested that the width of this band should be sensitive to the large amplitude motion of the distal heme environment. The temperature dependencies of the C-O bands of carboxycomplexes of horseradish peroxidase, myoglobin and hemoglobin at different pH were quantitatively interpreted taking into account electrostatic coupling of the band to the motion of the heme environment. The analysis of the parameters of the fitting procedure showed that upon heating in the liquid solvent water molecule enters the heme pocket of the proteins with the capacious pocket

* E-mail: stavrov@post.tau.ac.il; Fax: (972 3) 640 5168; Telephone: (972 3) 640 9859.

(horseradish peroxidase and "open" conformation of myoglobin and hemoglobin), this water molecule mainly contributing into the temperature dependence of the band. In the glassy matrix the large amplitude motions of the pocket amino acids are arrested and the disordered water cannot enter or leave the pocket. Appearance of the water in the heme pocket causes transition of the protein to another conformational substate, at room temperature almost all protein molecules exist in this conformational substate. To our best knowledge this is the first observed example of almost full transition of a protein from one conformational substate to another, caused by the temperature change.

INTRODUCTION

Heme proteins (HP) are widely used in the studies of the relationship between the structure, dynamics and function of proteins. Having the same active center, iron-porphyrin (Fe(P)), they manifest a wide variety of functions; this fact allows studying effect of the protein globule, in general, and heme environment, in particular, on the prosthetic group properties and activities [1]. The fact that HP's can be studied by using virtually all known spectroscopic techniques allows to obtain broad information about their dynamics by using time resolved studies [2-4]. HPs can be essentially affected by the state and nature of the solvent (see, for example [4]); this makes possible to examine the effect of the protein environment on their configuration and dynamics.

The resonance Raman scattering and infrared absorption spectra contain bands corresponding to the Fe-CO and C-O vibrations of the carbonyl complexes of heme proteins (HP(CO)). The related vibrational frequencies, ν_{FeC} and ν_{CO}, vary within a very wide range even in HP(CO)s, where Fe(P) is bound to the protein by the same histidine amino acid [5-8]. Hence, they are widely used to obtain information on the protein heme environment structure and dynamics.

In this paper the results of the theoretical study of the effect of the external electric field and distortions of the Fe(P) complex with CO and imidazole (Fe(P)(Im)CO, see Figure 1a) on ν_{FeC} and ν_{CO} are reviewed and applied to the interpretation of the experimental data on the structure and dynamics of HP(CO).

METHODS

Intermediate Neglect of Differential Overlap (INDO) version of MO LCAO approach (realized in ZINDO program [9-12]) was used for quantum chemical calculations of the Fe(P)(Im)(CO) electronic structure and orbital electron density transfers to and from CO upon its coordination, Δq_i. To calculate Δq_i the MO's obtained by the INDO calculations were rewritten in the basis of the eigenfunctions of the free CO and atomic functions of other atoms. Then the occupations of the CO eigenfunctions in the complex were calculated as the Mulliken population of the corresponding orbitals.

Structure of the Fe(P)(Im)(CO) complex is presented on Figure 1a. For the calculations the known structures of the porphyrin and imidazole rings [13] were used. The distances Fe-N_{Im} = 2.04 Å and Fe-C = 1.745 Å were taken from the X-ray diffraction studies of the model compounds [14-17] and for the C-O distance the free molecule value was used [18], 1.128 Å.

Effect of various distortions of the porphyrin ring and the FeCO unit, and differently located model point charges (±1 e$^-$) on the electronic structure of Fe(P)(Im)(CO), bond indexes and Δq_i was studied [19,20], see Figures (1-3).

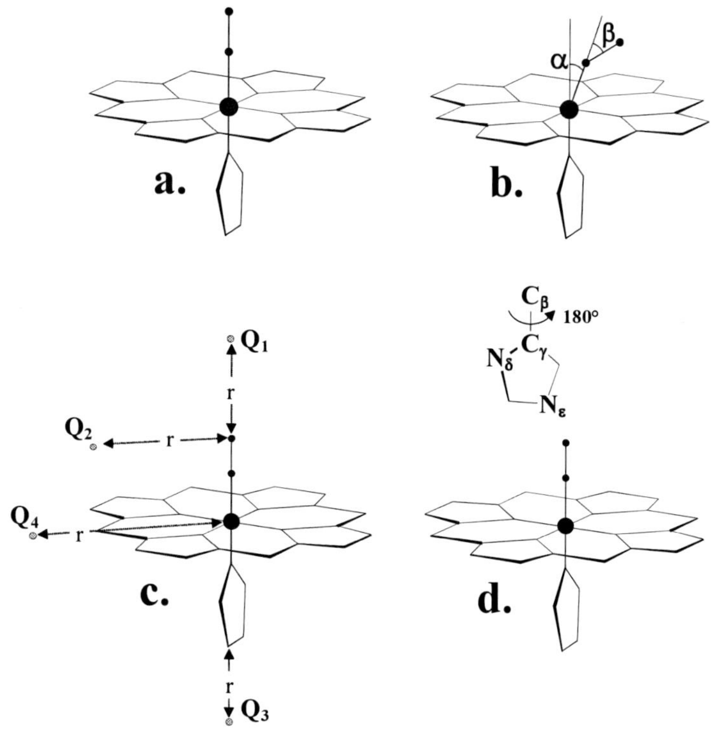

Figure 1. Schematic representation of the geometry of the complexes with linear perpendicular (a); bent and tilt CO (b); with model point charges (c); and distal histidine (d).

The vibronic theory of activation (VTA) bridges the MO description of the electronic structure and the coupling of electronic states with the nuclear configuration described by the vibronic coupling theory [21,22], and is described in detail elsewhere [21-23].

Similar to the integral linear diagonal vibronic constants

$$F^{(I)}=\langle I|(\partial H/\partial R)_{R=R_0}|I\rangle \tag{1}$$

orbital vibronic constants are introduced [21,22]

$$f_i=\langle i|(\partial H/\partial R)_{R=R_0}|i\rangle. \tag{2}$$

It is simply to show that

$$F^{(I)}=\sum_i q_i^{(I)} f_i. \tag{3}$$

In these formulae, i and I are the wave functions of the one-electron i-th MO and the I-th state of the diatomic, respectively; H is the Hamiltonian; R is the interatomic nuclear coordinate; R_0 is its equilibrium magnitude; $q_i^{(I)}$ is the i-th MO population number in the I-th electronic state of the diatomic; and the electronic configuration of different I states differ by the excitation of the electron from the i-th to the j-th MO. The expressions for the curvature constants, K and k_i, and anharmonicity constants, Γ and γ_i, can be written in full analogy with Eqs. (1) and (2).

By coordination q_i changes to $q_i + \Delta q_i$. In the first order approximation with respect to q_i a distorting force (ΔF), and changes in curvature (ΔK) and anharmonicity ($\Delta\Gamma$) at the equilibrium point of free CO R_0, occur (below the index of the electronic state I is omitted for simplicity):

$$\Delta F(R_0)=\sum_i \Delta q_i f_i$$
$$\Delta K(R_0)=\sum_i \Delta q_i k_i$$
$$\Delta\Gamma(R_0)=\sum_i \Delta q_i \gamma_i \quad (4)$$

Appearance of the distorting force and changes in the curvature and anharmonicity of the diatomic adiabatic curve directly affect the interatomic distance and vibrational frequency of the coordinated ligand. To estimate the changes of these parameters from Eq. (4) the diatomic adiabatic potential was assumed to be the Morse potential.

Eqs. (4) are valid for any rearrangement of the electronic structure of the molecule including excitation, ionization, reduction, and coordination, provided that Δq_i are sufficiently small as compared to the whole valence electron charge. This allowed to obtain f_i, k_i, and γ_i for the chemically active 5σ and $2\pi^*$ MOs of CO from the experimental data on different electronic states of free carbon monoxide [23]:

$$f(5\sigma) = -0.25\ mdyn,\ k(5\sigma) = 1.23\ mdyn/Å,\ \gamma(5\sigma) = 16.45\ mdyn/Å^2,$$
$$f(2\pi^*) = -1.71\ mdyn,\ k(2\pi^*) = 4.66\ mdyn/Å,\ \gamma(2\pi^*) = -21.04\ mdyn/Å^2. \quad (5)$$

The ground state values of K_0 = 19.00 mdyn/Å and Γ_0 = -131.22 mdyn/Å^2 were also used for the calculations. Note, that the $2\pi^*$ orbital vibronic constants are considerably larger then the 5σ ones, this being a result of the weak antibonding character of the latter.

Estimation of the relationship between ν_{CO} and ν_{FeC} using bond indexes. Note first of all, that at small changes of the force field constant K the change of the vibrational frequency ν equals

$$\Delta\nu = (\nu + \Delta\nu) - \nu = \sqrt{\frac{K}{M}}\left(\sqrt{1 + \frac{\Delta K}{K}} - 1\right) \approx \nu\frac{\Delta K}{2K}. \quad (6)$$

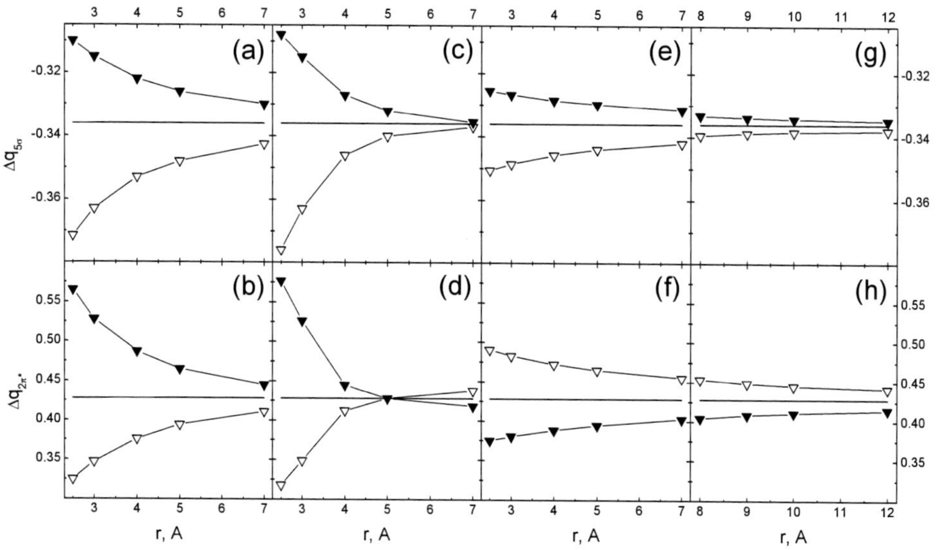

Figure 2. Dependence of the σ donation and π back-bonding on the magnitude and position of the Q_1 (a and b), Q_2 (c and d), Q_4 (e and f) and Q_3 (g and h) point charge (+1 e^-, solid triangles; -1 e^-, open triangles; and no charge - solid line).

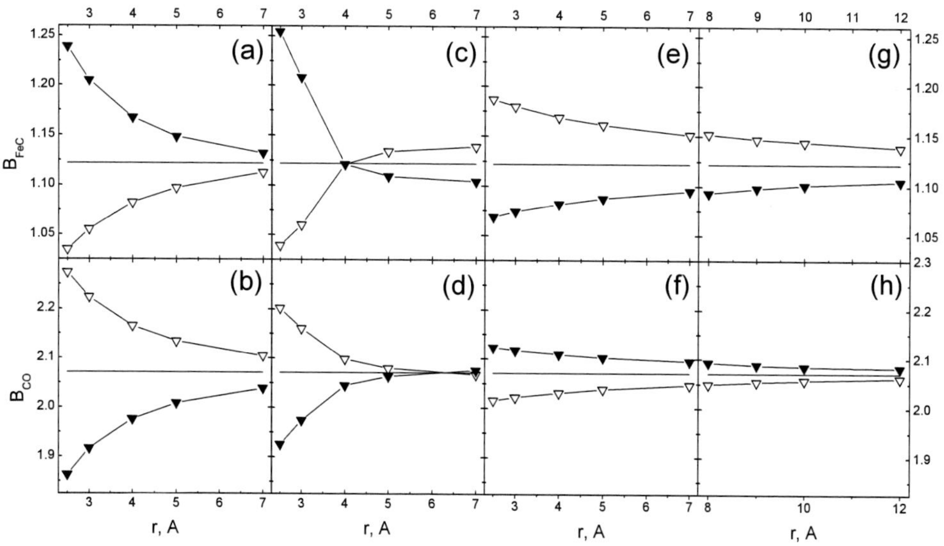

Figure 3. Dependence of the C-O (a and b) and Fe-C (c and d) bond indexes on the magnitude and position of the Q_1 (a and b), Q_2 (c and d), Q_3 (e and f) and Q_4 (g and h) point charge Notations for different charge magnitudes are the same as on Figure 2.

Then, taking into account that all the perturbations of the Fe(P)(Im)(CO) system (the heme geometry changes and the charges) affect the absolute values of B_{FeC} and B_{CO} relatively weakly (see Figure 3), we can state that the dependence of each bond index on the perturbation must be close to the linear one. Therefore, the relationship between the changes of this bond indexes (ΔB_{FeC} and ΔB_{CO}) is expected to be also linear. Assuming that the relative change of the force field constants is proportional to the corresponding relative change of the bond indexes, we can write for the relationship between the changes of the force field constants and bond indexes:

$$\frac{\Delta K}{K} = \zeta \frac{\Delta B}{B}, \qquad (7)$$

where ζ is a coefficient of proportionality. Using Eqs. (6) and (7) one obtains that

$$\frac{\Delta \nu_{FeC}}{\Delta \nu_{CO}} = \xi \frac{\nu_{FeC}}{\nu_{CO}} \frac{B_{CO}}{B_{FeC}} \frac{\Delta B_{FeC}}{\Delta B_{CO}}, \qquad (8)$$

where ξ takes into account that ζ_{FeC} and ζ_{CO} can be different.

Using the following values of the vibrational frequencies and bond orders $\nu_{CO} \approx 2000$ cm^{-1}, $\nu_{FeC} \approx 500$ cm^{-1}, $B_{CO} \approx 2$ and $B_{FeC} \approx 1$ and notations

$$\alpha_\nu = \Delta\nu_{FeC} / \Delta\nu_{CO}, \quad \alpha_B = \Delta B_{FeC} / \Delta B_{CO} \qquad (9)$$

one obtains for the relationship between α_ν and α_B

$$\alpha_\nu \approx \tfrac{1}{2} \xi \alpha_B \qquad (10)$$

Effect of pure dephasing on the shape of an infrared absorption band. We assume that the circular frequency of a molecular vibration, which absorbs infrared light (Ω) is much higher than the energy of thermal motion at room temperature. Thus, its excited vibrational states are hardly populated even at room temperature and barely contribute to the formation of the band. Therefore, as it was discussed [24,25], the temperature dependence of this band can be described in the framework of the Condon approximation of general theory of optical absorption bandshapes [26-29].

Assuming that the coupling of the vibration to the motions of the environment of the molecule is weak, these motions are harmonic and describing them by one effective coordinate, one can write down the following Hamiltonian, which takes into account the coupling using linear approximation,

$$H_0 = \frac{\mathbf{P}^2}{2M} + \frac{\mathbf{p}^2}{2m} + \tfrac{1}{2}(1+2\alpha q) M\Omega^2 Q^2 + \tfrac{1}{2} m\omega^2 q^2, \qquad (11)$$

where P and p, M and m, Ω and ω, and Q and q, are moments, reduced masses, initial circular frequencies, and displacements along the normal coordinates of the molecular vibration and the effective mode of the environment, respectively. α is linear constant of the coupling between the q and Q displacements, which describes the degree of the effect of the environment electric field on Ω.

The coupling under consideration rigidly shifts the adiabatic surface of the first excited molecular vibrational state along the q coordinate in respect to the ground state [30,31]. It follows from general theory [26-29] that in this case the intensity and position of the absorption band do not depend on the temperature. The second moment (M_2) manifests very specific temperature dependence

$$M_2(T) = A + \tfrac{1}{2} B \cdot \hbar\omega \cdot coth\left(\frac{\hbar\omega}{2k_B T}\right), \tag{12}$$

where

$$B = \frac{\alpha^2 (\hbar\Omega)^2}{m\omega^2}, \tag{13}$$

k_B is the Boltzmann constant, and $A \geq 0$ is the temperature independent contribution of inhomogeneous broadening or the high-frequency vibrations, which hardly change their amplitudes in the temperature interval under investigation.

In the classical limit

$$\hbar\omega << k_B T \tag{14}$$

the band has a Gaussian shape, second moment of which is

$$\sigma^2(T) = A + B \cdot k_B T. \tag{15}$$

Note that in the case of strong inhomogeneous contribution (usually this is the case in protein systems because of their distribution over different conformational substates) the bandshape is also close to Gaussian, second moment of which depends on the temperature according to Eq. (12).

It follows from Eq. (12) that in the case of harmonic motion along the q coordinate M_2 cannot increase steeper than proportional to T, Eq. (15).

Effect of External Electric Field on the Fe-C and C-O Chemical Bonds

The problem of the calculation of the force field constant and frequency of the stretching vibration of the coordinated CO in the Fe(P)(Im)(CO) complex and their dependence on the external electric field and the complex distortions for the first time was addressed theoretically by taking advantage of the combined use of ZINDO computations and VTA [19,20,23], as it was described in Methods.

The calculations confirmed that the σ-donation from 5σ CO orbital to Fe and back-bonding from Fe to $2\pi^*$ CO orbital mainly contribute into formation of the Fe-CO bond. It was shown [19,20] that $\Delta q_{5\sigma}$ and $\Delta q_{2\pi^*}$ are significantly affected by the changes in the Fe-CO distance and orientation of the FeCO unit and hardly influenced by the porphyrin distortions. Both axial homogeneous electric field and point charges located at different positions in respect to the complex were shown to considerably affect $\Delta q_{2\pi^*}$, $\Delta q_{5\sigma}$ being affected much weaker, see Figure 2. This result is easily understandable, because the charges create the electrostatic potential difference between Fe and CO. The latter causes shift of electron density of the mobile π electrons. Shift of the electron density to the $2\pi^*$ CO orbital weakens the C-O bond ($2\pi^*$ orbital is the CO strongly antibonding orbital) and strengthens the Fe-C bond reducing population of the antibonding $e_g(d_\pi)$-$2\pi^*$ orbital of the FeCO unit. This effect was invoked to interpret qualitatively the experimentally observed negative correlation between ν_{CO} and ν_{FeC} [5].

The calculations [19] showed that both the tilting and bending of the FeCO unit reduce both B_{FeC} and B_{CO}. Consequently (see Eqs (7)-(10)), the FeCO distortions are expected to reduce both ν_{CO} and ν_{FeC}, this conclusion contradicts to the experimentally observed negative correlation between these frequencies [7].

On contrary [19,20], presence of the axial and equatorial charges (Q_1, Q_2, and Q_4 on Figure 1c) change B_{CO} and B_{FeC} in opposite directions (Figure 3 and Figure 4) supporting the suggestion [5-7] that the variation in ν_{CO} and ν_{FeC} in different HPs stems mainly from the variation in the protein electric field and not from distortion of the FeCO unit. This conclusion was confirmed later by additional experimental [32-37] and theoretical [34,38-40] studies.

The Fe-CO distance variation substantially changes the bond indexes, these changes being also negatively correlated [19]. Consequently, this kind of distortion of the FeCO unit can additionally contribute to the negative correlation between B_{CO} and B_{FeC}.

The effect of different point charges on the CO force field constant K(C-O) was calculated using VTA (see Methods). The ν_{CO} values were evaluated utilizing the obtained K(C-O) and taking into account vibrational coupling in the FeCO unit, K(FeC) = 2.48 mdyn/Å, H(Fe-C-O) = 0.8 mdyn*Å/rad^2 and K(FeC,CO) = 0.8 mdyn/Å [5]. The results are presented on Figure 5, which shows that the charges strongly affect ν_{CO}.

The effect of the Q_2 charges (Figure 1c) deserves special consideration [19]. A displacement of a (for example) positive charge in the direction parallel to the porphyrin plane does not affect the Q_2-porphyrin distance and, consequently, practically does not affect the porphyrin potential. Therefore, attraction of the electron density to the porphyrin is practically not affected by the charge displacement, while its attraction to CO decreases even faster than in the previous case. Note, that since π electrons are more mobile than the σ ones, dependence of the back-bonding on the charge coordinate is steeper than the dependence of the σ donation (Figure 2c and 2d). For example, a positive charge located at 5 Å from the FeCO line still reduces σ donation, but practically does not affect back-bonding. Since both B_{CO} and ν_{CO} are controlled mainly by $\Delta q_{2\pi^*}$, the effect of such a charge on the B_{CO} (Figure 3b) and ν_{CO} (Figure 5b) is negligible. The charge located at $r > 5$ Å already attracts π electrons to the porphyrin ring stronger than to CO and, as a result, reduces the back-bonding in respect to its "no charge" value, the effect of this charge on the σ donation being very weak.

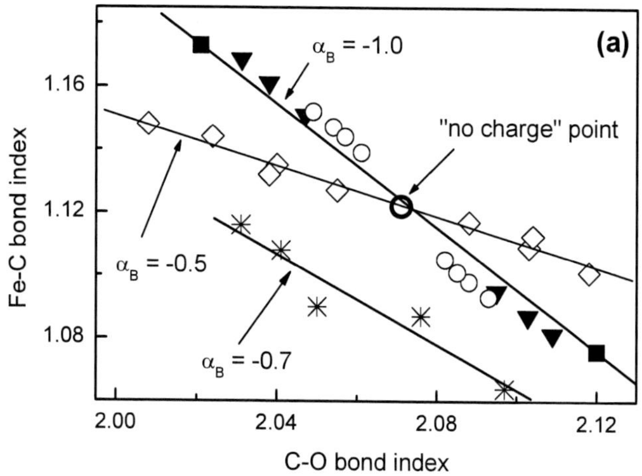

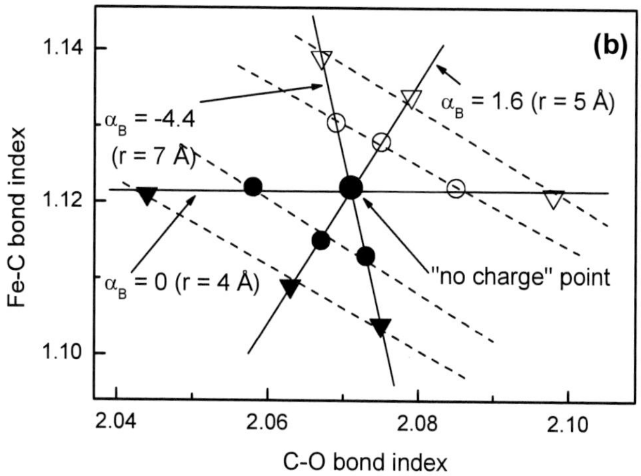

Figure 4. Theoretical data on the dependence of relationship between the Fe-C and C-O bond indexes on (a) the homogeneous electric field (squares) and Q_1 (diamonds), Q_3 (triangles) and Q_4 (circles) charges; (b) Q_2 charges (+1 e^-, solid triangles; +0.5 e^-, solid circles; -0.5 e^-, open circles; and -1 e^-, open triangles). Relationships between B_{FeC} and B_{CO} on the substitution for different charges at the same position (b, solid lines) and for the same charge at different positions (b, dashed lines) are represented.

Therefore, a positive charge located at the distance greater than 5 Å is expected to strengthen the C-O bond and increase respective vibrational frequency. It implies, for example, that the charges of opposite signs located closer and further than 5 Å from the FeCO line must cause changes in the same direction of $\Delta q_{2\pi}$ and, consequently, of ν_{CO}, see Figure 5b.

These interactions were calculated [19] to cause violation of the negative correlation between B_{CO} and B_{FeC} (and consequently, between ν_{CO} and ν_{FeC}), if different charged groups are located in the area of the porphyrin edge ($r \approx 4 - 6$ Å), see Figure 4b. It explained the lack of the linear negative correlation in model compounds containing different charged peripheral substitutes [19]. It also predicted the strong effect of the charged groups on the CO ^{17}O isomeric chemical shift (δ) and its poorer correlation with ν_{CO} in such models comparing to HPs. This conclusion was supported by the recent experimental data [41] (note that the authors erroneously interpreted these results as a prove of a weak contribution of the electrostatic effects into the vibrational and NMR parameters of the coordinated CO).

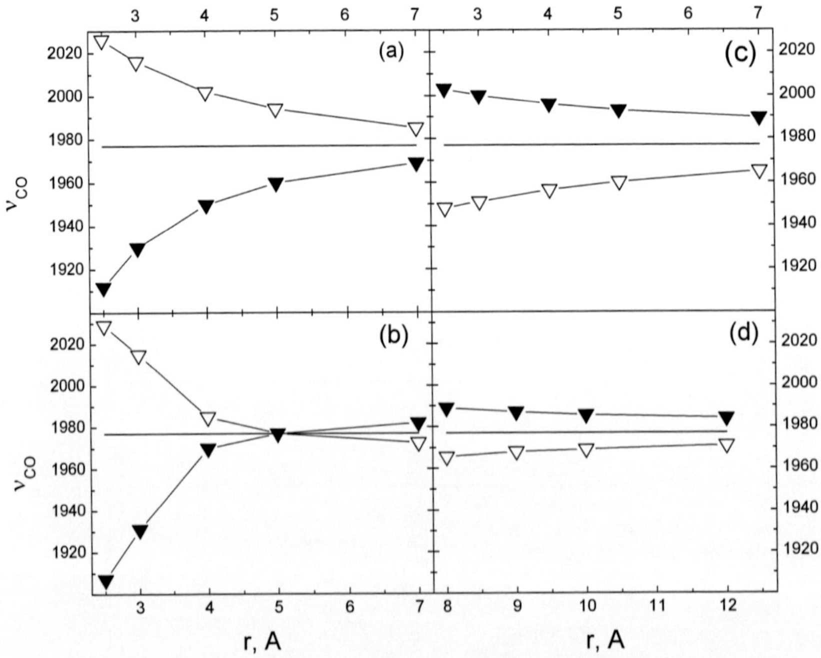

Figure 5. Dependence of ν_{CO} on the position and magnitude of the Q_1 (a), Q_2 (b), Q_3 (c) and Q_4 (d) point charges.

The effect of the electric field of the distal histidine on B_{CO}, B_{FeC}, ν_{CO} and δ in the carbonyl complex of myoglobin (Figure 1d) was computed [20]. The orientation of the imidazole of the distal histidine in respect to Fe(P)(Im)(CO) was obtained in the X-ray diffraction study of wild-type sperm whale myoglobin (Mb(CO)) [42] and used in these calculations. The effect of the neutral imidazole tautomerization (the proximal N_ε or distal N_δ nitrogen is protonated), 180° rotations around the histidine C-C bond (Figure 1d), and the protonation of both the nitrogens, was studied.

It was shown [20] that the four C-O vibrational bands of Mb(CO) (A_0, A_1, A_2, and A_3) [43] can be assigned to four different conformations of the distal histidine: histidine is displaced out of the pocket (A_0); N_δ histidine tautomer located in the pocket and 180° rotated around the C-C bond (A_2); and N_ε histidine tautomers without (A_1) and with (A_3) N_δ hydrogen-bonded to the solvent water. This assignment allowed [20] also to interpret quantitatively the experimentally observed pH dependence of δ [6]. The A_3 assignment naturally explained the experimentally unusually fast (~ 1 ns) interconversion of the protein between the A_1 and A_3 states [44].

CO Bandshape as a Probe of the Heme Environment Dynamics

It follows from Figure 5, that changes in the position of the point charge affect ν_{CO}. This implies that motion of the charged groups of the heme environment broadens the CO band, this broadening being often called pure dephasing [45]. However, it follows from the same figure, that the broadening is expected to be significant (~10 cm^{-1}) only if the amplitude of the motion is big enough (~1 Å) and the charged groups are located close enough (~2 – 4 Å) to the CO oxygen. Contribution of other charged groups (including those located on the imidazole side of the heme and equatorial) into the CO band broadening is considerably weaker. Therefore, the results presented on Figure 5 suggest the CO band as an effective probe of large amplitude motions of the heme environment, in general, and of the distal side of the HP(CO) heme pocket, in particular.

To test this conclusion the infrared absorption spectra of the samples of carbon monoxide complex of horseradish peroxidase (HRP(CO)) at pH 6.0 and 9.3, and its complex with benzohydroxamic acid (BHA) at pH 6.0 in 60% glycerol/water v/v solvent, were measured (Figure 6) [24,25]. Note that the glycerol/water solvent has glass liquid transition at $T_c \approx 180$ K. The bandshape and moments of the CO band can be analyzed by using the theoretical approach described in Methods, because ν_{CO} (~ 2000 cm^{-1}) is much larger than the energy of thermal motion at room temperature.

However M_2 of the CO band of HRP(CO) in liquid glycerol/water solvent manifests [24,25] the temperature dependence, which is much steeper than the dependence expected in the case of harmonic motion of the heme environment, Eq. (15). This very strong temperature dependence of M_2 points to the anharmonic character of the motion of heme environment.

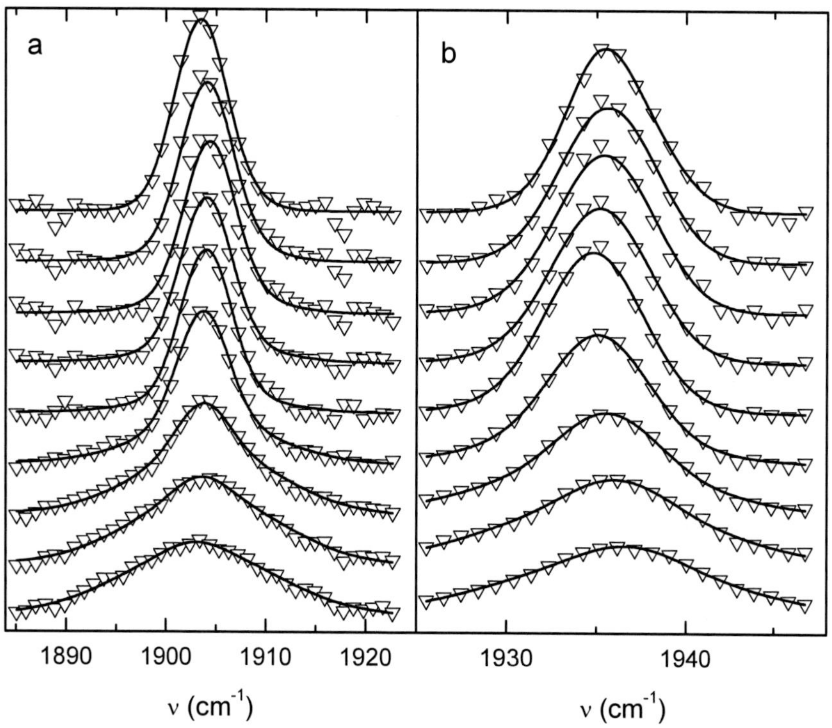

Figure 6. Temperature dependence of the CO infrared absorption band of HRP(CO) in glycerol/water solvent at pH 6.0 (a) and pH 9.3 (b) [24,25]. The spectra from top to bottom correspond to the temperatures 12(15), 50, 90, 130, 170, 200, 230, 260 and 290 K.

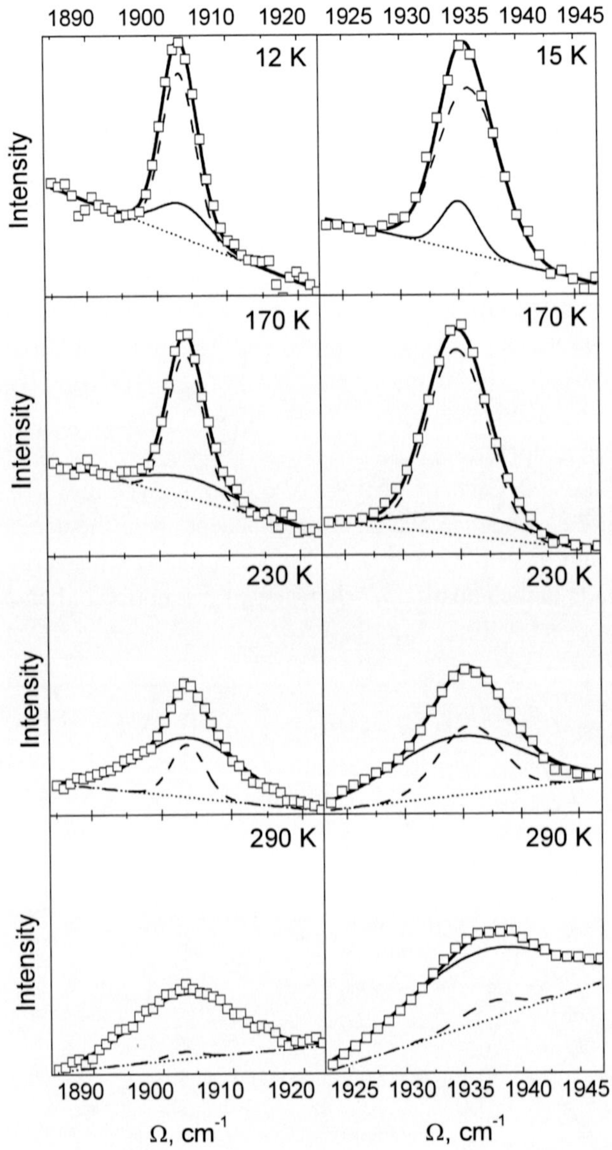

Figure 7. Temperature dependence of the contribution of CSS_l and CSS_h into the CO band (thick solid curve corresponds to the fit of the band; dashed and thin solid curves describe the contributions of CSS_l and CSS_h into the band, respectively; and the dotted line is background). The left and right columns correspond to the pH 6.0 and 9.3 samples, respectively.

The anharmonicity was interpreted in frameworks of the concept of conformational substates (CSS) [4,46,47], which postulates that protein exists in different CSSs, separated by energy barriers. Existence of CSSs was confirmed by the observation that proteins containing heme derivatives show inhomogeneously broadened optical spectra [48-51]. The CSSs are organized hierarchically being grouped in tiers of different energy, their population control the protein dynamics [47,52-55]. Consequently, one can describe protein dynamics as a superposition of two types of motions [56]. The first type is the "non-protein-specific motion" corresponding to harmonic vibrations of relatively small protein parts (for example, amino acid internal vibrations). The second is a large-amplitude "protein-specific motion", which corresponds to the protein molecule transition from one CSS to another; these transitions are affected by the protein surround and are strongly hindered in a glassy matrix [54,57-66].

To interpret the temperature dependence presented on Fig. 6 it was assumed [24,25,67] that the protein and the heme environment can exist in two CSSs, lower-energy (CSS_l) and higher-energy (CSS_h), the transition between them is linked to the protein surface and, consequently, is frozen in the glassy matrix. However in each of these states the heme environment motion is disconnected from the protein surface and is harmonic in the whole interval of temperatures.

In this case the CSS_h population can be written

$$p\left(T_{eff}\right)=\left\{1+exp\left[\frac{\Delta F\left(T_{eff}\right)}{k_B T_{eff}}\right]\right\}^{-1}, \tag{16}$$

where ΔF is a free energy difference between CSSh and CSSl

$$\Delta F\left(T\right)=\Delta E-\Delta S\cdot T_{eff}, \tag{17}$$

and T_{eff} reflects our assumption that the population of CSSs under consideration does not depend on the temperature in the glassy matrix, $T < T_c$

$$T_{eff}=\begin{Bmatrix} T, & T\geq T_c \\ T_{c,} & T<T_c \end{Bmatrix}. \tag{18}$$

In the case of the harmonic motion in each CSS and strong inhomogeneous contribution, the CO band is described by the expression

$$F(\Omega)=I_0\left\{p\left(T_{eff}\right)G\left[\Omega-\Omega_h,\sigma_h^2(T)\right]+\left[1-p\left(T_{eff}\right)\right]G\left[\Omega-\Omega_l,\sigma_l^2(T)\right]\right\}, \tag{19}$$

where I_0 is the band intensity, $G[\Omega-\Omega_{h(l)}, \sigma_{h(l)}^2(T)]$ is a normalized Gaussian with maximum at $\Omega_{h(l)}$ and second moment of $\sigma_{h(l)}^2(T)$, which corresponds to the absorption by $CSS_{h(l)}$. The temperature dependence of $\sigma_{h(l)}^2(T)$ is described by Eq. (15) with constants $A_{h(l)}$ and $B_{h(l)}$.

The temperature dependence of M_2 of the band can be easily obtained using Eq. (19)

$$M_2(\Phi)=p\left(T_{eff}\right)\cdot\sigma_h^2+\left[1-p\left(T_{eff}\right)\right]\sigma_l^2+p\left(T_{eff}\right)\left[1-p\left(T_{eff}\right)\right]\left(\Omega_h-\Omega_l\right)^2. \tag{20}$$

Eqs (16) – (19) allowed to fit the experimental spectra (see Figs. 6 and 7), the fitting parameters are presented in Table 1. Figure 8 shows the temperature dependence of the CSS_h population obtained using these parameters.

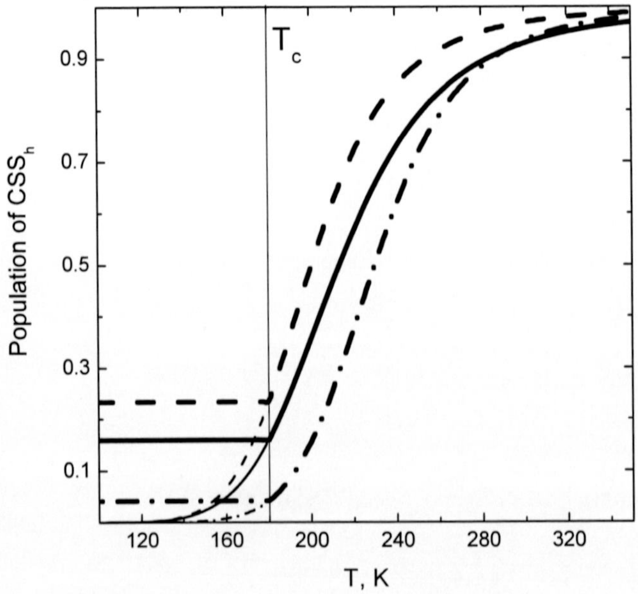

Figure 8. Temperature dependence of the population of the excited conformational substate in the HRP(CO) at pH 6.0 and 9.3 (thick dashed and solid curves, respectively) and of the "open" conformation of Mb(CO) at pH 5.0 (thick dotted curve). The thin curves represent the would be populations of CSS_h, if the samples were liquid in the whole interval of the temperatures.

Table 1. Spectroscopic and dynamic parameters of the studied proteins

	ΔE cm^{-1}	ΔS e. u.	A_h cm^{-2}	B_h cm^{-1}	A_l cm^{-2}	B_l cm^{-1}	Ω_h cm^{-1}	Ω_l cm^{-1}
HRP(CO) pH 9.3 (model 3)	1325± 120	6.2± 0.6	0	0.22± 0.02	6.9± 0.14	0	1935.6± 0.6	1934.8. ±0.6
HRP(CO) pH 6.0 (model 3)	1453± 145	7.2± 0.6	12.0± 4.2	0.24± 0.04	6.0± 0.2	0	1904.3± 0.5	1903.7± 0.5
A_0 band, Mb(CO) pH 5.0 (model 3)	1805± 315	7.8± 0.8	37.2± 8.3	0.04± 0.05	15.8± 1.2	0	1966.0	1966.0
HRP(CO)+BHA pH 6.0 (model 2)	–	–	–	–	0	0.07± 0.05	–	–
A_1 band, Mb(CO) pH 6.8 (model 2)	–	–	–	–	0	0.12± 0.03	–	–
A_1 band, Hb(CO) pH 6.8 (model 2)	–	–	–	–	7.35	0.04± 0.005	–	–

Analyzing the fitting parameters one should take into account that pK_α of the HRP(CO) distal histidine (His42) is reported to be 8.3 [68-71]. Increasing pH above this value causes the His42 deprotonation transforming its positively charged imidazolinium to neutral imidazole, which has moderate dipole moment. This change alters the structure of the distal part of the heme pocket [72]. The changes in the pocket structure and the His42 charge essentially affect the electrostatic interaction between the C-O dipole moment and the heme pocket amino acids. As a result the CO band [34,72] notably shifts upon the change in pH from 6.0 to 9.3, see Figure 6.

Consequently, if the thermal broadening of the CO band was caused by the motion of the heme pocket amino acids, then the parameter of the electrostatic interaction of this motion with the C-O dipole moment (B in Eq. (13)) would be essentially different at pH 6.0 and 9.3. However, it follows from Table 1, that B_h is weakly affected by the pH change, decreasing only by less than 10% upon the His42 deprotonation, whereas B_l is the same at both the pH values. Moreover, the CO band position hardly shifts upon the transition of the protein from CSS_l to CSS_h in both the pH 6.0 and 9.3 samples, this fact points to the weak change in the heme pocket structure upon the transition. Therefore one should conclude that most probably the CO band broadening is caused not by the increase in dynamics of the heme pocket amino acids in CSS_h, but by some other factors.

From our point of view, the best candidate for this role is a disordered water molecule, which appears in the heme pocket upon heating, and CSS_l and CSS_h correspond to the protein conformation without and with this molecule in the heme pocket. Indeed, HRP has a big pocket, which can accommodate not only water, but also much bigger substrate molecules and their analogs [73,74], and the analysis of the crystal structure of ferric HRP suggests presence of a disordered water molecule in the heme pocket [73]. The pocket is less polar than the solvent; therefore CSS_l corresponds to the protein conformation with the water molecule outside of the pocket. H_2O has strong dipole moment; it can weakly bind in different places of the heme pocket and move at these places and between them, affecting the CO band width without notably changing its position. H_2O entry to the pocket from the solvent increases the entropy of the whole system; this qualitative conclusion coincides with the results of the fitting procedure, Table 1. Moreover, at pH 6.0 (His42 is protonated) the heme pocket is more polar and has more places to bind H_2O, than at pH 9.3. Therefore both the increase in entropy upon the water entrance (ΔS) and the inhomogeneous broadening in CSS_h (A_h) are expected to be stronger at pH 6.0, than at pH 9.3. This qualitative conclusion also coincides with the results of the fitting procedure (note, however, the uncertainty of the ΔS evaluation), see Table 1. Water motion inside the heme pocket is hardly connected to the protein surface and is expected to take place in CSS_h even in the glassy environment. On contrary, the water entrance into the pocket ($CSS_l \rightarrow CSS_h$ transition) is coupled to the large amplitude motions of the heme pocket, their arrest by the glassy matrix making impossible the water entrance.

The appearance of the disordered water molecule in the heme pocket can be also caused by a cleavage upon heating of the hydrogen bond between one of the ordered heme pocket water molecule and the corresponding amino acid. Such ordered water molecule was observed in the HRP(CO) pocket at low temperature [75]. We cannot exclude this possibility, but note that presence of hydrogen bonded ordered water molecule in the pocket produces static electric field. This field is expected to notably shift the CO band position. At the same time the disordered water molecule only broadens the band. Consequently, one should expect not only broadening, but also notable shift of the CO band upon heating when the hydrogen bond is cleaved and the water moves in the pocket. This conclusion contradicts to the very close values of Ω_l and Ω_h and makes this mechanism less likely.

Additional room temperature X-ray diffraction experiments can be suggested to distinguish between these two mechanisms. If water molecule enters the heme pocket at higher temperatures (in liquid solvent) the number of ordered water molecules at low and room temperatures is expected to be the same. On the contrary, if heating of the sample liberates an ordered hydrogen bonded water molecule, the experiment must clearly show this.

It follows from the consideration presented above, that position of the CO band in HRP(CO) is controlled by the electrostatic interaction with heme pocket amino acids [72], whereas main contribution to the band broadening most probably stems from the interaction with the disordered water molecule in the pocket.

On Figure 9a the temperature dependences of M_2 of the CO bands of HRP(CO) at pH 6.0 and 9.3 (calculated using Eq. (20) and parameters from Table 1) and of the A_0 band of Mb(CO) at pH 5.0 [52] are presented. The A_0 CO infrared absorption band was shown [20,76-79] to correspond to the "open" protein conformation with the distal histidine located outside of the heme pocket. In this conformation the distal part of the heme pocket is big enough to accommodate nitrite [80] likewise water molecule. The myoglobin heme pocket is hydrophobic; consequently the CSS_l must correspond to the state with the water molecule out of the pocket. Upon heating the water molecule can enter the pocket forming CSS_h. This naturally explains why the same model, which involves an assumption about the presence of CSS_h, fits in the experimental data. Our interpretation of CSS_h as CSS with a disordered water in the pocket is also supported by the fact that the band position hardly depends on the temperature [52], suggesting small difference between Ω_h and Ω_l. Note, that this result reinstates an earlier proposition [53,81] that in the "open" conformation there is CSS_h, which was called by the authors A'_0. Population of this CSS_h strongly increases upon heating (see Figure 8), being lower than in HRP almost in the whole interval of the temperatures studied. This result is easily understandable, because the Mb heme pocket is much smaller and less polar than that of HRP.

Our conclusion about the presence of disordered water molecules in the heme pocket of HPs was confirmed lately by the molecular dynamics simulation of cytochrome P-450 [82]

In the "closed" conformation, which is mostly populated at pH 6.8, the distal histidine is located inside the heme pocket leaving much less free space. Consequently, the CSS_h energy is much higher and its population is expected to be much less. As a result, it hardly contributes to the CO band, and the thermal broadening temperature dependence of the related A_1 CO band is expected to behave harmonically in the liquid solvent. In the glassy environment the band is expected to depend on temperature harmonically if the heme environment is not linked to the protein surface and be temperature independent if the heme environment is strongly linked to the protein surface. Fitting shows (see Figure 9b) that in the "closed" conformation of Mb(CO) the latter situation takes place. The same is true for HRP(CO)+BHA and the "closed" conformation of Hb(CO), showing that the BHA binding in the HRP heme pocket displaces the disordered water even at room temperature [74]. Most probably in all these cases the CO broadening is caused by the electrostatic coupling to the motions of the heme pocket amino acids, which are linked to the protein surface. This cause of the broadening presumably exists also in CSS_h of HRP(CO) and Mb(CO) "open" conformation, but is masked by much stronger contribution of the disordered water.

Trehalose (T_c = 331 K) exists in glassy state in the whole interval of temperatures studied in this paper. The infrared absorption spectra of the pH 6.0 sample of HRP(CO) in trehalose consist of three clearly distinct (at least at low temperatures) peaks [25]. This is very different from other HRP(CO) spectra (Figure 6) , which manifest only one absorption peak in the interval 1900-2000 cm^{-1}. On Figure 9b the temperature dependence of M_2 of the most intense central peak is presented: despite the glassy environment it significantly depends on temperature. This dependence was fitted to the harmonic model, Eq. (12), which suggests that there is no $CSS_l \rightarrow CSS_h$ transition. This dependence is much stronger than that of the closed conformations of Mb(CO), Hb(CO) and HRP(CO)+BHA, where there is no $CSS_l \rightarrow CSS_h$ transition as well. Moreover, this is the only case, where the quantum effects are clearly seen and the effective frequency of the active vibration is found out, $\omega = 223\ cm^{-1}$. All these facts imply that in this case the heme pocket structure significantly differs from that in the glycerol/water mixture.

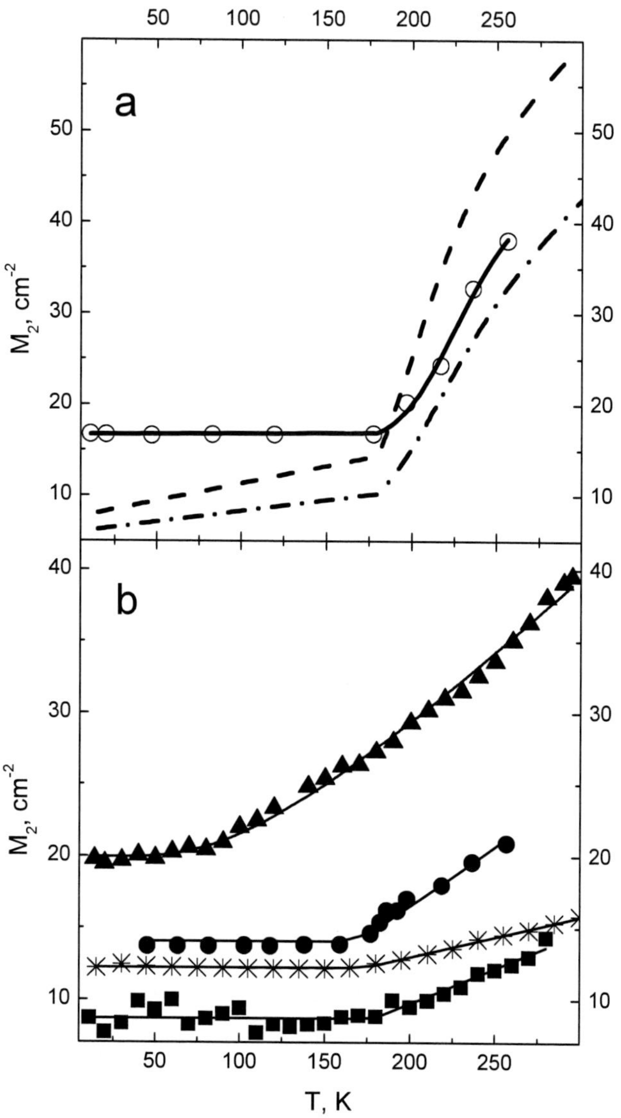

Figure 9. Temperature dependence of M_2 of the CO IR absorption band of different proteins. (a) HRP(CO) at pH 6.0 (dashed curve) and 9.3 (dashed-dotted curve), and A_0 component of the "open" conformation of Mb(CO) at pH 5.0 (empty circles) [52]; (b) central component of HRP(CO) in trehalose glass at pH 6.0 (solid triangles) [25]; A_1 component of the "closed" conformation of Mb(CO)

at pH 6.8 (solid circles) [52]; A_1 component of the "closed" conformation of Hb(CO) at pH 7.0 (asterisks) [85]; and HRP(CO)+BHA at pH 6.0 (solid squares) [24]. Solid curves represent fits to different models as described in the text.

Three hypotheses can be invoked to explain the experimentally observed temperature dependence under consideration. First, the HRP(CO) trehalose sample contains disordered water. This is possible, because the solid sample was prepared from the trehalose-water solution [25], and the probability for a water molecule to enter the pocket at $T > 300$ K in liquid solvent is close to 1, see Figure 8. Formation of the glassy matrix upon the sample drying can lock the water in the pocket keeping the system in CSS_h at all the temperatures studied. This assumption is supported by the close magnitudes of the coupling of the CO band to the heme environment, $B_h = 0.24 \pm 0.04$ (glycerol-water solvent) and 0.17 ± 0.01 (trehalose glass) cm^{-1}. The quantum effects in the temperature dependence can stem from the change in the water molecule motion, because in the glass under osmotic stress the protein in general and the heme environment in particular can become more compact.

Another possibility is that the reduction of the distances between the CO and the pocket amino acids upon the heme pocket contraction in the glassy matrix increases the coupling of the CO band to the internal vibrations of these amino acids. If this interpretation is correct, the 223 cm^{-1} is an effective frequency, which corresponds to a group of internal vibrations of the heme pocket amino acids.

And finally, the central peak of HRP(CO) in trehalose at pH 6.0 can correspond to the conformation with a trehalose molecule located in the pocket (despite its big size trehalose is flexible and can enter the pocket). In this case the stronger thermal broadening than in the cases of HRP(CO)+BHA, and "closed" conformations of Mb(CO) and Hb(CO) and the quantum effects can be interpreted as a manifestation of the coupling of the CO band to internal vibration of the trehalose molecule. This assumption is supported by the facts that trehalose has vibrations in the region of 223 cm^{-1} [83] and another peroxidase accommodates a molecule of co-solvent, glycerol, in its pocket [84].

Note that in this discussion we constrained ourselves to the simplest model of the protein dynamics, which invokes only one CSS_h. It is clear that it can be (and probably are) several higher energy conformational states. However, the fact that the simplest model allows interpreting the experimental data successfully shows that no conclusions about the larger number of CSS_h and their nature can be done on the basis of the experimental data under consideration.

CONCLUSION

The analysis of the effect of the Fe(P)(Im)(CO) complex distortion and model point charges on the bond indexes of the Fe-C and C-O bonds showed that the experimentally observed in HP(CO)s negative linear correlation between ν_{CO} and ν_{FeC} can be caused only by external electric fields and not by the complex distortions. At the same time it follows from the results, that some charges can violate this correlation; this shows that an absence of such a correlation does not necessarily points to the Fe(P)(Im)(CO) distortion.

It was obtained that appearance of charges on both sides of the porphyrin plane substantially affects ν_{CO}. However, motion of the charges located on the distal side of the porphyrin ring affects ν_{CO} much stronger than that of the proximal charges. Yet a movement of the distal charge does not affect ν_{CO} strongly. These results lead to a conclusion, that the width of the CO band is mainly affected by distal environment of the heme; its effect being strong only in the case of large-amplitude motions.

The theory of bandshape was applied to the interpretation of the temperature dependence of the CO-band of HRP(CO) in glycerol/water solvent at different pH with and without BHA. It was shown that in the liquid solvent the protein moves between two conformational substates. They correspond to the protein conformations without and with water molecule in the heme pocket; the higher is the temperature, the larger is the population of the latter. At room temperature almost all protein molecules exist in the higher energy CSS with the water in the pocket. The same transition takes place in the "open" conformation of Mb(CO), where the distal histidine is located out of the heme pocket leaving enough space for the water molecule. Note, that to our best knowledge this is a first example of an almost complete protein transition from a lower-energy conformational substate to the higher-energy one, caused by the temperature change.

In the glassy solvent the motions of the heme pocket amino acids are strongly hindered. Therefore water cannot enter or leave the pocket. Consequently, cooling the sample below the temperature of the glass liquid transition does not change the number of the protein molecules with the water molecule in the pocket. Hence, the temperature dependence of the band in glass stems from the temperature dependence of the motion of these water molecules.

In the HRP(CO)+BHA and "closed" conformations of Mb(CO) and Hb(CO) the disordered water is forced out of the pocket by BHA or the distal histidine. Thus the only contribution to the thermal broadening of the CO band stems from the electrostatic coupling of the CO vibration to the amino acids of the heme pocket. Their motions are linked to the motion of the protein surface. Therefore these motions are frozen in the glassy matrix, whereas in the liquid solvent they are well described in the harmonic approximation. As the result, the CO band second moment is temperature independent in the glassy matrix and is proportional to the temperature in a liquid solvent.

It follows from the consideration presented above, that position of the CO infrared absorption band of HP(CO)s is mainly affected by the structure of the heme pocket amino acids, whereas its width is essentially controlled by the dynamics of disordered water molecules, which populate the pocket.

References

[1] Eichorn, G. B., editor. 1973. *Inorganic Biochemistry*. Amsterdam: Elsevier.

[2] Rousseau, D. L. and Friedman, J. M. (1988). Transient and cryogenic studies of photodissociated hemoglobin and myoglobin. In T. G. Spiro (Ed.), *Biological Application of Raman Spectroscopy* (pp. 133-215). New York: Wiley and Sons.

[3] Rodgers, K. R. and Spiro, T. G. (1994). Nanosecond Dynamics of the R-T Transition in Hemoglobin - Ultraviolet Raman Studies. *Science, 265,* 1697-1699.

[4] Frauenfelder, H., Wolynes, P. G., and Austin, R. H. (1999). Biological physics. *Reviews of Modern Physics, 71,* S419-S430.

[5] Li, X. Y. and Spiro, T. G. (1988). Is Bound CO Linear or Bent in Heme-Proteins - Evidence from Resonance Raman and Infrared Spectroscopic Data. *J. Amer. Chem. Soc., 110,* 6024-6033.

[6] Park, K. D., Guo, K., Adebodun, F., Chiu, M. L., Sligar, S. G., and Oldfield, E. (1991). Distal and proximal ligand interactions in heme proteins: correlations between C-O and Fe-C vibrational frequencies, oxygen-17 and carbon-13 nuclear magnetic resonance. *Biochemistry, 30,* 2333-2347.

[7] Ray, G. B., Li, X. Y., Ibers, J. A., Sessler, J. L., and Spiro, T. G. (1994). How far can proteins bend the FeCO unit? Distal polar and steric effects in heme-proteins and models. *J. Amer. Chem. Soc., 116,* 162-176.

[8] Spiro, T. G., Zgierski, M. Z., and Kozlowski, P. M. (2001). Stereoelectronic factors in CO, NO and O-2 binding to heme from vibrational spectroscopy and DFT analysis. *Coord. Chem. Rev., 219,* 923-936.

[9] Ridley, J. and Zerner, M. (1973). An intermediate neglect of differential overlap technique for spectroscopy: pyrrole and the azines. *Theor. Chim. Acta, 32,* 111-134.

[10] Bacon, A. D. and Zerner, M. C. (1979). An intermediate neglect of differential overlap theory for transition metal complexes: Fe, Co and Cu chlorides. *Theor. Chim. Acta., 53,* 21-54.

[11] Zerner, M. C., Loew, G. H., Kirchner, R. E., and Mueller-Westerhoff, U. T. (1980). An intermediate neglect of differential overlap technique for spectroscopy of transition-metal complexes. Ferrocene. *J. Amer. Chem. Soc., 102,* 589-599.

[12] Anderson, W. P., Edwards, W. D., and Zerner, M. C. (1986). Calculated spectra of hydrated ions of the first transition-metal series. *Inorg. Chem., 25,* 2728-2732.

[13] Eaton, W. A., Hanson, L. K., Stephens, P. J., Sutherland, J. C., and Dunn, J. B. R. (1978). Optical spectra of oxy- and deoxyhemoglobin. *J. Amer. Chem. Soc., 100,* 4991-5003.

[14] Ricard, L., Weiss, R., and Momenteau, M. (1986). Crystal and molecular structure of a highly hindered iron(II) porphyrin complex. *Journal of the Chemical Society-Chemical Communications,* 818-820.

[15] Kim, K., Fettinger, J., Sessler, J. L., Cyr, M., Hugdahl, J., Collman, J. P., and Ibers, J. A. (1989). Structural characterization of a sterically encumbered iron(II) porphyrin CO complex. *J. Amer. Chem. Soc., 111,* 403-405.

[16] Kim, K. and Ibers, J. A. (1991). Structure of a carbon monoxide adduct of a "capped" porphyrin: Fe(C_2-Cap)(CO)(1-methylimidazole). *J. Amer. Chem. Soc., 113,* 6077-6081.

[17] Tetreau, C., Lavalette, D., Momenteau, M., Fischer, J., and Weiss, R. (1994). Structure-reactivity relationship in oxygen and carbon monoxide binding with some heme models. *J. Amer. Chem. Soc., 116,* 11840-11848.

[18] Huber, K. P. and Herzberg, G. (1979). Molecular Spectra and Molecular Structure. IV. Constants of diatomic Molecules. New York: Van Nostrand Reinhold.

[19] Kushkuley, B. and Stavrov, S. S. (1996). Theoretical study of the distal-side steric and electrostatic effects on the vibrational characteristics of the FeCO unit of the carbonylheme proteins and their models. *Biophys. J., 70,* 1214-1229.

[20] Kushkuley, B. and Stavrov, S. S. (1997). Theoretical study of the electrostatic and steric effects on the spectroscopic characteristics of the metal-ligand unit of heme proteins. 2. C-O vibrational frequencies, O-17 isotropic chemical shifts, and nuclear quadrupole coupling constants. *Biophys. J., 72,* 899-912.

[21] Bersuker, I. B. (1978). Mutual vibronic influence of weakly coordinated molecular systems in chemical reactions and catalysis. *Chem. Phys., 31,* 85-93.

[22] Bersuker, I. B. (1984). The Jahn-Teller Effect and Vibronic Interactions in Modern Chemistry. New York: Plenum Press.

[23] Stavrov, S. S., Decusar, I. P., and Bersuker, I. B. (1993). Chemical Activation of Oxygen and Carbon-Monoxide by Hemoproteins - the Vibronic Approach. *New Journal of Chemistry, 17,* 71-76.

[24] Kaposi, A. D., Vanderkooi, J. M., Wright, W. W., Fidy, J., and Stavrov, S. S. (2001). Influence of static and dynamic disorder on the visible and infrared absorption spectra of carbonmonoxy horseradish peroxidase. *Biophys. J., 81,* 3472-3482.

[25] Stavrov, S. S., Wright, W. W., Vanderkooi, J. M., Fidy, J., and Kaposi, A. D. (2002). Optical and IR absorption as probe of dynamics of heme proteins. *Biopolymers, 67,* 255-258.

[26] Lax, M. (1952). The Franck-Condon principle and its application to crystals. *J. Chem. Phys., 20,* 1752-1760.

[27] Markham, J. J. (1959). Interaction of normal modes with electron traps. *Reviews of Modern Physics, 31,* 956-989.

[28] Perlin, Y. E. (1964). Modern methods in the theory of many-phonon processes. *Uspekhi Fizicheskix Nauk, 80,* 553-595.

[29] Rebane, K. K. (1970). Impurity Spectra of Solids. Elementary Theory of Vibrational Structure. New York: Plenum Press.

[30] Bitler, A. and Stavrov, S. S. (1999). Iron-histidine resonance Raman band of deoxyheme proteins: Effects of anharmonic coupling and glass-liquid phase transition. *Biophys. J., 77,* 2764-2776.

[31] Rosenfeld, Y. B. and Stavrov, S. S. (1994). Anharmonic coupling of soft modes and its influence on the shape of the iron-histidine resonance Raman band of heme-proteins. *Chem. Phys. Lett., 229,* 457-464.

[32] Phillips, G. N., Teodoro, M. L., Li, T. S., Smith, B., and Olson, J. S. (1999). Bound CO is a molecular probe of electrostatic potential in the distal pocket of myoglobin. *J. Phys. Chem. B, 103,* 8817-8829.

[33] Spiro, T. G. and Wasbotten, I. H. (2005). CO as a vibrational probe of heme protein active sites. *J. Inorg. Biochem., 99,* 34-44.

[34] Kaposi, A. D., Wright, W. W., Fidy, J., Stavrov, S. S., Vanderkooi, J. M., and Rasnik, I. (2001). Carbonmonoxy horseradish peroxidase as a function of pH and substrate: Influence of local electric fields on the optical and infrared spectra. *Biochemistry, 40,* 3483-3491.

[35] Park, E. S. and Boxer, S. G. (2002). Origins of the sensitivity of molecular vibrations to electric fields: Carbonyl and nitrosyl stretches in model compounds and proteins. *J. Phys. Chem. B, 106,* 5800-5806.

[36] Springer, B. A., Sligar, S. G., Olson, J. S., and G. N. Phillips, J. (1994). Mechanism of ligand recognition in myoglobin. *Chem. Rev., 94,* 699-714.

[37] Lim, M., Jackson, T. A., and Anfinrud, P. A. (1995). Binding of CO to Myoglobin from a Heme Pocket Docking Site to Form Nearly Linear Fe-C-O. *Science, 269,* 962-966.

[38] McMahon, M. T., deDios, A. C., Godbout, N., Salzmann, R., Laws, D. D., Le, H. B., Havlin, R. H., and Oldfield, E. (1998). An experimental and quantum chemical investigation of CO binding to heme proteins and model systems: A unified model

based on C-13, O-17, and Fe-57 nuclear magnetic resonance and Fe-57 Mossbauer and infrared spectroscopies. *J. Amer. Chem. Soc., 120,* 4784-4797.

[39] Franzen, S. (2002). An electrostatic model for the frequency shifts in the carbonmonoxy stretching band of myoglobin: Correlation of hydrogen bonding and the Stark tuning rate. *J. Amer. Chem. Soc., 124,* 13271-13281.

[40] De Angelis, F., Jarzecki, A. A., Car, R., and Spiro, T. G. (2005). Quantum chemical evaluation of protein control over heme ligation: CO/O-2 discrimination in myoglobin. *J. Phys. Chem. B, 109,* 3065-3070.

[41] Kalodimos, C. G., Gerothanassis, I. P., Pierattelli, R., and Ancian, B. (1999). Carbon-13 and oxygen-17 chemical shifts, (O-16/O-18) isotope effects on C-13 chemical shifts, and vibrational frequencies of carbon monoxide in various solvents and of the Fe-C-O unit in carbonmonoxy heme proteins and synthetic model compounds. *Inorg. Chem., 38,* 4283-4293.

[42] Quillin, M. L., Arduini, R. M., Olson, J. S., and G. N. Phillips, J. (1993). High-resolution crystal structures of distal histidine mutants of sperm whale myoglobin. *J. Mol. Biol., 234,* 140-155.

[43] Potter, W. T., Hazzard, J. H., Choc, M. G., Tucker, M. P., and Caughey, W. S. (1990). Infrared spectra of carbonyl hemoglobins: Characterization of dynamic heme pocket conformers. *Biochemistry, 29,* 6283-6295.

[44] Johnson, J. B., Lamb, D. C., Frauenfelder, H., Muller, J. D., McMahon, B., Nienhaus, G. U., and Young, R. D. (1996). Ligand binding to heme proteins .6. Interconversion of taxonomic substates in carbonmonoxymyoglobin. *Biophys. J., 71,* 1563-1573.

[45] Fayer, M. D. (2001). Fast protein dynamics probed with infrared vibrational echo experiments. *Annual Review of Physical Chemistry, 52,* 315-356.

[46] Austin, R. H., Beeson, K. W., Eisenstein, L., Frauenfelder, H., and Gunsalus, I. C. (1975). Dynamics of ligand binding to myoglobin. *Biochemistry, 14,* 5355-5373.

[47] Frauenfelder, H., Nienhaus, G. U., and Young, R. D. (1994). Relaxation and disorders in proteins. In R. Richert andA. Blumen (Eds.), *Disorder Effects on Relaxation Processes* (pp. 592-614). Berlin: Springer-Verlag.

[48] Angiolillo, P. J., Leigh, J. S. J., and Vanderkooi, J. M. (1982). Resolved fluorescence emission spectra of iron-free cytochrome c. *Photochem. Photobiol., 36,* 133-1377.

[49] Kaposi, A. D., Wright, W. W., and Vanderkooi, J. M. (2003). Consequences of inhomogeneous broadening on fluorescence line narrowing spectra. *J. Fluoresc., 13,* 59-67.

[50] Fidy, J., Laberge, M., Kaposi, A. D., and Vanderkooi, J. M. (1998). Fluorescence line narrowing applied to the study of proteins. *Biochim. Biophys. Acta-Protein Struct. Molec. Enzym., 1386,* 331-351.

[51] Kaposi, A. D. and Vanderkooi, J. M. (1992). Vibronic energy map and excited-state vibrational characteristics of magnesium myoglobin determined by energy- selective fluorescence. *Proc. Natl. Acad. Sci. USA., 89,* 11371-11375.

[52] Ansari, A., Berendzen, J., Braunstein, D., Cowen, B. R., Frauenfelder, H., Hong, M. K., Iben, I. E. T., Johnson, J. B., Ormos, P., Sauke, T. B., Scholl, R., Schulte, A., Steinbach, P. J., Vittitow, J., and Young, R. D. (1987). Rebinding and relaxation in the myoglobin pocket. *Biophys. Chem., 26,* 337-355.

[53] Hong, M. K., Braunstein, D., Cowen, B. R., Frauenfelder, H., Iben, I. E. T., Mourant, J. R., Ormos, P., Scholl, R., Schulte, A., Steinbach, P. J., Xie, A. H., and Young, R. D.

(1990). Conformational Substates and Motions in Myoglobin - External Influences on Structure and Dynamics. *Biophys. J., 58,* 429-436.

[54] Huang, J., Ridsdale, A., Wang, J. Q., and Friedman, J. M. (1997). Kinetic hole burning, hole filling, and conformational relaxation in heme proteins: Direct evidence for the functional significance of a hierarchy of dynamical processes. *Biochemistry, 36,* 14353-14365.

[55] Prabhu, N. V., Dalosto, S. D., Sharp, K. A., Wright, W. W., and Vanderkooi, J. M. (2002). Optical spectra of Fe(II) cytochrome c interpreted using molecular dynamics simulations and quantum mechanical calculations. *J. Phys. Chem. B, 106,* 5561-5571.

[56] Parak, F. and Knapp, E. W. (1984). A consistent picture of protein dynamics. *Proc. Natl. Acad. Sci. U. S. A., 81,* 7088-7092.

[57] Cordone, L., Cupane, A., Leone, M., and Vitrano, E. (1986). Optical absorption spectra of deoxy- and oxyhemoglobin in the temperature range 300-20 K. Relation with protein dynamics. *Biophys. Chem., 24,* 259-275.

[58] Frauenfelder, H., Alberding, N. A., Ansari, A., Braunstein, D., Cowen, B. R., Hong, M. K., Iben, I. E. T., Johnson, J. B., Luck, S., Marden, M. C., Mourant, J. R., Ormos, P., Reinisch, L., Scholl, R., Schulte, A., Shyamsunder, E., Sorensen, L. B., Steinbach, P. J., Xie, A. H., Young, R. D., and Yue, K. T. (1990). Proteins and pressure. *Journal of Physical Chemistry, 94,* 1024-1037.

[59] Hagen, S. J., Hofrichter, J., and Eaton, W. A. (1996). Geminate rebinding and conformational dynamics of myoglobin embedded in a glass at room temperature. *Journal of Physical Chemistry, 100,* 12008-12021.

[60] Gottfried, D. S., Peterson, E. S., Sheikh, A. G., Wang, J. Q., Yang, M., and Friedman, J. M. (1996). Evidence for damped hemoglobin dynamics in a room temperature trehalose glass. *Journal of Physical Chemistry, 100,* 12034-12042.

[61] Cottone, G., Cordone, L., and Ciccotti, G. (2001). Molecular dynamics simulation of carboxy-myoglobin embedded in a trehalose-water matrix. *Biophys. J., 80,* 931-938.

[62] Cordone, L., Ferrand, M., Vitrano, E., and Zaccai, G. (1999). Harmonic behavior of trehalose-coated carbon-monoxy-myoglobin at high temperature. *Biophys. J., 76,* 1043-1047.

[63] Hagen, S. J., Hofrichter, J., and Eaton, W. A. (1995). Protein reaction-kinetics in a room-temperature glass. *Science, 269,* 959-962.

[64] Schlichter, J. and Friedrich, J. (2001). Glasses and proteins: Similarities and differences in their spectral diffusion dynamics. *J. Chem. Phys., 114,* 8718-8721.

[65] Gafert, J., Friedrich, J., and Parak, F. (1993). A comparative pressure tuning hole-burning study of protoporphyrin-IX in myoglobin and in a glassy host. *J. Chem. Phys., 99,* 2478-2486.

[66] Stavrov, S. S. (2001). Optical absorption band III of deoxyheme proteins as a probe of their structure and dynamics. *Chem. Phys., 271,* 145-154.

[67] Kaposi, A. D., Vanderkooi, J. M., and Stavrov, S. S. (2006). Infrared Absorption Study of the Heme Pocket Dynamics of Carbonmonoxyheme Proteins. *Biophys. J., 91,* 4191-4200.

[68] Evangelista-Kirkup, R., Smulevich, G., and Spiro, T. G. (1986). Alternative carbon monoxide binding modes for horseradish peroxidase studied by resonance Raman spectroscopy. *Biochemistry, 25,* 4420-4425.

[69] Smulevich, G., Paoli, M., DeSanctis, G., Mantini, A. R., Ascoli, F., and Coletta, M. (1997). Spectroscopic evidence for a conformational transition in horseradish peroxidase at very low pH. *Biochemistry, 36,* 640-649.

[70] Feis, A., Rodriguez-Lopez, J. N., Thorneley, R. N. F., and Smulevich, G. (1998). The distal cavity structure of carbonyl horseradish peroxidase as probed by the resonance Raman spectra of His 42 Leu and Arg 38 Leu mutants. *Biochemistry, 37,* 13575-13581.

[71] Uno, T., Nishimura, Y., Tsuboi, M., Makino, R., T.Iizuka, and Y.Ishimura. (1987). Two types of conformers with distinct Fe-C-O configuration in the ferrous CO complex of hoseradish peroxidase. Resonance Raman and infrared spectroscopic studies with native and deuteroheme-substituted enzymes. *Biochemistry, 262,* 4549-4556.

[72] Kaposi, A. D., Prabhu, N. V., Dalosto, S. D., Sharp, K. A., Wright, W. W., Stavrov, S. S., and Vanderkooi, J. M. (2003). Solvent dependent and independent motions of CO-horseradish peroxidase examined by infrared spectroscopy and molecular dynamics calculations. *Biophys. Chem., 106,* 1-14.

[73] Gajhede, M., Schuller, D. J., Henriksen, A., Smith, A. T., and Poulos, T. L. (1997). Crystal structure of horseradish peroxidase C at 2.15 angstrom resolution. *Nature Structural Biology, 4,* 1032-1038.

[74] Henriksen, A., Schuller, D. J., Meno, K., Welinder, K. G., Smith, A. T., and Gajhede, M. (1998). Structural interactions between horseradish peroxidase C and the substrate benzhydroxamic acid determined by X-ray crystallography. *Biochemistry, 37,* 8054-8060.

[75] Carlsson, G. H., Nicholls, P., Svistunenko, D., Berglund, G. I., and Hajdu, J. (2005). Complexes of horseradish peroxidase with formate, acetate, and carbon monoxide. *Biochemistry, 44,* 635-642.

[76] Morikis, D., Champion, P. M., Springer, B. A., and Sligar, S. G. (1989). Resonance Raman Investigations of Site-Directed Mutants of Myoglobin: Effects of Distal Histidine Replacement. *Biochemistry, 28,* 4791-4800.

[77] Zhu, L., Sage, J. T., Rigos, A. A., Morikis, D., and Champion, P. M. (1992). Conformational Interconversion in Protein Crystals. *J. Mol. Biol., 224,* 207-215.

[78] Yang, F. and G. N. Phillips, J. (1996). Crystal structure of CO-, deoxy- and met-myoglobins at various pH values. *Journal of Molecular Biology, 256,* 762-774.

[79] Muller, J. D., McMahon, B. H., Chien, E. Y. T., Sligar, S. G., and Nienhaus, G. U. (1999). Connection between the taxonomic substates and protonation of histidines 64 and 97 in carbonmonoxy myoglobin. *Biophys. J., 77,* 1036-1051.

[80] Frauenfelder, H., McMahon, B. H., Austin, R. H., Chu, K., and Groves, J. T. (2001). The role of structure, energy landscape, dynamics, and allostery in the enzymatic function of myoglobin. *Proc. Natl. Acad. Sci. USA., 98,* 2370-2374.

[81] Iben, I. E. T. 1988. The kinetics of protein conformational relaxation in sperm whale myoglobin following a pressure jump. [Ph. D.]. Urbana-Champaign: University of Illinois at Urbana-Champaign. 91 p.

[82] Rydberg, P., Rod, T. H., Olsen, L., and Ryde, U. (2007). Dynamics of water molecules in the active-site cavity of human cytochromes P450. *J. Phys. Chem. B, 111,* 5445-5457.

[83] Ballone, P., Marchi, M., Branca, C., and Magazu, S. (2000). Structural and vibrational properties of trehalose: A density functional study. *J. Phys. Chem. B, 104,* 6313-6317.

[84] Gupta, K., Selinsky, B. S., Kaub, C. J., Katz, A. K., and Loll, P. J. (2004). The 2.0 A resolution crystal structure of prostaglandin H-2 synthase-1: Structural insights into an unusual peroxidase. *J. Mol. Biol., 335,* 503-518.

[85] Cupane, A., Leone, M., and Militello, V. (2003). Conformational substates and dynamic properties of carbonmonoxy hemoglobin. *Biophys. Chem., 104,* 335-344.

In: Biopolymer Research Trends
Editor: Tamas S. Nemeth, pp. 145-163

ISBN: 978-1-60021-983-2

Chapter 5

VOLATILE GENERAL ANESTHETIC INTERACTIONS WITH FOUR-α-HELIX BUNDLE PROTEINS

***Tao Zhang*[1] *and Jonas S. Johansson*[*1,2]**

[1]Departments of Anesthesiology and Critical Care,
[2]Biochemistry and Biophysics, and the Johnson Research Foundation,
University of Pennsylvania, Philadelphia, PA 19104, USA

ABSTRACT

Although volatile general anesthetics are administered to over 20 million patients in the United States each year for a broad range of surgical procedures, their mechanisms of action remain poorly understood. Volatile general anesthetics are believed to exert their clinical effects by modulating the activity of neuronal plasma membrane ligand-gated ion channels, including the γ-aminobutyric acid type A receptor and the glycine receptor. The structures of these membrane proteins are currently unknown but their transmembrane domains are thought to be composed of the commonly occurring four-α-helix bundle protein fold, based upon homology modeling with the related nicotinic acetylcholine receptor Cys-loop ligand-gated ion channel. Site-directed mutagenesis studies on intact ligand-gated ion channels expressed in different cell types implicate the transmembrane domains of the protein as constituting volatile general anesthetic sites of action. A similar conclusion has been drawn using photoaffinity labeling of the nicotinic acetylcholine receptor with halothane, followed by microsequencing to identify volatile general anesthetic binding sites. Experimental studies with synthetic four-α-helix bundle proteins reveal that this folding motif is capable of binding several contemporary clinically used volatile general anesthetics with dissociation constants that correlate closely with their respective EC_{50} values (effective concentration in 50% of test subjects) in humans for maintaining the anesthetic state. Detailed biophysical studies on these synthetic four-α-helix bundle proteins provide insight into how volatile general anesthetic binding can lead to altered protein activity, by modulating the structure, flexibility and overall stability of the system. In addition, molecular dynamics simulations on both synthetic-

[*] Corresponding author. Mailing address: 319C, John Morgan Building, Department of Anesthesiology and Critical Care, University of Pennsylvania, 3620 Hamilton Walk, Philadelphia, PA 19104, USA. Telephone: 215-349-5472. Fax: 215-349-5078. E-mail: JohanssJ@uphs.upenn.edu

and natural four-α-helix bundle protein domains provide further evidence for how biomolecular function can be modulated in the presence of bound volatile general anesthetic molecules. This chapter will present recent advances gained into the fundamental mechanisms of volatile general anesthetic action based upon studies with a number of different four-α-helix bundle motifs.

INTRODUCTION

Volatile general anesthetics are administered to more than 20 million surgical patients in the United States each year [1]. Although widely and successfully used in the clinical arena, the basic mechanisms of action of these agents remain largely unknown. The majority of investigators in the field believe that the volatile general anesthetics act by altering the function of specific central nervous system membrane proteins involved in neuronal information transfer. In particular, membrane proteins located at central nervous system synapses have been identified as likely targets for the volatile general anesthetics [2-6].

These central nervous system plasma membrane proteins include members of the Cys-loop ligand-gated ion channel family such as the γ-aminobutyric acid type A receptor and the glycine receptor. Both of these ion channels allow the passage of chloride ions from the extracellular medium into the neuronal cytoplasm upon the binding of their respective ligands (γ-aminobutyric acid or glycine), which leads to hyperpolarization of the neuronal plasma membrane. Volatile general anesthetics are able to enhance the activity of both the γ-aminobutyric acid type A receptor and the glycine receptor, which results in the passage of additional chloride ions across the neuronal plasma membrane, which in turn decreases the likelihood of action potential generation. The structures of the γ-aminobutyric acid type A receptor and the glycine receptor are currently unknown. However, a 4 Å resolution cryoelectron microscopy structure of the related Cys-loop ligand-gated ion channel nicotinic acetylcholine receptor from *Torpedo marmorata* has been published recently [7, 8]. This work reveals that the transmembrane domain of each of the five subunits that constitute the intact ligand-gated ion channel is composed of a four-α-helix bundle motif (Figure 1). The four transmembrane α-helices of each of the Cys-loop ligand-gated ion channel subunits are designated M1-M4.

Another type of central nervous system plasma membrane protein that is sensitive to volatile general anesthetics is the excitatory glutamatergic ion channel known as the *N*-methyl-D-aspartate receptor, which can also be located at synapses between neurons. This ligand-gated ion channel is normally activated by the binding of both glutamate and glycine, which leads to the opening of a cation-selective pore that allows the conduction of both calcium- and sodium ions down their respective electrochemical gradients into the cytoplasm, resulting in neuronal plasma membrane depolarization via an excitatory post-synaptic current [9-11]. In addition, the elevation in the neuronal cytoplasmic calcium concentration acts as a biochemical stimulus that initiates various physiological signaling cascades involved in, for example, synaptic plasticity. Volatile general anesthetics act on the *N*-methyl-D-aspartate receptor to inhibit its activity [2-6], which again would have the net effect of tending to hyperpolarize the neuronal plasma membrane, favoring electrical quiescence.

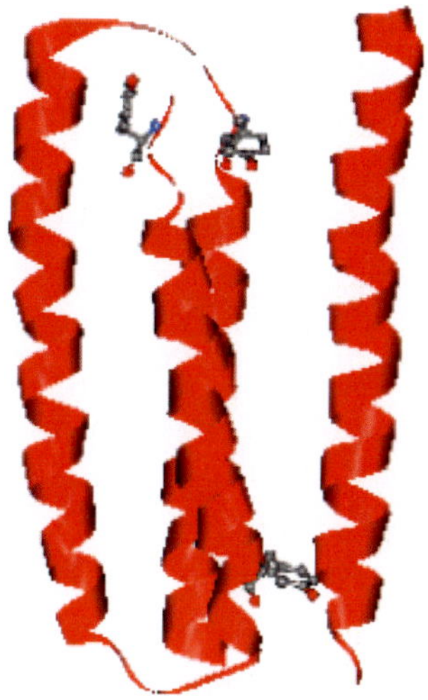

Figure 1. Structure of the transmembrane domain of the nAChR α subunit (Protein Data Bank access code 1OED), showing the three tyrosine residues at positions 213, 234, and 277.

Structurally, the *N*-methyl-D-aspartate receptor is a heteromeric protein complex that is by consensus composed of a total of four subunits, of which there are three different types: NR1 (which binds glycine), NR2 (which binds glutamate) and NR3 [11]. There are eight different NR1 subunits, four different NR2 subunits and two NR3 subunits, with most intact central nervous system *N*-methyl-D-aspartate receptors being composed of two NR1- and two NR2 subunits [10, 11]. The NR3 subunits can also combine with NR1 subunits to form functional *excitatory* glycine receptors that are insensitive to glutamate [12]. From an ultrastructural perspective, the *N*-methyl-D-aspartate receptor is less well understood than the Cys-loop ligand-gated ion channel family. An electron microscopy study on the related ionotropic glutamate AMPA (α-amino-3-hydroxy-5-methyl-4-isoxazole propionic acid) receptor from rat brain at a resolution of approximately 40 Å has been published, which indeed indicates that the intact receptor is a tetramer [13]. In addition, each subunit has three complete transmembrane segments (M1, M3 and M4) and a partial transmembrane sequence that loops back on the cytoplasmic domain of the protein (M2, "re-entrant loop").

This chapter will review the interaction of volatile general anesthetics with four-α-helix bundle folds from both natural- and designed synthetic proteins, based upon experimental and theoretical approaches. These types of investigations suggest plausible fundamental mechanisms whereby volatile general anesthetics may ultimately reversibly alter central nervous system protein function.

The Transmembrane Domains of the γ-Aminobutyric Acid Type A Receptor Are Important for the Effects of Volatile General Anesthetics

The chloride ion conducting properties of the γ-aminobutyric acid type A receptor in response to its natural neurotransmitter are enhanced by several volatile general anesthetics including halothane (Figure 2), isoflurane, enflurane, sevoflurane and desflurane [14]. Using chimeric receptors expressed in *Xenopus laevis* oocytes that were composed of the glycine receptor α1 subunit sequence and the γ-aminobutyric acid ρ1 subunit receptor sequence, a 45 amino acid residue segment was identified which included parts of both M2 and M3 that was responsible for the potentiating effects of enflurane on glycine receptor α1 subunit chloride ion conduction [15]. In particular, mutations of the residues Ser270 on M2 and Ala291 on M3 to bulkier residues such as isoleucine altered potentiation by enflurane.

Further studies with the γ-aminobutyric acid type A receptor α1 and α2 subunits identified the same residues as being of importance for the potentiating effects of enflurane and isoflurane on these Cys-loop ligand-gated ion channels. Volatile general anesthetic sensistivity in the case of the γ-aminobutyric acid type A receptor α2 subunit could be abolished by making the Ala291Trp mutation. Based upon homology modeling with the known cryoelectron microscopy structure of the nicotinic acetylcholine receptor [7], the residues Ser270 and Ala291 of the γ-aminobutyric acid type A receptor α1 subunit are predicted to be part of the hydrophobic core of the transmembrane four-α-helix bundle. These two residues are postulated to form part of a volatile general anesthetic binding site, or alternatively may play a role in transducing the free energy associated with anesthetic binding to cause the resulting modulation of chloride ion conduction across the membrane through the open channel [4].

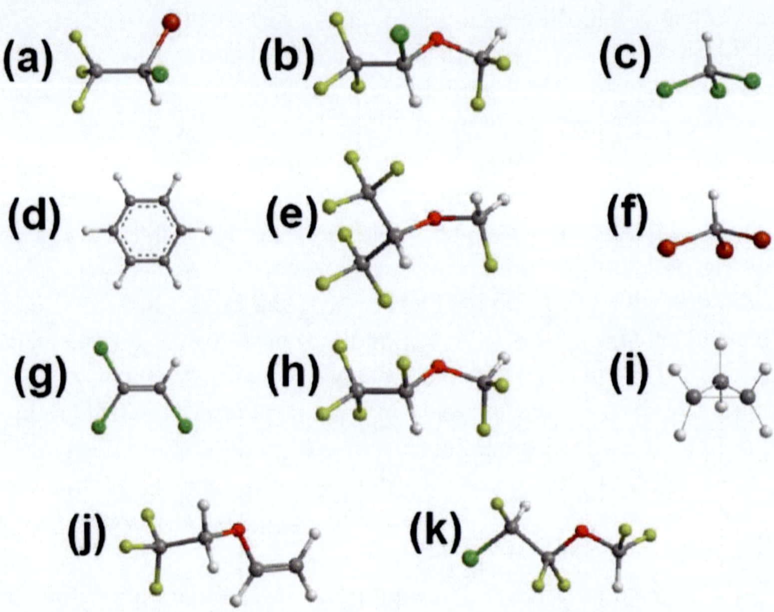

Figure 2. Ball-and-stick chemical structures of eleven inhaled general anesthetics. (a) Halothane (2-bromo-2-chloro-1,1,1-trifluoroethane), (b) isoflurane (1-chloro-2,2,2-trifluoroethyl difluoromethyl ether), (c) chloroform, (d) benzene, (e) sevoflurane (fluoromethyl 2,2,2-trifluoro-1-(trifluoromethyl) ethyl ether), (f) bromoform, (g) trichloroethylene, (h) desflurane (1,2,2,2-tetrafluoroethyl difluoromethyl ether), (i) cyclopropane, (j) fluroxene (2,2,2-trifluoroethyl vinyl ether), and (k) enflurane (2-chloro-1,1,2-trifluoroethyl difluoromethyl ether). Fluorine atoms are in light green, carbon atoms in gray, oxygen atoms in red, hydrogen atoms in white, chlorine atoms in dark green, and bromine atoms in purple.

Electrophysiological studies on a series of γ-aminobutyric acid type A receptor α2 subunit Ser270 mutants expressed in human embryonic kidney 293 cells indicated that the ability of isoflurane to potentiate the chloride ion conduction through the channel correlated with the volume of the amino acid residue present at this position. A small residue such as an alanine at position 270 resulted in a 160% enhancement of chloride ion conduction in response to a combination of 500 μM isoflurane and 5 μM γ-aminobutyric acid, while potentiation of charge transfer by isoflurane was essentially abolished by introducing the bulkier aromatic residues histidine, phenylalanine, tyrosine or tryptophan at position 270 [16].

The potentiating effects of sevoflurane and desflurane on chloride ion conduction by γ-aminobutyric acid type A α1β2γ2s receptors in response to 5 μM γ-aminobutyric acid expressed in human embryonic kidney 293 cells can also be abolished by substituting Ser270 in M2 of the α1 subunit by the larger side-chains of isoleucine or tryptophan [17]. These findings suggest that these two volatile general anesthetics might share common binding sites with enflurane and isoflurane on the γ-aminobutyric acid type A receptors.

The site-directed mutagenesis work on expressed γ-aminobutyric acid type A receptors indicates that volatile general anesthetics interact with the transmembrane domains of the protein subunits. These transmembrane domains are thought to be composed of four-α-helix bundle folds based upon homology modeling with the related nicotinic acetylcholine receptor Cys-loop ligand-gated ion channel.

Volatile General Anesthetic Interactions with Four-α-Helix Bundle Proteins

Volatile General Anesthetic Interactions with Designed Synthetic Four-α-Helix Bundle Proteins

Over the past fifteen years, new possibilities for studying ligand-receptor interactions from structural and energetic perspectives have arisen from the introduction of designed synthetic proteins [18]. This approach has been extended to small, highly α-helical synthetic proteins that can incorporate cofactors such as porphyrins and metals [19, 20] into a four-α-helix bundle framework. These water-soluble bundle proteins exhibit a high degree of secondary structure and thermodynamic stability, and have a hydrophobic core consisting of both aliphatic and aromatic residues that is amenable to modification using standard protein engineering techniques.

The utility of the hydrophobic core of these water-soluble four-α-helix bundles as a laboratory for understanding the structural features of volatile general anesthetic binding sites on proteins has been explored over the past ten years in this laboratory [21-29]. The relative importance of protein cavity volume and polar contributions to anesthetic binding can therefore, in principle, be determined. This information provides the foundations and guidelines necessary for recognizing the structural nature of potential native central nervous system target sites, allowing, for example, for a focused search of the Protein Data Bank [30] with regard to both currently existing- and future entries.

Apart from the nicotinic acetylcholine receptor which occurs in a naturally concentrated form (representing 25% of the total protein) in the electric organs of the marine rays of the genus *Torpedo* [31], other likely *in vivo* central nervous system targets (such as the γ-aminobutyric acid type A receptor) cannot currently be obtained in sufficient quantities, because of technical limitations, to allow direct binding studies to be performed. Investigators have therefore examined volatile general anesthetic binding to various model systems, which at least structurally share some features with the transmembrane domains of Cys-loop ligand-gated ion channels, which as noted above, are composed of four-α-helix bundles (Figure 1).

The first study directly addressing volatile general anesthetic interactions with four-α-helix bundle proteins focused on halothane binding to H10A24 [21], a biomolecule initially designed to incorporate heme molecules in its hydrophobic core, which consisted of alanine, leucine, phenylalanine, arginine, and histidine residues [32]. This H10A24 four-α-helix bundle protein is composed of four identical, highly α-helical 27-residue segments that are tethered by two disulfide linkages $(GGGC)_2$ at the N-termini and by hydrophobic interactions in the protein interior. The α-helical segments were designed using the heptad repeat as a guide, where positions *a* and *d* constitute the hydrophobic core of the protein, with additional contributions from heptad positions *e* and *g,* while heptad positions *b*, *c* and *f* are solvent exposed and consist of lysines and glutamates. The H10A24 di-α-helix protein had the following primary structure, with hydrophobic heptad *a* and *d* position residues indicated in bold:

Ac-**L**KK**L**R**E**E **A**LK**L**LEE **F**KK**L**LEE **H**LKW**L**E GGGC

CGGG E**L**WKL**H** EE**L**LKK**F** EE**L**LKL**A** EE**R**LKK**L**-$CONH_2$.

The N-terminus was acetylated and the C-terminus carboxyamidated in order to enhance the α-helical content of the H10A24 four-α-helix bundle protein [33]. The fluorescence quantum yield of Trp25 was used as an indicator of the binding of halothane to the hydrophobic core of the H10A24 four-α-helix bundle protein. Because the halothane molecule contains a bromine atom, this volatile general anesthetic will quench the tryptophan fluorescence provided that the binding occurs in close proximity to the indole ring [24, 34]. Halothane bound to the hydrophobic core of the H10A24 four-α-helix bundle protein with a dissociation constant (K_d) of 2.3 ± 0.4 mM. Time-resolved fluorescence measurements were used to corroborate the steady-state fluorescence measurements, and confirmed that halothane was indeed binding to the hydrophobic core of the H10A24 four-α-helix bundle protein. The free energy change (ΔG°) associated with the binding of halothane to the hydrophobic core of this four-α-helix bundle protein at 25 °C is –3.6 kcal/mol, which is almost exclusively accounted for by the hydrophobic effect, since partitioning of halothane into *n*-hexane is energetically quite comparable (-3.4 kcal/mol) [35].

Cavities, or packing defects, in the protein matrix represent sites where small molecules can bind if they are of the appropriate size, shape, and polarity [36-38]. By increasing macromolecular flexibility, these packing defects are thought to play a crucial role in normal physiological protein function [39, 40]. A cavity was introduced into the hydrophobic core of the four-α-helix bundle protein $(A\alpha_2)_2$ by replacing three larger leucine residues with smaller alanine residues at positions 19, 44 and 48 [22]. The resulting volume of this packing defect in the hydrophobic core of the four-α-helix bundle protein $(A\alpha_2)_2$ of 171 Å^3 was predicted to be sufficiently large to accommodate a halothane molecule, which has a calculated van der Waals volume of 123 Å^3 [41]. The tryptophan was moved to position 15 (a hydrophobic heptad *a* position) in the four-α-helix bundle protein $(A\alpha_2)_2$ design, where it occupied a location lining the proposed volatile general anesthetic binding pocket. $(A\alpha_2)_2$ is a dimeric four-α-helix bundle protein consisting of two 62-residue sequences that in turn comprise two 27-residue α-helical segments joined by a $(gly)_8$ linker. The primary structure of the di-α-helical $A\alpha_2$ with hydrophobic heptad *a* and *d* position residues shown in bold is:

Ac-**L**KK**L**R**E**E **A**AK**L**F**E**E **W**KK**L**A**E**E **A**AK**L**L**E** GGGG

GGGG E**L**LK**L**C EE**A**AKK**A** EE**L**FK**L**A EE**R**LKK**L**-$CONH_2$.

Halothane now bound to the hydrophobic core of the four-α-helix bundle protein $(A\alpha_2)_2$ with a K_d = 0.71 ± 0.04 mM as monitored by the concentration-dependent quenching of the Trp15 fluorescence quantum yield. The $\Delta G°$ associated with halothane binding to the hydrophobic core of the four-α-helix bundle protein $(A\alpha_2)_2$ at 25 °C is –4.3 kcal/mol, indicating that a preexisting binding pocket can contribute on the order of -0.7 kcal/mol of binding energy for a prototypical volatile general anesthetic. Preexisting packing defects, or cavities, in the protein matrix are thus predicted to represent likely sites where small molecules such as volatile general anesthetics, may bind. In addition, by filling such packing defects, volatile general anesthetics would be predicted to limit macromolecular flexibility, at least locally, and therefore might alter protein function.

A methionine residue was introduced at position 38 in the four-α-helix bundle protein to create $(A\alpha_2\text{-L38M})_2$, in order to begin to explore the importance of potential polar interactions on volatile general anesthetic binding to a macromolecular target [23]. This substitution was based on the findings [35] that halothane, enflurane, isoflurane and desflurane all partitioned better into ethyl methyl sulfide (a model for the methionine side-chain) than into *n*-hexane (a model for the leucine side-chain) by $\Delta\Delta G°$ values at 25 °C of –0.7-1.0 kcal/mol. Halothane bound to the four-α-helix bundle protein $(A\alpha_2\text{-L38M})_2$, with a K_d = 200 ± 10 μM as assessed by its ability to quench Trp15 fluorescence, corresponding to a 3.5-fold improved affinity compared to the interaction with the $(A\alpha_2)_2$ four-α-helix bundle protein. Although methionine is similar to leucine in terms of its hydrophobicity [42-44], size (both amino acids have side-chain volumes of 124 Å^3) [45] and helix-propensity [46], it does contain the more polarizable sulfur atom. Compared to the corresponding $–CH_2–$ group on leucine, the sulfur atom on methionine is 1.7-times as polarizable [47], suggesting that more favorable van der Waals interactions would be achieved with a bound hydrophobic ligand such as a volatile general anesthetic.

Global protein stability can also be modulated by a bound volatile general anesthetic. Binding of halothane to the hydrophobic core of the four-α-helix bundle protein $(A\alpha_2\text{-L38M})_2$ stabilized the folded conformation of the macromolecule by –0.9 kcal/mol as assessed by amide hydrogen exchange rates [23]. By extension then, volatile general anesthetic binding might stabilize the open conformation of the γ-aminobutyric acid type A receptor, or the closed conformation of the *N*-methyl-D-aspartate receptor.

The role played by Trp15 in the binding of halothane to the four-α-helix bundle protein $(A\alpha_2\text{-L38M})_2$ was investigated by replacing this bulky aromatic residue with a smaller tyrosine residue to form the four-α-helix bundle protein $(A\alpha_2\text{-L38M/W15Y})_2$ [25]. Halothane was shown to quench the fluorescence of tyrosine with the same efficiency with which it quenches tryptophan fluorescence. Halothane bound to the hydrophobic core of the four-α-helix bundle protein $(A\alpha_2\text{-L38M/W15Y})_2$ with a $K_d = 660 \pm 70$ μM, or with an approximately three-fold lower affinity compared to the binding to the four-α-helix bundle protein $(A\alpha_2\text{-L38M})_2$. The relative binding affinities reflect the electron richness of the π-systems of the two residues (tryptophan > tyrosine), indicating that halothane is interacting directly with the aromatic rings of Trp15 and Tyr15 in the two four-α-helix bundle systems. The results suggest that the acidic hydrogen on the number two carbon of halothane forms a CH/π interaction with the aromatic ring systems of the two amino acids (Figure 3) in the two four-α-helix bundle proteins $(A\alpha_2\text{-L38M/W15Y})_2$ and $(A\alpha_2\text{-L38M})_2$. The indole moiety of tryptophan is the most efficient π-acceptor group found in proteins [48]. Although the CH/π bond is the weakest type of hydrogen bond with an enthalpic contribution of approximately 1 kcal/mol [49], it is considerably stronger than what is expected from a dispersion mechanism (≤ 0.1 kcal/mol) [50]. The results of this study [25] predict that the side-chain of tryptophan will form stronger interactions with a bound halothane molecule on natural proteins than will that of tyrosine.

Isothermal titration calorimetry [51] has been used to measure the binding of ten volatile general anesthetics to the hydrophobic core of the four-α-helix bundle protein $(A\alpha_2\text{-L38M})_2$, which has to date proven to be the design that binds these apolar compounds with the highest affinity [52, 53]. In an isothermal titration calorimetry experiment the amount of heat taken up or given off as a function of time with each successive injection of ligand into a protein solution is precisely measured [54, 55]. Because the binding sites on the protein become progressively occupied with each addition of ligand to the protein solution, less and less heat is absorbed or released.

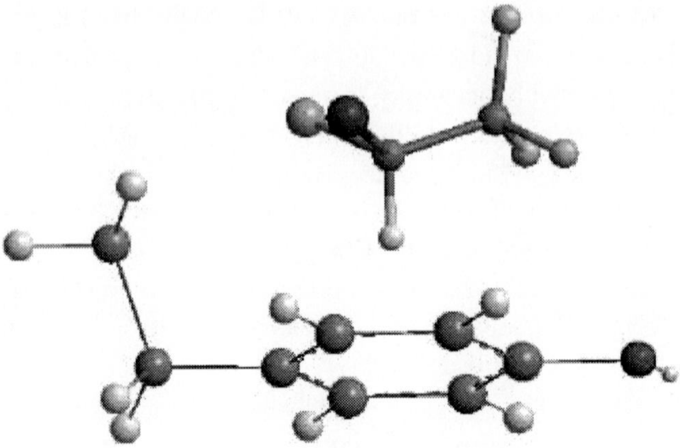

Figure 3. Proposed halothane interaction with the Tyr15 side-chain in the four-α-helix bundle protein $(A\alpha_2\text{-L38M/W15Y})_2$. Hydrogens are in white, carbons in gray, oxygen in red, fluorines in light green, bromine in purple, and chlorine in dark green. The lone hydrogen (soft acid) on halothane is forming a CH/π interaction with the π electrons (soft base) of the aromatic ring.

This principle allows a direct determination of the affinity (the association constant $K_a = 1/K_d$) of the interaction between the ligand and the protein during an isothermal titration calorimetry experiment. Further, the enthalpy change ($\Delta H°$) underlying ligand binding to the protein is directly measured from the heat that is absorbed or released during the course of the titration. The free energy change ($\Delta G°$) associated with ligand binding to the protein can then be calculated from the relationship $\Delta G° = R{\cdot}T{\cdot}\ln K_d$, where R is the gas constant and T is the temperature of the experiment in kelvins. Finally, the entropy change ($\Delta S°$) associated with ligand binding to the protein can be calculated from the Gibbs-Helmholtz equation, where $\Delta G° = \Delta H° - T{\cdot}\Delta S°$.

The advantage of isothermal titration calorimetry is that it allows the determination of all three thermodynamic parameters ($\Delta G°$, $\Delta H°$ and $\Delta S°$) from a single experiment, circumventing the need for the more time-consuming van't Hoff analysis. In the case of ligands like the volatile general anesthetics, isothermal titration calorimetry has also proven very useful because measuring the interaction of these compounds with proteins has historically proven technically difficult. For example, approaches such as ^{19}F-nuclear magnetic resonance spectroscopy [56, 57], halothane photoaffinity labeling [58] and fluorescence spectroscopy [34] have many drawbacks, not the least of which are their suitability for only a limited range of anesthetic molecules.

Figure 4 shows a calorimetric titration of a pH 7.0 solution of the four-α-helix bundle protein $(A\alpha_2\text{-L38M})_2$ with sevoflurane. Each separate peak during the course of the binding isotherm (upper panel, Figure 4) reflects the amount of heat that is evolved during a single injection of sevoflurane. The fact that this is an exothermic reaction (heat is evolved) is reflected by the negative deflections from baseline. The enthalpy change associated with the sequential injections of sevoflurane are plotted as a function of the sevoflurane/$(A\alpha_2\text{-L38M})_2$ molar ratio in the lower panel of Figure 4, and the values of $\Delta H°$, K_d, $\Delta G°$, and $\Delta S°$ are determined from the plot. The dissociation constant for sevoflurane binding to the hydrophobic core of the four-α-helix bundle protein $(A\alpha_2\text{-L38M})_2$ is 140 ± 10 μM as determined using isothermal titration calorimetry [52], which is comparable to the EC_{50} value in humans for maintaining the anesthetic state of 260 μM [14]. For sevoflurane binding to the hydrophobic core of the four-α-helix bundle protein $(A\alpha_2\text{-L38M})_2$, the $\Delta G° = -5.2 \pm 0.1$ kcal/mol, the $\Delta H° = -8.2 \pm 0.2$ kcal/mol, and the $\Delta S° = -8.0$ cal/mol K. The $\Delta S°$ for this interaction is therefore unfavorable and reflects the fact that binding is of relatively high affinity, which results in decreases in both rotational- and translational degrees of freedom of the volatile general anesthetic and the adjacent amino acid side-chains at the binding site in the hydrophobic core of the four-α-helix bundle protein [59].

A Meyer-Overton plot of the affinity of the four-α-helix bundle protein $(A\alpha_2\text{-L38M})_2$ for ten different volatile general anesthetics compared to their respective whole animal- or human EC_{50} values for maintaining the anesthetic state is shown in Figure 5. The plot reveals a good correlation with a least squares linear regression correlation coefficient (r) of 0.962 [60], indicating that the hydrophobic core of the four-α-helix bundle protein $(A\alpha_2\text{-L38M})_2$ represents a good model for the *in vivo* central nervous system sites of volatile general anesthetic action. For comparison, the partitioning of seven general anesthetics (halothane, isoflurane, enflurane, sevoflurane, fluroxene, desflurane and cyclopropane) into lecithin, olive oil or benzene, generated least squares linear regression correlation coefficients of 0.972-0.988 [61].

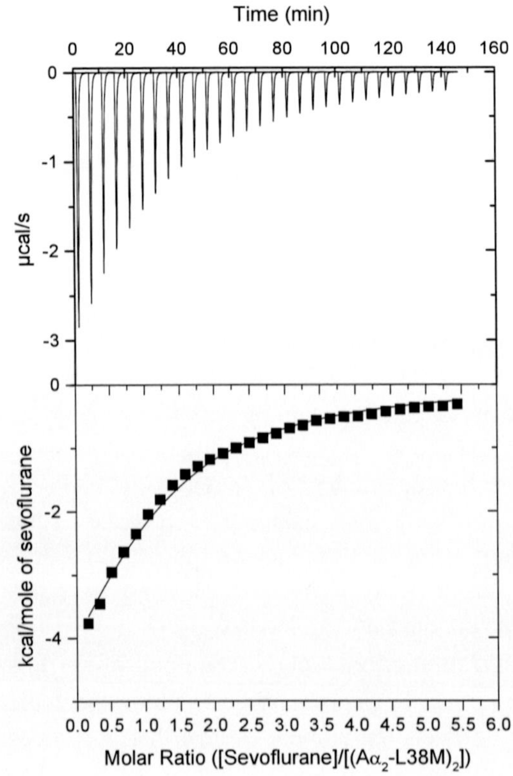

Figure 4. Titration of the four-α-helix bundle protein $(A\alpha_2\text{-L38M})_2$ with sevoflurane, showing the calorimetric response as successive injections of volatile general anesthetic are added to the reaction cell [52]. The lower panel depicts the binding isotherm of the calorimetric titration shown in the upper panel. The continuous line represents the least-squares fit of the data to a single-site binding model. Reprinted with permission from the Biophysical Society.

In addition, the partitioning of 21 general anesthetics into phosphatidylcholine lipid bilayers was associated with a least squares linear regression correlation coefficient of 0.965 [62]. The hydrophobic core of the four-α-helix bundle protein $(A\alpha_2\text{-L38M})_2$ therefore obeys the Meyer-Overton correlation to the same degree as that exhibited by a lipid bilayer.

Studies with the expressed four-α-helix bundle protein $(A\alpha_2\text{-L1M/L38M})_2$ have provided evidence that volatile general anesthetics can alter the structure of the target upon complex formation [27, 28]. Figure 6 shows the ^1H-^{15}N heteronuclear single quantum coherence nuclear magnetic resonance spectra of the four-α-helix bundle protein $(A\alpha_2\text{-L1M/L38M})_2$ with and without bound sevoflurane. Binding of the volatile general anesthetic to the hydrophobic core of the four-α-helix bundle protein $(A\alpha_2\text{-L1M/L38M})_2$ alters the chemical shifts of at least 15 of the cross-peaks present in the ^1H-^{15}N heteronuclear single quantum coherence nuclear magnetic resonance spectrum, indicating widespread structural changes. Such changes in protein architecture following volatile general anesthetic binding may represent a fundamental mechanism of action whereby macromolecular activity is modulated.

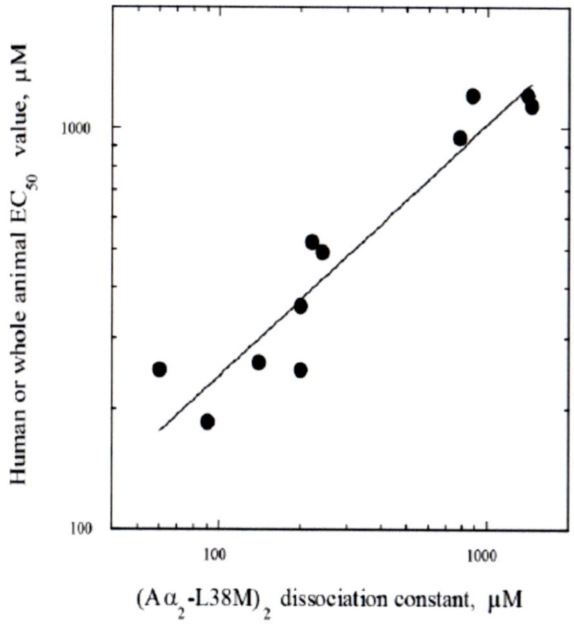

Figure 5. Least-squares linear regression correlation of human- or whole animal potency data for ten volatile general anesthetics with their respective dissociation constants for binding to the hydrophobic core of the four-α-helix bundle protein ($A\alpha_2$-L38M)$_2$. The ten volatile general anesthetics are halothane, isoflurane, enflurane, desflurane, sevoflurane, chloroform, bromoform, trichloroethylene, benzene and fluroxene. Reprinted from [60] with permission from Elsevier B.V.

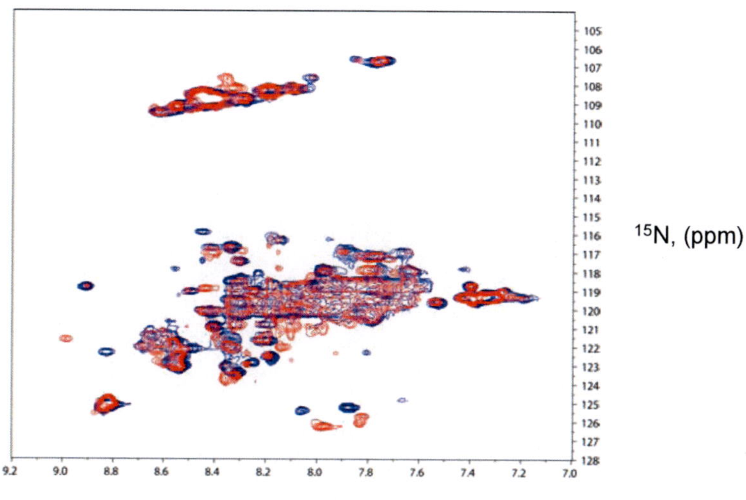

Figure 6. ^{15}N-^{1}H heteronuclear single quantum coherence nuclear magnetic resonance spectrum of ($A\alpha_2$-L1M/L38M)$_2$ in 20 mM phosphate buffer at a temperature of 25 °C (blue). The red ^{15}N-^{1}H heteronuclear single quantum coherence nuclear magnetic resonance spectrum was recorded following the addition of 1 mM sevoflurane [28]. Reprinted with permission from the American Chemical Society.

Volatile General Anesthetic Interactions with Naturally Occurring Four-α-Helix Bundle Motifs

The SNARE complex is a naturally occurring four-α-helix bundle composed of three different proteins termed syntaxin-1A, synaptobrevin-II and SNAP-25B (synaptosome-associated protein of relative molecular mass 25 kDa) that is involved in facilitating the fusion of neurotransmitter-containing synaptic vesicles with the neuronal plasma membrane [63]. Using ^{19}F-nuclear magnetic resonance spectroscopy, isoflurane has been shown to interact directly with the SNARE complex proteins in the concentration range of 120-460 μM, in an apparently saturable fashion, indicating specific binding [64]. For comparison, the clinical EC_{50} value of isoflurane in humans for maintaining the anesthetic state is 260 μM [14]. Whether binding of a volatile general anesthetics alters the activity of the SNARE complex has not been addressed to date, but it is clearly a study that needs to be carried out, since this would represent an attractive widespread site of action where central nervous system function could be perturbed.

Halothane may be used as a photoaffinity labeling agent because the C-Br bond can be broken using 254 nm radiation [58] to yield a chlorotrifluoroethyl radical that in turn attaches covalently to the protein in the vicinity of the native volatile general anesthetic binding site. By using ^{14}C-labeled halothane it is possible to use standard scintillation counting in order to identify either intact proteins or fragments of proteins that have been irreversibly labeled with the volatile general anesthetic. Halothane photoaffinity labeling has been used to identify binding sites on the nicotinic acetylcholine receptor from *Torpedo californica* electric organ [65]. Photoaffinity labeling with halothane was followed by protein microsequencing to allow identification of residues with incorporated volatile general anesthetic. The four transmembrane segments (M1-M4) of each subunit form a four-α-helix bundle motif as revealed by the 4 Å cryoelectron microscopy structure of this Cys-loop ligand-gated ion channel [7]. Halothane was found to photoaffinity label Tyr213 in the α subunit M1 α-helix and Tyr228 in the δ subunit M1 α-helix in a protein conformation-dependent manner (Figure 7), with anesthetic binding being enhanced in the case of the desensitized channel state compared to the closed channel state. This study represents the first experimental evidence that volatile general anesthetics are indeed capable of interacting directly with a member of the Cys-loop ligand-gated ion channel family of receptors.

Both isoflurane and halothane have been shown to bind to equine apoferritin, which is a protein composed of 24 four-α-helix bundle subunits [66]. Using isothermal titration calorimetry, halothane and isoflurane bound to apoferritin with dissociation constants of approximately 5 μM, which is remarkable since this represents only $1/50^{th}$ of their respective clinical EC_{50} values in humans for maintaining the anesthetic state [14], and are the highest affinity interactions for volatile general anesthetic binding to protein that have been described to date. X-ray crystal structures of apoferritin with bound volatile general anesthetics were solved to a resolution of 1.7 Å, and revealed that both halothane and isoflurane bound in preexisting cavities in the protein and formed favorable van der Waals interactions with Leu24, Ser27 and Tyr28. This study was the first to show a high resolution structure of a bound ether anesthetic (the ones currently in clinical use in the United States) in any protein, and provided much-needed insight into the structural features of volatile general anesthetic binding sites on protein targets.

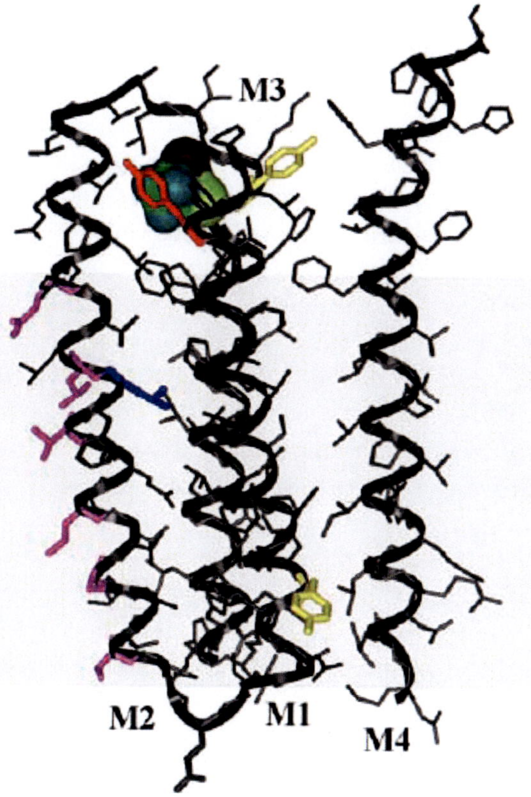

Figure 7. A stick and ribbon representation of the transmembrane domain of the nicotinic acetylcholine receptor δ subunit in the closed state with a Connolly surface depiction of halothane bound next to Tyr228 on M1 (in red) [65]. The extracellular surface is up and the lipid bilayer is represented in gray. Unlabeled Tyr248 on M1 and Tyr291 on M3 are shown in yellow. M2 residues that line the ion channel pore are shown in magenta. Reprinted with permission from the American Chemical Society.

Studies with designed synthetic- and natural four-α-helix bundle motifs indicate that this particular protein fold can bind a number of different volatile general anesthetics with affinities that correlate with their respective clinical EC_{50} values in humans for maintaining the anesthetic state. Biophysical studies on these proteins are providing fundamental mechanistic insight into how the binding of a volatile general anesthetic results in altered protein activity.

MOLECULAR DYNAMICS SIMULATIONS

Molecular dynamics simulations provide detailed atomic level information on protein structure and flexibility on the nanosecond on up to the microsecond time-scale [67-69]. In addition, the simulations yield insight into how a bound ligand, such as a volatile general anesthetic, might interact with a biomolecular target, and in turn how such binding might alter protein structure and motions.

A one nanosecond all-atom molecular dynamics simulation at a temperature of 298 K and at a constant pressure of one atmosphere of the 124-residue designed synthetic four-α-helix bundle protein $(A\alpha_2)_2$ in the presence of a halothane molecule bound in its hydrophobic core has been reported [70]. Halothane localized in the designed binding cavity engineered into the hydrophobic core of the four-α-helix bundle protein $(A\alpha_2)_2$ and caused the α-helical scaffold to become more compact. Although the halothane molecule is fully free to both rotate and to translate in the binding pocket in the hydrophobic core of $(A\alpha_2)_2$ it tended to preferentially orient itself towards the Trp15 residue in such a way that the bromine atom on the halothane approaches the indole ring. This finding is in good agreement with the experimental work that shows marked quenching of the Trp15 fluorescence in the four-α-helix bundle protein $(A\alpha_2)_2$ following halothane binding [22], since this effect requires close proximity (orbital overlap) between the bromine atom on the halothane molecule and the fluorophore [71-73].

A follow-up study describing a one nanosecond all-atom molecular dynamics simulation at a temperature of 298 K and at a constant pressure of one atmosphere of the 124-residue designed synthetic four-α-helix bundle protein $(A\alpha_2\text{-L38M})_2$ with a bound halothane molecule in its hydrophobic core has also been published [74]. This molecular dynamics trajectory was carried out in the presence of a fully hydrated (approximately 6600 water molecules) four-α-helix bundle protein. The existence of two symmetrical cavities of sufficient volume where anesthetic molecules could localize in the hydrophobic core of the four-α-helix bundle was apparent, consistent with the original design of this protein [23]. Halothane again bound preferentially with its bromine atom pointing towards the indole ring of Trp15 in good agreement with the photoaffinity labeling results on this four-α-helix bundle protein [23], and the presence of the anesthetic in the hydrophobic core of the protein was found to decrease the flexibility of this aromatic side-chain. The latter finding is in accord with polarized fluorescence results obtained with bovine serum albumin with either bound halothane or isoflurane that suggest that volatile general anesthetic binding can attenuate tryptophan side-chain mobility [75]. In addition, the binding of halothane to the hydrophobic core of the four-α-helix bundle protein $(A\alpha_2\text{-L38M})_2$ resulted in some local distortion of the α-helical scaffold (Figure 8), indicating that a bound anesthetic molecule can alter the structure of biomolecules, which may be a basic mechanism of action underlying some of their clinical effects.

Molecular dynamics simulations investigating the interaction of halothane with the transmembrane domains (four-α-helix bundle motifs, Figure 1) of the α and δ subunits of the nicotinic acetylcholine receptor from *Torpedo marmorata* have recently been reported [76]. The transmembrane domains were embedded in a fully hydrated 1,2-dioleoyl-*sn*-glycero-3-phosphocholine lipid bilayer and simulations were carried out at one atmosphere of pressure and a temperature of 305 K for 16 nanoseconds in the presence of halothane, which was introduced initially into the aqueous phase. Halothane molecules tended to concentrate at the lipid-water interface, in agreement with experimental results obtained with ^{19}F-nuclear magnetic resonance spectroscopy examining the location of isoflurane in phosphatidylcholine vesicles [77]. In the case of the α-subunit transmembrane domain of the nicotinic acetylcholine receptor, one halothane molecule partitioned into the interior of the four-α-helix bundle and localized close to Tyr277. The presence of the volatile general anesthetic resulted in a decrease in the mobility of the adjacent residues in the binding pocket inside the four-α-helix bundle, and also attenuated the flexibility of the M2-M3 loop.

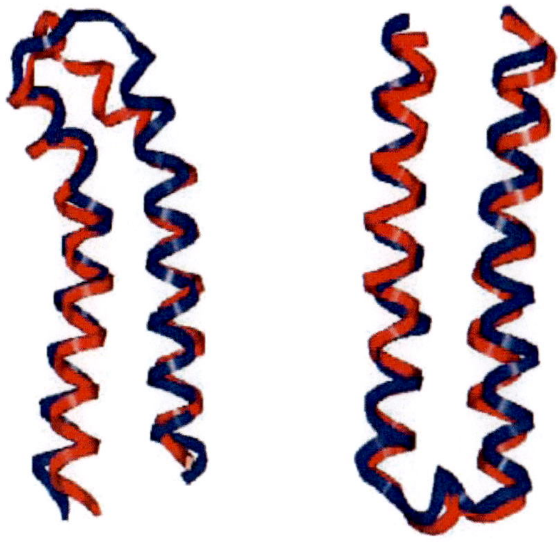

Figure 8. Ribbon representations of the 1 ns configuration of the α-helical backbones of the $A\alpha_2$-L38M monomers with amino acid side-chains omitted for clarity [74]. Red is in the presence of halothane and blue is in its absence. Panel (a) is the $A\alpha_2$-L38M monomer with which halothane interacts more tightly resulting in some distortion of the protein structure. Reprinted with permission of the Federation of European Biochemical Societies.

The latter finding is of interest because the M2-M3 loop has been implicated in playing an important role in the gating mechanism of the nicotinic acetylcholine receptor [78], and changes in the dynamics of this region of the transmembrane domain of the α-subunit therefore provides insight into how a bound volatile general anesthetic might modify the ion conducting properties of this structurally best understood member of the Cys-loop ligand-gated ion channel receptor family of proteins. Apart from modulating protein flexibility, halothane binding to the transmembrane domain of the α-subunit also had the effect of making the four-α-helix bundle more compact by increasing the correlation of the motions between the helices M3 and M4.

Molecular dynamics simulations are providing basic insights into how volatile general anesthetics might alter protein flexibility and structure, which may underlie many of their clinical effects. As the field evolves, studies on ever more complex systems all the way up to fully hydrated intact ligand-gated ion channels embedded in physiological lipid membranes are anticipated. The results of molecular dynamics simulations in many cases corroborate experimental findings, thereby validating the underlying theoretical foundations. However, novel findings based on molecular dynamics simulations can also serve as a guide for future experimental efforts.

Conclusion

Basic mechanisms of volatile general anesthetic action are becoming clearer based upon studies with expressed neuronal plasma membrane ligand-gated ion channels and simpler model systems using both experimental and theoretical approaches. Binding of volatile general anesthetics can alter the stability, flexibility and the structure of the protein target, all of which are expected to alter the function of the latter. Understanding fundamental mechanisms of volatile general anesthetic action should ultimately pave the way for the introduction of more selective agents with fewer side-effects.

ACKNOWLEDGEMENT

Work supported by National Institutes of Health grants GM55876 and GM65218, Bethesda, MD 20892, USA.

REFERENCES

[1] Sebel, PS; Bowdle, TA; Ghoneim, MM; Rampil, IJ; Padilla, RE; Gan, TJ; Domino, KB. The incidence of awareness during anesthesia: A multicenter United States study. *Anesth. Analg.* 2004 99, 833-839.

[2] Campagna, JA; Miller, KW; Forman, SA. Mechanisms of actions of inhaled anesthetics. *N. Engl. J. Med.* 2003 348, 2110-2124.

[3] Rudolph, U; Antkowiak, B. Molecular and neuronal substrates for general anaesthetics. *Nat. Rev. Neurosci.* 2004 5, 709-720.

[4] Hemmings, HC; Akabas, MH; Goldstein, PA; Trudell, JR; Orser, BA; Harrison, NL. Emerging molecular mechanisms of general anesthetic action. *Trends Pharmacol Sci.* 2005 26, 503-510.

[5] Franks, NP. Molecular targets underlying general anaesthesia. *Br. J. Pharmacol.* 2006 147, S72-S81.

[6] Orser, BA. Lifting the fog around anesthesia. *Sci. Am.* 2007 296, 54-61.

[7] Miyazawa, A; Fujiyoshi, Y; Unwin, N. Structure and gating mechanism of the acetylcholine receptor pore. *Nature* 2003 423, 949-955.

[8] Unwin, N. Refined structure of the nicotinic acetylcholine receptor at 4 Å resolution. *J Mol. Biol.* 2005 346, 967-989.

[9] Popescu, G. Mechanism-based targeting of NMDA receptor functions. *Cell Mol. Life Sci.* 2005 62, 2100-2111.

[10] Waxman, EA; Lynch, DR. *N*-methyl-D-aspartate receptor subtypes: Multiple roles in excitotoxicity and neurological disease. *Neuroscientist* 2005 11, 37-49.

[11] Paoletti, P; Neyton, J. NMDA receptor subunits: function and pharmacology. *Curr Opin Pharmacol.* 2007 7, 39-47.

[12] Chatterton, JE; Awobuluyi, M; Premkumar, LS; Takahashi, H; Talantova, M; Shin, Y; Cui, J; Tu, S; Sevarino, KA; Nakanishi, N; Tong, G; Lipton, SA; Zhang, D. Excitatory glycine receptors containing the NR3 family of NMDA receptor subunits. *Nature* 2002 415, 793-798.

[13] Nakagawa, T; Cheng, Y; Sheng, M; Walz, T. Structure and different conformational states of native AMPA receptor complexes. *Nature* 2005 433, 545-549.

[14] Krasowski, MD; Harrison, NL. General anaesthetic actions on ligand-gated ion channels. *Cell Mol Life Sci* 1999 55, 1278-1303.

[15] Mihic, SJ; Ye, Q; Wick, MJ; Koltchine, VV; Krasowski, MD; Finn, SE; Mascia, MP; Valenzuela, CF; Hanson, KK; Greenblatt, EP; Harris, RA; Harrison, NL. Sites of alcohol and volatile anaesthetic action on $GABA_A$ and glycine receptors. *Nature* 1997 389, 385-389.

[16] Koltchine, VV; Finn, SE; Jenkins, A; Nikolaeva, N; Lin, A; Harrison, NL. Agonist gating and isoflurane potentiation in the human γ-aminobutyric acid type A receptor determined by the volume of a second transmembrane domain residue. *Mol. Pharmacol* 1999 56, 1087-1093.

[17] Nishikawa, K; Harrison, NL. The actions of sevoflurane and desflurane on the γ-aminobutyric acid receptor type A. Effects of TM2 mutations in the α and β subunits. *Anesthesiology* 2003 99, 678-684.

[18] DeGrado, WF; Woolfson, DN. Engineering and design. *Curr. Opin. Struct. Biol.* 2006 16, 505-507.

[19] Cochran, FV; Wu, SP; Wang, W; Nanda, V; Saven, JG; Therien, MJ; DeGrado, WF. Computational de novo design and characterization of a four-helix bundle protein that selectively binds a nonbiological cofactor. *J. Am. Chem. Soc.* 2005 127, 1346-1347.

[20] Wade, H; Stayrook, SE; DeGrado, WF. The structure of a designed diiron(III) protein: Implications for cofactor stabilization and catalysis. *Angew Chem. Int. Ed.* 2006 45, 4951-4954.

[21] Johansson, JS; Rabanal, F; Dutton, PL. Binding of the volatile anesthetic halothane to the hydrophobic core of a tetra-α-helix bundle protein. *J. Pharmacol. Exp. Ther.* 1996 279, 56-61.

[22] Johansson, JS; Gibney, BR; Rabanal, F; Reddy, KS; Dutton, PL. A designed cavity in the hydrophobic core of a four-α-helix bundle improves volatile anesthetic binding affinity. *Biochemistry* 1998 37, 1421-1429.

[23] Johansson, JS; Scharf, D; Davies, LA; Reddy, KS; Eckenhoff, RG. A designed four-α-helix bundle that binds the volatile general anesthetic halothane with high affinity. *Biophys J* 2000 78, 982-993.

[24] Johansson, JS; Solt, K; Reddy, KS. Binding of the general anesthetics chloroform and 2,2,2-trichloroethanol to the hydrophobic core of a four-α-helix bundle protein. *Photochem. Photobiol.* 2003 77, 89-95.

[25] Manderson, GA; Johansson, JS. Role of aromatic side chains in the binding of volatile general anesthetics to a four-α-helix bundle. *Biochemistry* 2002 41, 4080-4087.

[26] Manderson, GA; Michalsky, SJ; Johansson, JS. Effect of four-α-helix bundle cavity size on volatile anesthetic binding energetics. *Biochemistry* 2003 42, 11203-11213.

[27] Pidikiti, R; Shamim, M; Mallela, KMG; Reddy, KS; Johansson, JS. Expression and characterization of a four-α-helix bundle protein that binds the volatile general anesthetic halothane. *Biomacromolecules* 2005 6, 1516-1523.

[28] Pidikiti, R; Zhang, T; Mallela, KMG; Shamim, M; Reddy, KS; Johansson, JS. Sevoflurane-induced structural changes in a four-α-helix bundle protein. *Biochemistry* 2005 44, 12128-12135.

[29] Solt, K; Johansson, JS; Raines, DE. Kinetics of anesthetic-induced conformational transitions in a four-α-helix bundle protein. *Biochemistry* 2006 45, 1435-1441.

[30] http://www.rcsb.org/pdb/home/home.do Accessed 5/5/2007.

[31] Ellena, JF; Blazing, MA; McNamee, MG. Lipid-protein interactions in reconstituted membranes containing acetylcholine receptor. *Biochemistry* 1983 22, 5523-5535.

[32] Robertson, DE; Farid, RS; Moser, CC; Urbauer, JL; Mulholland, SE; Pidikiti, R; Lear, JD; Wand, AJ; DeGrado, WF; Dutton, PL. Design and synthesis of multi-haem proteins. *Nature* 1994 368, 425-432.

[33] Choma, CT; Lear, JD; Nelson, MJ; Dutton, PL; Robertson, DE; DeGrado, WF. Design of a heme-binding four-helix bundle. *J. Am. Chem. Soc.* 1994 116, 856-865.

[34] Johansson, JS; Eckenhoff, RG; Dutton, PL. Binding of halothane to serum albumin demonstrated using tryptophan fluorescence. *Anesthesiology* 1995 83, 316-324.

[35] Johansson, JS; Zou, H. Partitioning of four modern volatile general anesthetics into solvents that model buried amino acid side-chains. *Biophys. Chem.* 1999 79, 107-116.

[36] Liang, J; Edelsbrunner, H; Fu, P; Sudhakar, PV; Subramaniam, S. Analytical shape computation of macromolecules: II. Inaccessible cavities in proteins. *Proteins* 1998 33, 18-29.

[37] Whittington, DA; Rosenzweig, AC; Frederick, CA; Lippard, SJ. Xenon and halogenated alkanes track putative binding cavities in the soluble methane monooxygenase hydrolase. *Biochemistry* 2001 40, 3476-3482.

[38] Musah, RA; Jensen, GM; Bunte, SW; Rosenfeld, RJ; Goodin, DB. Artificial protein cavities as specific ligand-binding templates: Characterization of an engineered heterocyclic cation-binding site that preserves the evolved specificity of the parent protein. *J. Mol. Biol.* 2002 315, 845-857.

[39] Hubbard, SJ; Argos, P. A functional role for protein cavities in domain:domain motions. *J. Mol. Biol.* 1996 261, 289-300.

[40] Wang, H; Lin, S; Allen, JP; Williams, JC; Blankert, S; Laser, C; Woodbury, NW. Protein dynamics control the kinetics of initial electron transfer in photosynthesis. *Science* 2007 316, 747-750.

[41] Abraham, MH; McGowan, JC. The use of characteristic volumes to measure cavity terms in reversed phase liquid chromatography. *Chromatographia* 1987 23, 243-246.

[42] Nozaki, Y; Tanford, C. The solubility of amino acids and two glycine peptides in aqueous ethanol and dioxane solutions. *J. Biol. Chem.* 1971 246, 2211-2217.

[43] Urry, DW; Gowda, DC; Parker, TM; Luan, CH; Reid, MC; Harris, CM; Pattanaik, A; Harris, RD. Hydrophobicity scale for proteins based on inverse temperature transitions. *Biopolymers* 1992 39, 1243-1250.

[44] Wimbley, WC; White, SH. Experimentally determined hydrophobicity scale for proteins at membrane interfaces. *Nature Struct. Biol.* 1996 3, 842-848.

[45] Richards, FM. The interpretation of protein structures: total volume, group volume distributions and packing density. *J. Mol. Biol.* 1974 82, 1-14.

[46] O'Neil, KT; DeGrado, WF. A thermodynamic scale for the helix-forming tendencies of the commonly occurring amino acids. *Science* 1990 250, 646-651.

[47] Fersht, AR; Dingwall, C. Cysteinyl-tRNA synthetase from *Escherichia coli* does not need an editing mechanism to reject serine and alanine. High binding energy of small groups in specific molecular interactions. *Biochemistry* 1979 18, 1245-1249.

[48] Steiner, T; Koellner, G. Hydrogen bonds with π-acceptors in proteins: Frequencies and role in stabilizing local 3D structures. *J. Mol. Biol.* 2001 305, 535-557.

[49] Levitt, M; Perutz, MF. Aromatic rings as hydrogen bond acceptors. *J. Mol. Biol.* 1988 201, 751-754.

[50] Cornell, WD; Cieplak, P; Bayly, CI; Gould, IR; Merz, KM; Ferguson, DM; Spellmeyer, DC; Fox, T; Caldwell, JW; Kollman, PA. A second generation force field for the simulation of proteins, nucleic acids, and organic molecules. *J. Am. Chem. Soc.* 1995 117, 5179-5197.

[51] Wiseman, T; Williston, S; Brandts, JF; Lin, L-N. Rapid measurement of binding constants and heats of binding using a new titration calorimeter. *Anal. Biochem.* 1989 179, 131-137.

[52] Zhang, T; Johansson, JS. An isothermal titration calorimetry study on the binding of four volatile general anesthetics to the hydrophobic core of a four-α-helix bundle protein. *Biophys. J.* 2003 85, 3279-3285.

[53] Zhang, T; Johansson, JS. A calorimetric study on the binding of six general anesthetics to the hydrophobic core of a model protein. *Biophys. Chem.* 2005 113, 169-174.

[54] Leavitt, S; Friere, E. Direct measurement of protein binding energetics by isothermal titration calorimetry. *Curr. Opin. Struct. Biol.* 2001 11, 560-566.

[55] Ababou, A; Ladbury, JE. Survey of the year 2004: Literature on applications of isothermal titration calorimetry. *J. Mol. Recognit.* 2006 19, 79-89.

[56] Dubois, BW; Evers, AS. ^{19}F-NMR spin-spin relaxation (T_2) method for characterizing anesthetic binding to proteins: Analysis of isoflurane binding to albumin. *Biochemistry* 1992 31, 7069-7076.

[57] Dubois, BW; Cherian, SF; Evers, AS. Volatile anesthetics compete for common binding sites on bovine serum albumin: A ^{19}F-NMR study. *Proc. Natl. Acad. Sci. USA* 1993 90, 6478-6482.

[58] Eckenhoff, RG; Shuman, H. Halothane binding to soluble proteins determined by photoaffinity labeling. *Anesthesiology* 1993 79, 96-106.

[59] Smithrud, DB; Wyman, TB; Diederich, F. Enthalpically driven cyclophane-arene inclusion complexation: Solvent-dependent calorimetric studies. *J. Am. Chem. Soc.* 1991 113, 5420-5426.

[60] Zhang, T; Johansson, JS. The binding of ten general anesthetics to a four-alpha-helix bundle protein displays enthalpy-entropy compensation. *International Congress Series* 2005 1283, 219-222.

[61] Taheri, S; Halsey, MJ; Liu, J; Eger, EI; Koblin, DD; Laster, MJ. What solvent best represents the site of action of inhaled anesthetics in humans, rats, and dogs? *Anesth. Analg.* 1991 72, 627-634.

[62] Janoff, AS; Pringle, MJ; Miller, KW. Correlation of general anesthetic potency with solubility in membranes. *Biochim. Biophys. Acta* 1981 649, 125-128.

[63] Sutton, RB; Fasshauer, D; Jahn, R; Brunger, AT. Crystal structure of a SNARE complex involved in synaptic exocytosis at 2.4 Å resolution. *Nature* 1998 395, 347-353.

[64] Crowder, CM; Berilgen, J. Isoflurane binds the rat synaptic protein SNAP-25 at clinical concentrations. *Anesthesiology* 2000 93, A806.

[65] Chiara, DC; Dangott, LJ; Eckenhoff, RG; Cohen, JB. Identification of nicotinic acetylcholine receptor amino acids photolabeled by the volatile anesthetic halothane. *Biochemistry* 2003 42,13457-13467.

[66] Liu, R; Loll, PJ; Eckenhoff, RG. Structural basis for high-affinity volatile anesthetic binding in a natural 4-helix bundle protein. *FASEB J.* 2005 19, 567-576.

[67] Grubmüller, H; Schulten, K. Special issue: Advances in molecular dynamics simulations. *J. Struct. Biol.* 2007 157, 443.

[68] Ayton, GS; Voth, GA. Multiscale simulation of transmembrane proteins. *J. Struct. Biol.* 2007 157, 570-578.

[69] Sanbonmatsu, KY; Tung, C-S. High performance computing in biology: Multimillion atom simulations of nanoscale systems. *J. Struct. Biol.* 2007 157, 470-480.

[70] Davies, LA; Klein, ML; Scharf, D. Molecular dynamics simulation of a synthetic four-α-helix bundle that binds the anesthetic halothane. *FEBS Lett.*1999 455, 332-338.

[71] Lam, W-C; Tsao, DHH; Maki, AH; Maegley, KA; Reich, NO. Spectroscopic studies of arsenic(III) binding to *Escherichia coli* RI methyltransferase and to two mutants, C223S and W183F. *Biochemistry* 1992 31, 10438-10442.

[72] Gonnelli, M; Strambini, GB. Phosphorescence lifetime of tryptophan in proteins. *Biochemistry* 1995 34, 13847-13857.

[73] Schauerte, JA; Steel, DG; Gafni, A. Time-resolved room temperature phosphorescence in proteins. *Methods Enzymol.* 1997 278, 49-71.

[74] Davies, LA; Zhong, Q; Klein, ML; Scharf, D. Molecular dynamics simulation of four-α-helix bundles that bind the anesthetic halothane. *FEBS Lett.* 2000 478, 61-66.

[75] Johansson, JS; Zou, H; Tanner, JW. Bound volatile general anesthetics alter both local protein dynamics and global protein stability. *Anesthesiology* 1999 90, 235-245.

[76] Vemparala, S; Saiz, L; Eckenhoff, RG; Klein, ML. Partitioning of anesthetics into a lipid bilayer and their interaction with membrane-bound peptide bundles. *Biophys J.* 2006 91, 2815-2825.

[77] Tang, P; Yan, B; Xu, Y. Different distribution of fluorinated anesthetics and nonanesthetics in model membrane: A ^{19}F-NMR study. *Biophys J.* 1997 72, 1676-1682.

[78] Law, RJ; Henchman, RH; McCammon, JA. A gating mechanism proposed from a simulation of a human $\alpha 7$ nicotinic acetylcholine receptor. *Proc. Natl. Acad. Sci. USA* 2005 102, 6813-6818.

In: Biopolymer Research Trends
Editor: Tamas S. Nemeth, pp. 165-188

ISBN: 978-1-60021-983-2

Chapter 6

EFFECTS OF MOLAR MASS ON THE COIL TO DOUBLE-HELIX TRANSITION OF POLYSACCHARIDE GELLAN GUMS IN AQUEOUS SOLUTIONS

Etsuyo Ogawa
Showagakuin Jr. College

ABSTRACT

Using 6 samples of well-purified sodium-type gellan gums with different molar masses (Na-gellan, G1-G6, M_w =120x10^3–17x10^3 at 40°C), the effects of molar mass on the coil to double-helix transition in aqueous solutions with and without 25 mmol NaCl were studied by light scattering and circular dichroism measurements, viscometry, and differential scanning calorimetry.

In aqueous solutions with 25 mmol NaCl, the temperature dependences of M_w, molar ellipticity at 201nm $[\theta]_{201}$, intrinsic viscosity $[\eta]$, and DSC exothermic curves of G1-G6 samples were measured from 5 to 90°C. It was found that the coil to double-helix transitions for G1-G5 samples (M_w =120x10^3 –32x10^3) took place at almost the same temperature, and the coil to double-helix transition accelerated with increase of M_w. The G6 sample (M_w =17x10^3) did not form a double-helix at 25 °C suggesting that the lowest molar mass, below which no helix is formed, lies between M_w = 32x10^3 and M_w =17x10^3. The $[\eta]$ and M_w obtained in the temperature range from 40 to 25 °C can be explained by a simple coil/double-helix equilibrium model using the double-helix contents determined from circular dichroism data. The van't Hoff's transition enthalpy ΔH_{vH} of Na-gellans depended on M_w. It is concluded that the coil to double-helix transitions of Na-gellans are all-or-none type transitions, and accelerated with increasing M_w.

In aqueous solutions without NaCl, the coil to double-helix transitions of G1-G5 samples were investigated from the temperature dependences of viscosity number η_{sp}/c and $[\theta]_{201}$ within the temperature range from 5 to 90 °C. It was found that the coil to double-helix transition temperatures for G1-G5 samples are almost the same, irrespective of M_w, which increase with increasing polymer concentrations. ΔH_{vH} values of G1-G5 samples, which are nearly the same as the values obtained in aqueous 25 mmol NaCl solutions, depend markedly on M_w. These results suggest that, in the same way as the results obtained in aqueous solutions with 25 mmol NaCl, the coil to double-helix transition without NaCl accelerated with increasing M_w. From the results of viscometry and CD measurements in aqueous solutions without NaCl, the conformational behavior of G6 is considered to be different from those of G1-G5.

INTRODUCTION

Gellan gum, an anionic polysaccharide produced by *Sphingomonas elodea* has complex tetrasaccharide repeating sequences of β-D-glucose, β-D-glucuronic acid, β-D-glucose, α-L-rhamnose and a carboxyl side group (Figure 1). [1,2] Gellan gum is widely used in the food industry and biomedical fields since a solution of gellan gum forms transparent and thermoreversible gels in the presence of a sufficient amount of cations, when it is cooled below the gelation temperature [3].

The physical nature of the gel states is controlled by the cross-linking structure and the cross-linking structure is related to the gelation process. Many biopolymers, e.g. κ-carrageenan, agarose, gelatin, and gellan gum, form gels in aqueous solutions. It is now accepted that the gelation of these biopolymers has a common feature. [4] These biopolymers show specific intermolecular interactions and form helical structures. The helical structures play an important role in biopolymer systems such as gelation. Gelation is preceded by the coil to helix conformational transition, and helices aggregate and form junction zones in gels. Therefore, gelation of polysaccharides is strongly coupled with polymer conformational transition. The detailed gelation mechanism of these biopolymers, however, has not been clarified yet. For example, the gelation process of the κ-carrageenan solution has been explained by two mechanisms. In the first mechanism, suggested by Morris et al. [5], junction zones are formed by segments of a double-helix. In the other mechanism, proposed by Smidsrød and Grasdalen [6,7] monohelices aggregate and form junction zones. Further discussion concerning the gelation mechanism has continued [8,9].

From a scientific point of view, gellan gum is not only an industrially important polysaccharide but a good model system for investigating gelation mechanisms of helix-forming polysaccharides in general, and it is useful to understand the structure and properties of other polysaccharide gels. Therefore, the investigation of the gelation mechanism of gellan gum in aqueous solutions is very important.

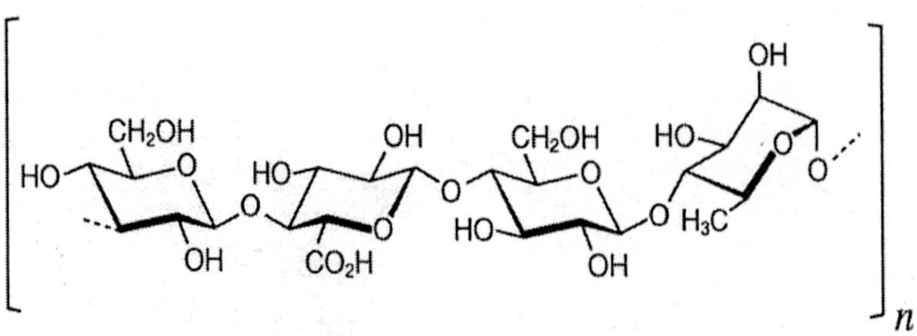

Figure 1. Repeating units of sodium-type gellan gum.

Many studies on the initial stage of the gelation mechanism of gellan gum molecules in aqueous solutions have been carried out with various experimental techniques, including rheology [10-13], small-angle x-ray scattering [14], light scattering (LS)[14], osmotic pressure [16-18], circular dichroism (CD)[18-20], optical rotation [21], nuclear magnetic resonance (NMR)[19-23], differential scanning calorimetry (DSC)[10-13], and viscometry.[17,18,22,24] In the above studies, it has been shown that the gelation mechanism of gellan gum aqueous solutions is as follows: gellan gum molecules change from the disordered state (single coil) to the ordered state (double-helix) with decreasing temperature, and helical molecules aggregate, leading to gelation. It is sometimes suggested that the helical parts aggregate to form cross-linking zones joined by flexible random coiled polymer chains.[25,26] The morphological change of gellan gum polymer during sol-gel transition has been directly visualized by transmission electron microscopy (TEM)[27] and atomic force microscopy (AFM).[28-30] These microscopic images of gellan gums[27-30] have provided new insights into the gellan gum gels supporting a fibrous model; the entire network strands and cross-links are continuously composed of thick fibers formed by the side-by-side aggregation of double-helices. Theoretical studies have been carried out for thermoreversible gelation in helix-forming polymer solutions. Tanaka [31] discussed the associations of multiple helices on the basis of the model that helices aggregate into network junction zones. Two possible mechanisms of network formation in helical gels are proposed by Viebke et al. [4] One is the gelation on the helical level and the other is the gelation on super helical level. On the helical level, junction zones are linked by flexible random coiled chains. This is the classical model of network formation. On the superhelical level, the network formation is where fully developed (single or multiple) helices aggregate to form a gel. This model seems similar to the fibrous model.

The coil to helix conformational transition is a prerequisite for gel formation. The gelation of gellan gums was strongly coupled with polymer conformational transition. The first step is the double-helix formation, and the second step is aggregation of helices. Therefore, the information on the precise mechanism of the conformational transition provides a better understanding of the second step and the ultimate gelation of gellan gums. However, the precise mechanism of coil to double-helix transition has not been clarified sufficiently. It is unknown whether the coil to double-helix transition proceeds directly or indirectly through intermediate steps, including single-helices. The possibility that the gellan gum molecules form single-helices which pair and form junction zones with decreasing temperature can not be ignored completely. It has been shown that the coil to double-helix transition and aggregation of double-helices proceeds in the presence of metal cations and the order of the effectiveness reported by many research groups [12,18,19] is as follows: K^{+}>>Na^{+}, Li^{+}. Therefore, commercially available gellan gum is produced predominantly in the potassium salt form.

The double-helix formation, however, is influenced strongly not only by the presence of metal cations but also by the characteristics of gellan gum molecules, such as molar mass. The molar mass dependence of the double-helix formation is unclear. Particularly, it is unknown how high molar mass gellan gum form double-helices. This is the most fundamental question for the helical structure of biopolymers. Only a few studies concerning the effects of molar mass on the double-helix formation have been carried out, partly due to the difficulty of sample preparation. Takahashi et al. [32] studied the conformational change of sodium-type gellan gums in 25 mmol aqueous NaCl solutions from LS and viscosity measurements. The molar mass dependence of the intrinsic viscosities, radii of gyration, and hydrodynamic radii in both the disordered and ordered states were analyzed on the basis of a wormlike chain model.

In this study, using 6 samples of well-purified sodium-type gellan gums (Na-gellans) with different molar masses, we investigated the conformational transition in aqueous solutions with and without 25 mmol dm^{-3} NaCl by LS, viscosity, and CD measurements and DSC. [33-36] We investigated the effect of molar mass on the conformational transition in aqueous solutions with 25 mmol NaCl from two angles. The first, described in sections 1-1 and 1-2, explains how high molar mass Na-gellans form double-helix. The second, described in sections 1-1 and 1-3, consider whether the coil to double-helix transition proceeds directly or indirectly, and examines the effect of molar mass on the transition. In section 2, we elucidated the conformational transition of Na-gellan samples in aqueous solutions without NaCl, comparing them with those obtained in aqueous solutions with 25 mmol NaCl.

EXPERIMENTAL

Materials

Na-gellan samples (G1-G6) were prepared from deacetylated gellan gum powder (San-Eigen F.F.I) by vibrational grinding and ion-exchange. Ground gellan gum samples were prepared as follows. [37,38] 50g of deacetylated gellan gum powder was placed in a grinding pot. After freezing at –60 ^{0}C, the pot was connected to a vibrational grinder (HEIKO Sample Mill Model TI-300) and the samples were ground for 15 min. Then, the pot and ground samples were frozen at –60 ^{0}C again. This freeze-grind cycle was repeated and 3g of ground gellan gum was taken out after the fixed grinding times: 0, 75, 120, 180, 300, and 360 minutes for G1, G2, G3, G4, G5, and G6, respectively.

Table 1. Metal Contents in the Gellan Gum Samples. (10^{-2}g/g)

	Original	G1	G2	G3	G4	G5	G6
Na	0.378	3.00	3.33	3.10	3.20	3.29	3.27
	(10)	(80)	(89)	(83)	(85)	(88)	(87)
K	1.72	0.07	0.00345	0.00251	0.00322	0.00849	0.00417
	(27)	(1.1)	(0.05)	(0.04)	(0.05)	(0.13)	(0.07)
Ca	0.93	0.005	0.00251	0.00101	0.00245	0.00154	0.00347
	(14)	(0.08)	(0.04)	(0.02)	(0.04)	(0.02)	(0.05)
Mg	0.408	0.0006	0.00019	0.00021	0.00017	0.00019	0.00024
	(10)	(0.02)	(0.005)	(0.005)	(0.004)	(0.005)	(0.006)
Fe	0.00757		0.00176	0.00243	0.00385	0.00213	0.00232
	(0.08)		(0.02)	(0.03)	(0.04)	(0.02)	(0.03)

The contents of carboxyl groups in dry gellan gum are indicated in parentheses (%).

A ground gellan sample (1.0g) was immersed in 0.5 dm^{3} deionized water with sodium citrate overnight at 40 ^{0}C. The mixture was stirred for 8h at 80 ^{0}C and filtered through a 0.45μm Millipore filter at 60 ^{0}C. This solution was dropped into deionized water/isopropanol mixture and precipitates were recovered by filtration. These precipitates were immersed in aqueous NaCl/isopropanol mixture for 2h. The precipitates were recovered by filtration and washed with deionized water/isopropanol mixture. The conversion to Na salt was checked by measuring the ionic contents of the Na-gellan samples. The ion contents of the purified Na-gellan samples are shown in Table 1.

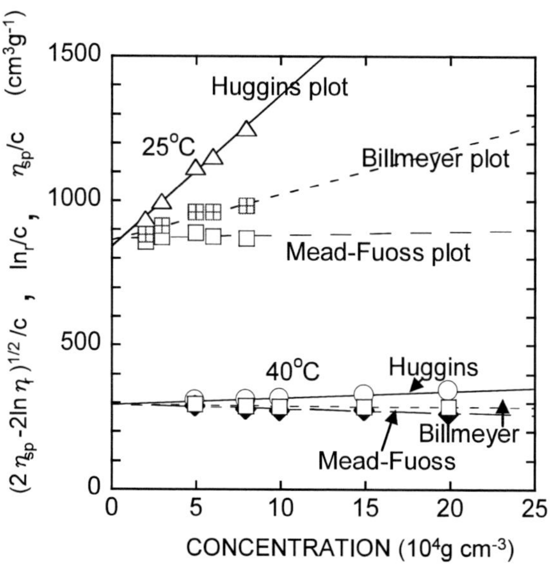

Figure 2. Typical viscosity data analyzed by Huggins, Billmeyer, and Mead-Fuoss plots.

Measurements

LS and CD measurements, viscometry, and DSC for Na-gellan were carried out in aqueous solutions with 25 mmol dm^{-3} NaCl. LS measurements were performed at 15, 25, and 40 °C, using a purpose-built spectrometer. [15] The scattering angle was 20-150°. The vertically polarized 488.0 nm line of an Ar ion laser was used as an incident beam, and vertically polarized scattered light was detected using the photon counting method. Intrinsic viscosities [η] were measured in 25 mmol dm^{-3} aqueous NaCl solutions over the temperature range from 45 to 6 °C using an Ubbelohde-type low-shear capillary viscometer. The flow time of the solvent was ca. 600 s at 45 °C. Viscosity data were analyzed by the Huggins [39], Billmeyer [40], and Mead-Fuoss [41] plots simultaneously to obtain reliable extrapolation to an infinite dilution. Typical examples of the Huggins, Billmeyer, and Mead-Fuoss plots for the Na-gellan aqueous solutions with 25 mmol NaCl are shown in Figure 2. CD measurements were performed using a Jasco J-75 Spectropolarimeter (Japan). The light-pass length was 1 mm. The CD spectra were measured at 10, 15, 20, 25, 30, 40, and 60 °C over a wave length range from 195 to 250 nm with a scanning rate of 10 nm/min. The molar ellipticity at 201 nm $[\theta]_{201}$ was also measured in the temperature range of from 60 to 5 °C. The temperature was lowered at 0.5 °C /min from 60 to 5 °C, and raised at the same rate up to 60 °C. DSC was performed using a Setaram micro DSC III calorimeter (Caluire, France). The measurements were carried out in a temperature range from 90 to 5 °C at a cooling rate of 0.5 °C /min.

The molar concentration in the values of molar ellipticity $[\theta]$ (deg cm^2 dmol^{-1}) and calorimetric enthalpy ΔH(kJ mol^{-1}) is expressed in terms of the repeating units.

RESULTS AND DISCUSSION

1. Conformational Properties of Na-Gellans in Aqueous Salt Solutions

1.1. Temperature Dependence of the Conformational Properties of Na-Gellans

In this study, we prepared 6 Na-gellan samples with different molar mass by vibrational grinding and ion exchange methods. Previously [37,38], we prepared 6 samples of Na-type gellan gums having different molar mass using the same methods as this study, and the values of number-average molar mass M_n and second virial coefficients A_2 were determined by osmometry. [19,20] From the results, it was found that the M_n values of those samples were lowered with increasing grinding time due to the cutting of principal chains of gellan gum molecules [19,20].

Table 2 shows the weight-average molar mass (M_w) for Na-gellan samples obtained in aqueous solutions with 25 mmol NaCl. These values were obtained by LS measurements. M_w obtained at 40 °C (M_w^{40}) for the 6 samples ranged from 17x10^3 to120x10^3. The ratios of M_w obtained at 25 and 40 °C (M_w^{25}/M_w^{40}), and those at 15 and 40□ (M_w^{15}/M_w^{40}) are shown in the third and fifth columns of Table 2, respectively. Takahashi et al. [32] studied molar mass dependence of [η] of Na-gellans at 40 and 25 °C in aqueous solutions with 25 mmol NaCl, using 9 Na-gellan samples ranging in M_w from 115x10^3 to 34.7x10^3 at 40 °C. It was reported that the ratios of M_w^{25}/M_w^{40} for Na- gellans were around 2, supporting the model of the conformational transition of gellan gum between a dissociated single chain and associated chain composed of two molecules (double-helix). The double-helix formation for Na-gellans in aqueous solutions with NaCl was reported by some authors using LS [14] and osmotic pressure measurements. [16-18] Recently, Atkin et al. [27] provided TEM images of laterally forming duplex aggregates of Na-gellans in aqueous solutions without NaCl.

Table 2. Weight-Average Molar Masses of Na-Gellan Samples[a]

Sample	$10^{-3}M_w^{40}$	M_w^{25} / M_w^{40}	$M_w^{25(\text{Eq 4})} / M_w^{40\,b}$	M_w^{15} / M_w^{40}	$M_w^{15(\text{Eq.4})} / M_w^{40\,b}$
G1	120	2.0	2.0	2.2	
G2	71	2.1	1.9	3.3	
G3	62	1.7	1.7		
G4	57	1.5	1.6	1.8	
G5	32	1.3	1.3	1.7	1.8
G6	17	1.0		1.3	

[a]Superscripts of 15, 25, and 40 denote the temperature at which data were obtained.
[b]Calculated values from Equation 4. Details are shown in the text.

As shown in Table 2, the values of M_w^{25}/M_w^{40} were ca. 2 for G1 and G2 samples, suggesting that the double-helix was formed at 25 °C. The values of M_w^{25}/M_w^{40} for the samples of G3, G4, and G5 were 1.7-1.3, suggesting that at 25 °C a double-helix formation took place partially; in these solutions disordered single coils and ordered double-helices coexist below the transition temperatures. However, the value of M_w^{25}/M_w^{40} for the sample of G6 was 1.0. This result suggests that the double-helix formation did not occur at 25 °C. The values of M_w^{15}/M_w^{40} for G1-G5 were larger than those of M_w^{25}/M_w^{40}, suggesting that on lowering the temperature, the aggregation of double-helices occurred for G1 and G2, and the amount of double-helices increased for G3-G5 with the progress of the double-helix formation. M_w^{15}/M_w^{40} for G2 was larger than that of G1. It is thought that this difference is partly due to a fairly wide molar mass distribution in the G2 sample. To elucidate the effect of the molar mass distribution further study is necessary. For the sample of G6, the values of M_w^{15}/M_w^{40} was 1.3. It seems that the partial double-helix formations took place at 15 °C.

The temperature dependencies of [η] for the Na-gellan aqueous solutions with 25 mmol NaCl are shown in Figure 3. The [η] of G1-G5 were almost constant at higher temperature regions but, with lowering temperature, these [η] increased rapidly in the temperature region of 32-25 °C. Below ca. 25 °C, the increase in the [η] values became gradual. The slope of [η] vs temperature plots in the region of 32-25 °C increases with increasing M_w. The variation of [η] can be interpreted as follows. On lowering the temperature, the coil to double-helix formation takes place for G1-G5 below ca. 32 °C and the aggregation of double-helices occurs below ca. 25 °C at least partly.[17] In addition, the double-helix formation is accelerated with increasing molar mass. While, as shown in Figure 3, the values of [η] for G6 show a slight linear increase and did not show any steep increase in the temperature region from 45 to 10 °C. It seems that for G6 the conformational transition such as the coil to double-helix transition did not occur in the temperature region of 32-25 °C. However, at 15°C, M_w^{15}/M_w^{40} for G6 was 1.3 suggesting that the partial double-helix formation took place. This partial double-helix formation could not be observed from the temperature dependence of [η] for G6 in Figure 3. This result can be explained as follows. Takahashi et al. [32] studied the molar mass dependence of [η] of Na-gellans at 40 and 25 °C in aqueous solutions with 25 mmol NaCl, using 9 Na-gellan samples ranging in M_w from 34.7x10^3 to 11.5x10^3. Data were analyzed on the basis of unperturbed wormlike chain models according to the Yamakawa-Fujii-Yoshizaki (TFY) theory.[42,43] The wormlike chain parameters reported were q=9.4 nm, M_L=355nm^{-1}, and d=1.0nm at 40 °C (single coil), and q=98 nm, M_L=650nm^{-1}, and d=2.4 nm at 25 °C (double-helix). Here q is the value of persistence length, M_L the molar mass per unit contour length of the chain, and d the chain diameter. From these results, it can be estimated that, for G6 solutions, the difference of [η] values for single coil and for double-helix may be small, and the partial double-helix formation at 15 °C could not be observed clearly from the [η] vs temperature plots shown in Figure 3.

In this study, the temperature at which the coil to double-helix transition takes place is defined as the transition temperature $T_{ch.}$ The CD spectra for the 6 Na-gellan aqueous solutions with 25 mmol NaCl (polymer concentration c_p= 0.5 wt%) were measured at 10, 15, 20, 25, 30, 40, and 60 °C over the wavelength range from 195 to 250 nm. The typical CD spectra of G1, G3, G5, and G6 solutions are shown in Figure 4. Absorption peaks were observed around 201 nm in the CD spectra of G1-G6. The peak around 201 nm in the CD spectra of G1-G5 shifted to lower wavelengths with decreasing temperature (a red shift) suggesting that the conformational change of Na-gellan molecules took place.

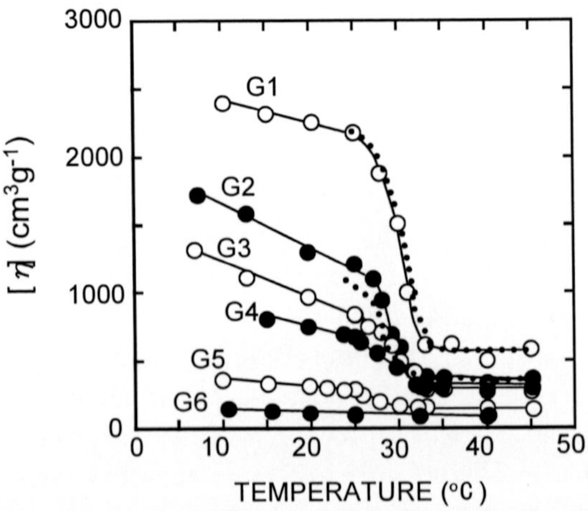

Figure 3. Temperature dependence of intrinsic viscosity for Na-gellan aqueous solutions with 25mmol NaCl. Experimental fit of the data ( ); estimated values using Equation 5( ). Details are shown in the text.

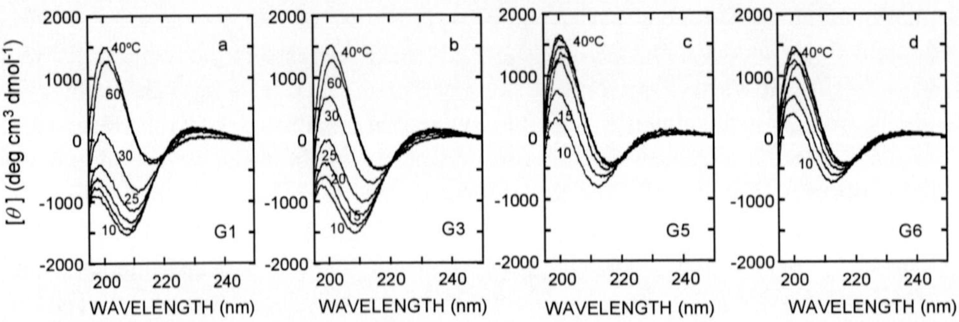

Figure 4. CD spectra of Na-gellan aqueous solutions with 25 mmol NaCl (c_p=0.5wt%). (a)G1, (b) G3, (c) G5, (d) G6.

On the other hand, the peak around 201 nm in G6 sample did not shift to lower wavelength with decreasing temperature. The peak at around 201 nm reflects the optically active chemical structure of the glucuronic acid in random coil form of the gellan gum.[18-20,46,47] Bosco et al.[22] reported that the transition detected by CD methods in dilute solutions seems mostly due to the chair-boat equilibrium. The variation of molar ellipticity of the peak at around 201 nm should correspond to a conformational change of the gellan molecules. On reducing the temperature below T_{ch}, the optically active high-order structure of the gellan chain, double-helical structure is formed, and the CD spectrum overlaps with the spectrum for the random coil and that for the double-helix. Therefore, the change in molar ellipticity at 201 nm $[\theta]_{201}$ is proportional to the change in the amounts of the random coil (or helix).

Temperature dependencies of $[\theta]_{201}$ for G1-G6 Na-gellan solutions with c_p= 0.25, 0.5, and 1.0 wt% were measured. The temperature was lowered from 60 to 5 °C (cooling process).

The plots with c_p= 0.25, 0.5, and 1.0 wt% show a similar behavior on the whole. Typical examples are shown in Figures 5a and 5b.

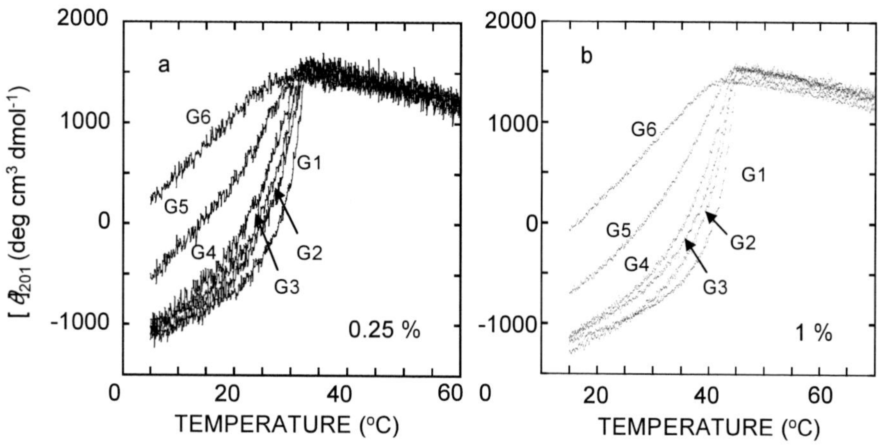

Figure 5. Temperature dependence of molar ellipticity at 201 nm, $[\theta]_{201}$, for Na-gellan aqueous solutions with 25 mmol NaCl (a) c_p=0.25 wt%; (b) c_p=1.0 wt%.

On lowering the temperature, the values of $[\theta]_{201}$ for G1-G6 increased slightly between the temperature range from 60 to T_{ch} (34-32 °C). It is considered that above T_{ch}, only the intrinsic optical activity of gellan chains in random coil form contributes to $[\theta]_{201}$ values. For G1-G5 Na-gellan solutions, the values of $[\theta]_{201}$ decreased steeply below T_{ch} and after that, the values of $[\theta]_{201}$ showed gradual downward curvature until 5 °C. The slopes of $[\theta]_{201}$ vs temperature plots in the temperature regions of 32 to ca. 25 °C increase with increasing M_w. Results for the temperature dependences for G1-G5 are explained as follows. Below the transition temperatures, optical activity of gellan chains due to the double-helical structure and aggregation of double-helices contributed to the CD spectra in addition to the intrinsic optical activity in coil form. The coil to double-helix conformational changes are accelerated with increasing M_w of Na-gellans. It is noted that T_{ch} for G1-G5 were almost coincident within experimental errors which suggest that the coil to double-helical transition for G1-G5 took place at the same temperature. These results are in good agreement with those observed by viscometry. On the other hand, as shown in Figures 5a and 5b, the values of $[\theta]_{201}$ for G6 showed a gradual decrease in the whole temperature range between 30 and 5°C, and T_{ch} for G6 was lower than those for G1-G5. It seems that the conformational behavior of the G6 sample in the temperature range 30 to 5°C in aqueous solutions was different from that of G1-G5 samples.

In order to examine thermal hysteresis of the conformational transition, the temperature was lowered from 60 to 5 °C (cooling process), and increased up to 60 °C (heating process). As typical examples, $[\theta]_{201}$ vs temperature plots of G1and G6 for the cooling process are shown in Figures 6a and 6b, respectively. For the 0.5 wt % solution, $[\theta]_{201}$ vs temperature plots for the heating processes are also shown in Figures 6a and 6b. In the cooling process, the plots with c_p= 0.25, 0.5, and 1.0 wt% for G1-G5 become higher with increasing polymer concentration and T_{ch} shifts towards slightly higher temperature regions, suggesting that the coil to double-helix transition and aggregation of double-helices proceeded with increasing polymer concentration.

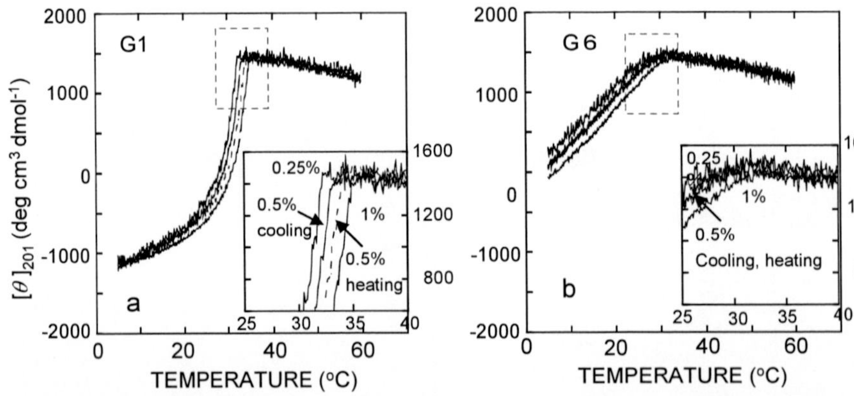

Figure 6. Temperature dependence of molar ellipticity at 201 nm, [θ]201, for Na-gellan aqueous solutions with 25 mmol NaCl. (a)G1, (b)G6. The portion of the curve surrounded by the dashed line is enlarged in the inset. Cooling process: cp=0.25, 0.5, and 1.0wt% ( ); heating process:0.5wt% ( ).

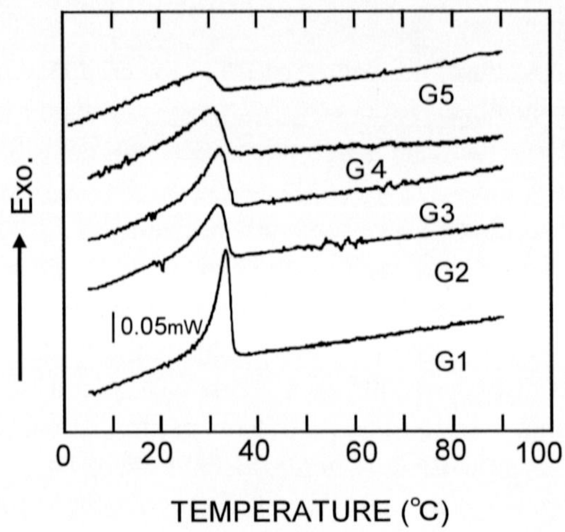

Figure 7. Cooling DSC curves of 1 wt% Na-gellan aqueous solutions with 25 mmol NaCl.

The double-helix to coil transition temperatures T_{hc} observed in the heating process are slightly higher than those of the coil to double-helix transition temperatures T_{ch} observed in the cooling process. These experimental findings support the above-mentioned conclusion that, for G1-G5, the coil to double-helix transition took place and the aggregation of double-helices was followed by decreasing temperature. Similar results of Na-gellans in aqueous solutions with NaCl were found in DSC measurements[12], CD measurements[20] and osmometry. [16,17] On the other hand, as shown in Figure 6b, $[\theta]_{201}$ for G6 showed a gradual decrease in the whole temperature range between 30 and 5°C. From the comparison between cooling process and heating process for G6 solutions, it is seen that $[\theta]_{201}$ for G6 in the cooling process coincided with that in the heating process and no thermal hysteresis was observed. These CD results indicate that the conformational behavior of G6 samples in the temperature range of 30 and 5°C in aqueous solutions differ from that of G1- G5 samples. It seems that the double-helix formation and/or aggregation of helices scarcely occurred for G6 in the temperature range between ca 32 and 25 °C, as mentioned above from the results of M_w^{25}/M_w^{40} and $[\eta]$ vs temperature plots.

DSC curves for G1-G5 sample solutions (c_p= 1.0 wt%) at cooling rate of 0.5 °C /min are shown in Figure 7. A clear peak for G6 could not be observed in a cooling DSC curve. Cooling curves for G1-G5 showed a single exothermic peak and the peak widths became broader with decreasing molar mass. From the previous studies [10,12,44], we judged that the exothermic peaks in the DSC curves were attributed to the coil to double-helix transition of Na-gellan molecules.

The transition temperature T_{ch} was defined at the intersection of the extrapolated lines from data points of $[\theta]_{201}$ in the higher temperature range and from the data points in the range of steep decrease at lower temperature in Figures 5 and 6. The values of T_{ch} determined from temperature dependence of $[\theta]_{201}$ for the cooling process are plotted against polymer concentration in Figure 8 together with the values of T_{ch} obtained by viscometry and DSC in Figures 3 and 7, respectively. As shown in Figure 8, T_{ch} for G1-G5 samples obtained by CD measurements, DSC and viscometry are almost the same and the values of T_{ch} shift to higher temperatures with increasing polymer concentration. We have shown that the double-helix formation was accelerated by increasing M_w. However, the results shown in Figure 8 indicate that the values of T_{ch} do not depend so much on M_w when M_w is higher than 32x10^3. The coil to double-helix transition of G1-G5 occurs at almost the same temperature. The transition, however, becomes sharper with increasing M_w. In contrast, for G6 sample, the coil to double-helix transition could not be detected by viscometry and DSC. The values of T_{ch} for G6 determined by CD measurements were much lower than those of G1-G5.

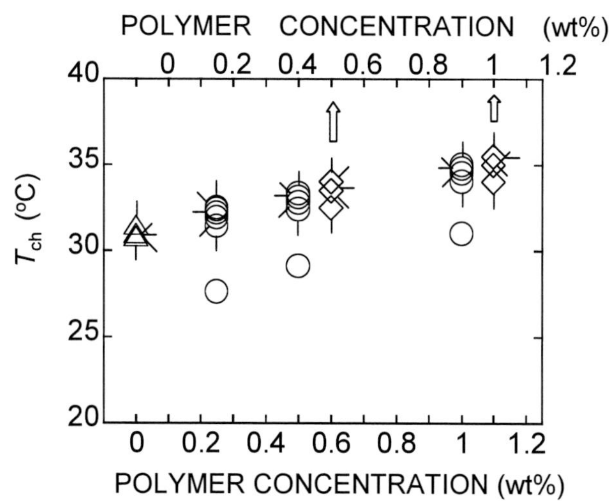

Figure 8 Polymer concentration dependence of coil to double-helix transition temperature, T_{ch}, for Na-gellan aqueous solutions with 25 mmol NaCl obtained from CD data: G1( ),G2( ),G3( ),G4( ),G5( ), and G6( ). T_{ch} obtained from DSC data:G1( ),G2( ),G3( ),G4( ), and G5( ). T_{ch} obtained from viscosity data: G1( ),G2( ),G3( ),G4( ), and G5( ).

Assuming a simple coil/double-helix equilibrium model for the Na-gellan solutions, the double-helix contents f can be written as [45];

$$[\theta]_{201} = f[\theta]_{201\mathrm{h}} + (1-f)[\theta]_{201\mathrm{c}} \tag{1}$$

This equation is set up between $f = 0$ (full coil states) and $f = 1$ (full double-helix states). Here $[\theta]_{201}$ is the molar ellipticity $[\theta]_{201}$ at a given temperature, $[\theta]_{201\mathrm{c}}$ and $[\theta]_{201\mathrm{h}}$ are the values of $[\theta]_{201}$ in the fully coil and fully double-helix states, respectively. We evaluated f for G1-G5 samples from Equation 1 using experimental $[\theta]_{201}$ for G1 obtained at T_{ch} and 25 °C as $[\theta]_{201\mathrm{c}}$ and $[\theta]_{201\mathrm{h}}$, respectively. Temperature dependence of f for G1-G6 samples with c_p= 0.25 and 1.0 wt% is shown in Figures 9a and 9b, respectively.

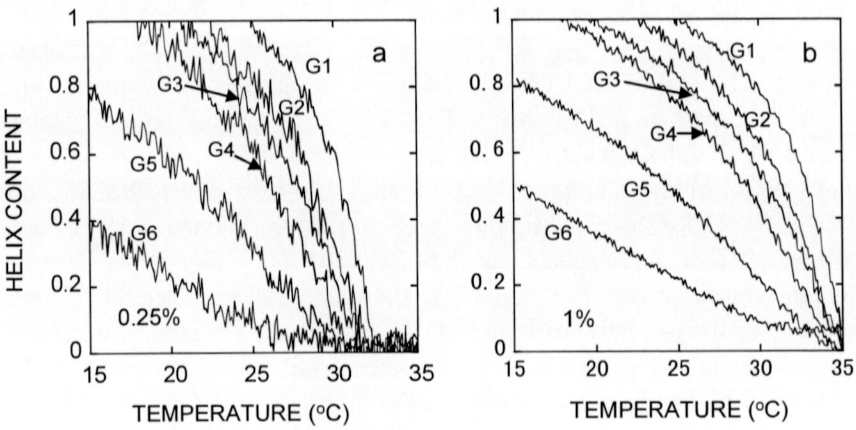

Figure 9. Temperature dependence of helix content for Na-gellans in aqueous solutions with 25 mmol NaCl. (a) c_p=0.25 wt%; (b) c_p=1.0 wt%.

As shown in Figures 9a and 9b, f for the Na-gellan molecules of G1-G6 in aqueous solutions with 25 mmol NaCl increased rapidly below T_{ch} with decreasing temperature, and the slopes of these plots increased with increasing M_{w}. These results of f can be explained as follows. On lowering temperature, coil to double-helix transitions took place and the coil to double-helix transitions were expedited with increasing M_{w}. From the comparison between the values of f in Figures 9a and 9b, values of f for the Na-gellan molecules of G1-G6 in aqueous solutions with 25 mmol NaCl in Figure 9b (c_p=1 wt%) are higher than those in Figure 9a (c_p= 0.25 wt%), indicating that the coil to double-helix transitions were expedited with increasing polymer concentration. It is noted that the f values for G6 in Figure 9a (c_p= 0.25 wt%) were very small (less than 0.08 at 25 °C), suggesting that the coil to double-helix transitions scarcely occurred at 25 °C, which supports the conclusion obtained from $M_{\mathrm{w}}^{25}/M_{\mathrm{w}}^{40}$ and $[\eta]$ data.

We analyzed the CD data using van't Hoff plots and calculated van't Hoff's enthalpy ΔH_{vH}. When single-coils and double-helices coexist in an equilibrium state in the solution at a given temperature, the equilibrium constant K and van't Hoff's enthalpy ΔH_{vH} can be written as follows.

$$K(T) = f/(1-f)^2 \tag{2}$$

$$d\, ln\, K / (d(1/T)) = -\Delta H_{vH} / R \quad (3)$$

Here, R is the gas constant. Typical examples of van't Hoff plots are shown in Figures 10a -10d. The $ln\ K$ values decreased abruptly at T_{ch}. Below T_{ch}, van't Hoff plots ($ln\ K$ vs $1/T$) for G1-G5 showed a two-step decrease, steep and then gradual (slopes 1 and 2 in Figures10a-10c). In order to calculate theΔH_{vH}, it is necessary to decide which slope should be selected. Christensen and his coworkers [46] reported that van't Hoff plots of xanthan solutions showed two slopes below T_{ch} , as found in the present study. They estimated ΔH_{vH} from the data just below T_{ch} (steep slope). In the present study, from the results of M_w and $[\eta]$, it is concluded that, on lowering temperature, the coil to double-helix transition occurs below T_{ch} and is followed by the aggregation of double-helices. Therefore, we considered that ΔH_{vH} for the coil to double-helix transition should be calculated from the $[\theta]_{201}$ data obtained just below T_{ch}. The values of ΔH_{vH} (kJ/mol) determined from slope 1 are plotted against M_w^{40} together with the values obtained from slope 2 in Figure 10.

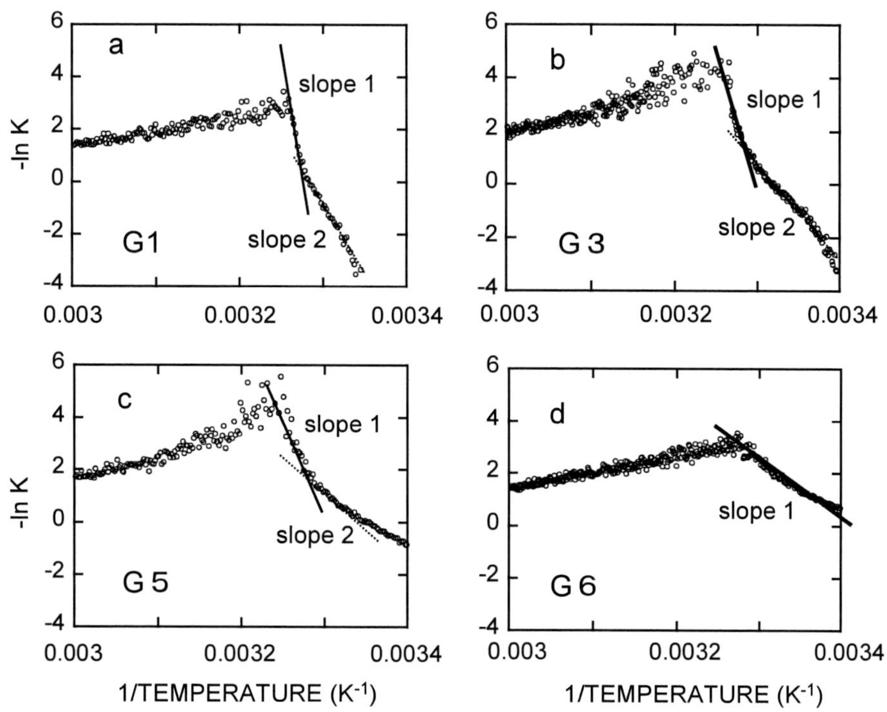

Figure 10. The van't Hoff plots of molar ellipticity, $[\theta]_{201}$, for Na-gellan aqueous solutions (c_p=1.0 wt%) with 25 mmol NaCl. (a) G1, (b)G3, (c) G5, (d) G6.

The values of ΔH_{vH} (kJ/mol) determined from slope 1 are plotted against M_w^{40} in Figure 11. Values of ΔH_{vH} shown in Table 3, are slightly dependent on c_p and depend markedly on M_w^{40}. These results support the conclusion that the double-helix formation is accelerated with the increase of molar mass. However, van't Hoff plots ($ln\ K$ vs $1/T$) for G6 in Figure10d showed a gradual one step decrease up to 25 °C (1/K=0.00335), andΔH_{vH} for G6 in Figure 11 is much lower than those for G1-G5. These results suggest that the conformational behavior of G6 in the temperature range from ca 32 to 25 °C is different from that of G1-G5.

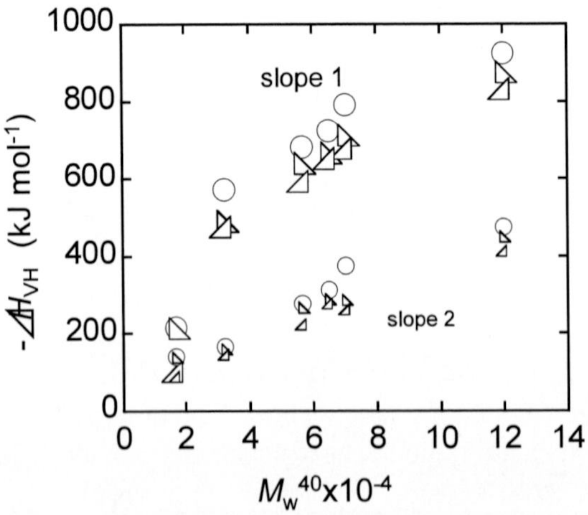

Figure 11. Dependence of van't Hoff enthalpy, ΔH_{vH}, of Na-gellan aqueous solutions with 25 mmol NaCl on weight-average molar mass, $M_W{}^{40}$. The values of $M_W{}^{40}$ are determined at 40°C. Determined from slope 1; c_p=0.25 wt% ( ), c_p=0.5 wt%( ), c_p=1.0 wt% ( ); Determined from slope 2; c_p=0.25 wt% (), c_p=0.5 wt%( ), c_p=1.0 wt% ( ).

1.2. Lowest Molar Mass of Na-Gellan for Double-Helix Formation

The gelation is influenced strongly by the molar mass of gellan molecules. However, only a small number of studies concerning molar mass effects on the initial stage of gelation such as double-helix formation have been carried out, partly due to the difficulty of sample preparation.

Table 3. Characteristics of the Coil to Double-Helix Transition of 1 wt% Na-Gellan Aqueous Solutions with 25 mmol dm^{-3} NaCl

	M_w x10^{-3} (40°C)	$-\Delta H_{vH}$[a] (kJ/mol)	$-\Delta H_{cal}$[b] (kJ/mol)	$\Delta H_{vH}/\Delta H_{cal}$	n[c]
G1	120	924	6.05	153	82
G2	71	792	5.43	146	48
G3	62	730	5.70	128	42
G4	57	670	5.40	125	39
G5	32	573			21
G6	17	214			11

[a] ΔH_{vH} was determined from slope 1 in the van't Hoff plots.

[b] ΔH_{cal} was determined from the DSC curve in Figure 7.

[c] The number of repeating units in a molecular chain. Details are shown in the text.

Therefore, molar mass dependence of the double-helix formation is unclear. Particularly, it is unknown how high molar mass gellan gums form double-helices. Gidley et al. [47] examined the mit chain length below which no double-helix formation occurs, using α-D-glucan oligomers as a model compound of starch. They reported that the minimum chain-length required for formation of the double-helix (crystallization) was 10 repeating units.

From the results of M_w^{25}/ M_w^{40} (Table 2) and $[\eta]$ vs temperature plots for G6 sample solutions (Figure 3), we conclude that at 25°C the coil to double-helix transition of G6 did not occur. However, this conclusion seems to be unclear in terms of the CD data, i.e. the transition temperature T_{ch} and transition enthalpy ΔH_{vH} for G6 molecules can be obtained from the CD data, even though the values for G6 molecules were much smaller than those of G1-G5 molecules. The above facts can be explained as follows. $[\theta]_{201}$ were measured in the solutions with c_p= 0.25-1.0 wt%, on the other hand, M_w and $[\eta]$ were extrapolated values to zero concentration. The double-helix formation and aggregation of double-helices were accelerated with lowering temperature and also with increasing polymer concentration. The partial double-helix formation could be observed in the CD spectra. However, as shown in Figure 4, the conformational behavior of G6 molecules observed by CD measurements in this study is undoubtedly different from that of G1-G5 molecules. The experimental findings for G6 molecules from CD measurements are consistent with the conclusion derived from LS and $[\eta]$ measurements that G6 molecules did not form a double-helix at 25 °C in aqueous solutions with 25 mmol NaCl.

Within the experimental conditions in this study, G6 molecules (M_w^{40} = 17x10^3) did not form double-helices at 25 °C, while G5 molecules (M_w^{40} = 32x10^3) formed double-helices. This means that the value of the smallest M_w, at which it is possible to form a double-helix at 25 °C, lies between 17x10^3 (G6) and 32x10^3 (G5). Chandrasekaran et al. [48,49] reported that potassium type gellan gum formed a left-handed 3 fold double-helix from the X-ray diffraction analysis of oriented fibers. Applying this report to Na-gellan samples in this study, we tried to analyze the double-helical structure of G5 and G6 molecules. Okamoto et al. [50] reported that the ratio of M_w/M_n for Na-gellan was 2.2 at 40 °C. Using this value, as shown in the sixth column of Table 3, the numbers of repeating units (molar mass of the repeating unit is ca.668) in a G5 molecule and a G6 molecule were roughly calculated as around 21 and 11, respectively. G5 chains have ca.7 turns in double-helical structure. On the other hand, if G6 chains formed double-helices, they are assumed to have 3-4 turns. From these considerations, it seems that the double-helix formation could not occur due to insufficient number of helices (repeating units) of G6 molecules; the dividing line between whether the chain forms a double-helix or not lies in between ca.7 and 3-4 turns. Recently [35], it has been found that the coil to double-helix transitions of Na-gellans in aqueous solutions with 25 mmol NaCl are all-or-none type transitions and the coil to double-helix transition proceeds directly. From the present study, it is concluded that G6 molecules in aqueous solutions with 25 mmol NaCl with zero concentration at 25 °C could not proceed in all-or-none transition due to insufficient number of repeating units in a chain. However, the coil to double-helix transition proceeds with increasing polymer concentration and/or with lowering temperature. In the solutions at lower temperature such as 15 °C and/or higher polymer concentrations such as c_p= 0.25-1 wt%, partial double-helix formations may occur. Further studies, using several low molar mass samples are necessary to confirm these considerations.

1.3. Coil to Double-Helix Transition of Na-Gellans

Many studies concerning the coil to helix transition of Na-gellans in aqueous solutions have been carried out. However, the mechanism of the transition is not sufficiently clear. In this section, from the results of molar mass, intrinsic viscosity, CD spectra, and DSC curves of 5 Na-gellans (G1-G5), we elucidated the mechanism of the transition such as whether the coil to double-helix transition proceeds directly or indirectly, and examined the effect of molar mass on the transition.

As shown in Figures 9a and 9b, the slopes of f vs temperature plots below the T_{ch} increase with increasing M_w, indicating that the double-helix formation occurs faster in high molar mass samples. From the comparison between the values of f in Figures 9a and 9b, it is found that f for the G1-G5 samples increased with increasing c_p. Using a coil/double-helix equilibrium model (single coils and double-helices coexist in the solution), the measured M_w and $[\eta]$ at a given temperature may be expressed as; [51]

$$M_w = fM_h + (1-f)M_c \tag{4}$$

$$[\eta] = f[\eta]_h + (1-f)\,[\eta]_c \tag{5}$$

Here subscripts h and c are used to denote the values of the molar mass and intrinsic viscosity of the double-helix and single coil, respectively (M_h, for example, is the molar mass of the double-helix). f denotes the helix content, and the values in Figure 9a (c_p= 0.25 wt%) were used for the following calculations. The ratios of M_w^{25}/M_w^{40} (G1-G5 samples) and M_w^{15}/M_w^{40} (G1, G2, G4, and G5 samples) determined at 25 and 15 °C by LS are shown in the third and fifth columns of Table 2, respectively. $M_w^{25\,(Eq.4)}$ (G1-G5 samples) and $M_w^{15(Eq.4)}$ (the G5 sample) at 25 and 15 °C were estimated from Equation 4 and Figure 9a using M_w^{40} experimentally determined by LS as M_c and $2M_w^{40}$ as M_h. The ratios of $M_w^{25\,(Eq.4)}/M_w^{40}$ and $M_w^{15(Eq.4)}/M_w^{40}$ evaluated in this way are shown in the fourth and sixth columns of Table 2, respectively. These ratios, estimated by Equation 4 (columns 4 and 6), showed fairly good agreement with the experimental values of M_w^{25}/M_w^{40} and M_w^{15}/M_w^{40} (columns 3 and 5, respectively), although the helix content f used for this estimation was determined at c_p=0.25% (Figure 9a) and not based on the value extrapolated to zero concentration. The dashed curves in Figure 3 show $[\eta]$ calculated from Equation 5, using experimental $[\eta]$ obtained at 40 and 25 °C as $[\eta]_c$ and $[\eta]_h$, respectively. The calculated $[\eta]$ values fit the observed data points well. These results of M_w and $[\eta]$ substantiate the validity of the assumption that only single chains and double-helices coexist in aqueous salt solutions. In other words, upon cooling, single-coil Na-gellan molecules change directly into double-helices below the transition temperature without taking an intermediate conformation such as a single-helix or partially broken double-helix between these two conformations.

The values of ΔH_{vH} (kJ/mol) for solutions of c_p= 1 wt% shown in Table 3 are plotted against M_w^{40} together with the values obtained from slope 2 in Figure 11. As shown in Figure 11, ΔH_{vH} are slightly dependent on the polymer concentration and depend markedly on M_w^{40}. These results support the conclusion that the double-helix formation is accelerated with the increase of molar mass. ΔH is the change in molar enthalpy for the coil to helix transition. Experimentally, ΔH can be determined from integration of the DSC curves. The ratio $\Delta H_{vH}/\Delta H$ [52-54] represents the cooperative unit size, which is related to the cooperativity parameter σ in the Zimm-Bragg theory[55] for the coil to single-helix transition [54].

$$\sigma^{-1/2} = \Delta H_{vH}/\Delta H \tag{6}$$

$\Delta H_{vH}/\Delta H$ is the average number of repeating units that undergoes the coil to helix transition in concert in a strictly two-state process, such as completely ordered (helix) and completely disordered (coil) states.[54] For the coil to double-helix transition of Na-gellans, some qualitative information can be obtained from Equation 6. The values of enthalpy change (ΔH_{cal}) for G1 to G4 samples were obtained by integration of the DSC curves shown in Figure 7. ΔH_{cal} are listed in Table 3. Determination of ΔH_{cal} for G5 is difficult since the peak width of G5 in the DSC spectra was broad. The values of $\Delta H_{vH} / \Delta H_{cal}$ were evaluated from Equation 6 using ΔH_{cal} as ΔH and listed in the fifth column of Table 3. In aqueous solutions with 25 mmol dm^{-3} NaCl, the coil to double-helix transition of Na-gellans took place with decreasing temperature following by the aggregation of helices. Therefore, the value of ΔH_{cal} contains the enthalpy change corresponding to the double-helix formation ΔH and other enthalpy changes such as the aggregation of double-helices. The value of ΔH_{vH} determined from slope 1 in van't Hoff plots corresponds to the enthalpy change for the coil to double-helix transition. Consequently, the cooperative unit sizes $\Delta H_{vH} / \Delta H$ in Equation 6 for Na-gellan molecules should be larger than the apparent cooperative unit sizes $\Delta H_{vH} / \Delta H_{cal}$ in Table 3. The number of repeating units in a Na-gellan molecule are listed in the sixth column of Table 3. From the comparison of $\Delta H_{vH} / \Delta H_{cal}$ (column 5) and n (column 6), it is found that $\Delta H_{vH} / \Delta H_{cal}$ are larger than n, respectively. Accordingly, $\Delta H_{vH} / \Delta H$ for G1 to G4 are much larger than n for G1 to G4, respectively. Theses results mean that the coil to double-helix transition is a highly cooperative phenomenon. Schellman [57,58] studied the coil to helix transitions for polypeptides using one-sequence approximation. His work leads us to consider the helix to coil transition as an "all-or-none" transition between complete helix and complete random coil. In a transition of the all-or-none type, all the units on a given helix are brought simultaneously from one state to another and no intermediate state exists. Further, the all-or-none approach was extended by Applequest and Damle [59] to states with only species A and B_N, where B_N represents a completely bonded species. Thereafter, transitions of the all-or-none type were reported for several biopolymers in aqueous solutions. For example, Nakanishi and Norisuye [51] reported that double-helical succinoglycans changed directly to single-coiled molecules which did not proceed through intermediate steps such as partial breaking of helical portions or formation of single helices. In the present study, the single coils of Na-gellan change directly into double-helices. These transitions are highly cooperative phenomena and proceed with increasing M_w. Therefore, the coil to double-helix transitions of Na-gellans can be considered as all-or-none type transitions. Takahashi et al.[32] reported that the value of persistence length for the Na-gellan chain was 9.4 nm and the Na-gellan chain behaves as a semi-flexible chain at 40 °C.[32,44] It is considered that the substantial stiffness of the single stranded Na-gellan chain facilitates the coil to double-helix transition in the all-or-none type. In this study, even at the highest polymer concentration (c_p=1.0 wt%), the Na-gellan solution did not form gels due to an insufficient number of helices to form a three-dimentional network [12]. However, a solution with much higher polymer concentration should form gels. Na-gellan gels composed of double-helices with few coiled portions seem favor the fibrous model obtained by recent microscopic images [27-30] than the classical model in which junction zones are linked by flexible random coiled chains.

2. Conformational Properties of Na-Gellans in Aqueous Solutions

The temperature dependence of viscosity number η_{sp}/c_p (= $(\eta_s - \eta_o)/\eta_o c_p$) for the Na-gellans was determined in aqueous solutions (polymer concentration c_p= 0.25, 0.5, and 1.0

wt%). Here, η_s and η_o are the viscosities of the solution and the solvent, respectively. Typical examples of the η_{sp}/c_p versus temperature plots for the Na-gellans in aqueous solutions are shown in Figures 12a-12c. On lowering the temperature, the η_{sp}/c_p of G1-G5 sample solutions with various polymer concentrations showed an almost constant or only slight increase at higher temperature regions, but increased rapidly below the coil to double-helix transition temperatures (T_{ch}). The transition temperatures of the Na-gellan solutions increased with increasing polymer concentrations.

The variation of η_{sp}/c_p is a reflection of the conformational change of Na-gellan molecules and can be interpreted as follows. In aqueous solutions, on lowering temperature, the coil to double-helix transition takes place below the transition temperature T_{ch} and the solutions of higher polymer concentrations have a greater tendency to form double-helices than those of lower polymer concentrations. At the temperatures, above T_{ch}, where polymer chains take a coil form, the values of η_{sp}/c_p increase with decreasing polymer concentrations, which are commonly observed for a flexible polyelectrolyte in aqueous solutions in the absence of added salt. This effect can be attributed to the fact that with decreasing polyelectrolyte concentration, the condensed counterions escape into the surrounding solution and hence the ionic strength increases, so that the polymer chain is expanded by the electrostatic repulsive interactions. However, as shown in Figure 12c, the values of η_{sp}/c_p for G6 show a slight linear increase and did not show any steep increase in the temperature region from 45 to 10 °C. It seems that for G6 the conformational transition such as the coil to double-helix transition did not occur in the temperature region of 30-10 °C.

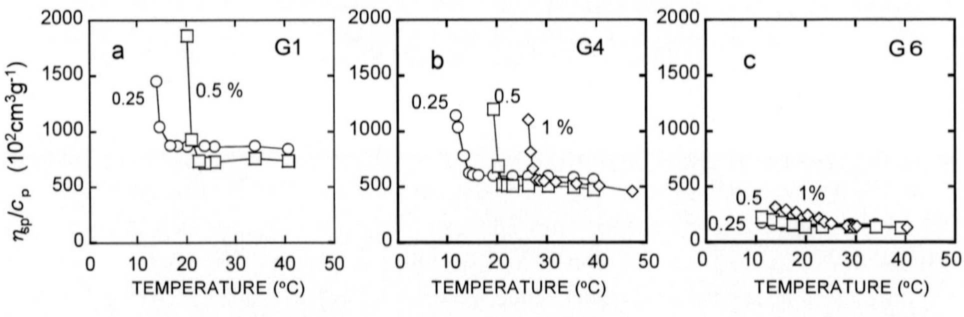

Figure 12. Temperature dependence of η_{sp}/c_p for Na-gellans in aqueous solutions. (a) G1, (b) G4, (c) G6.

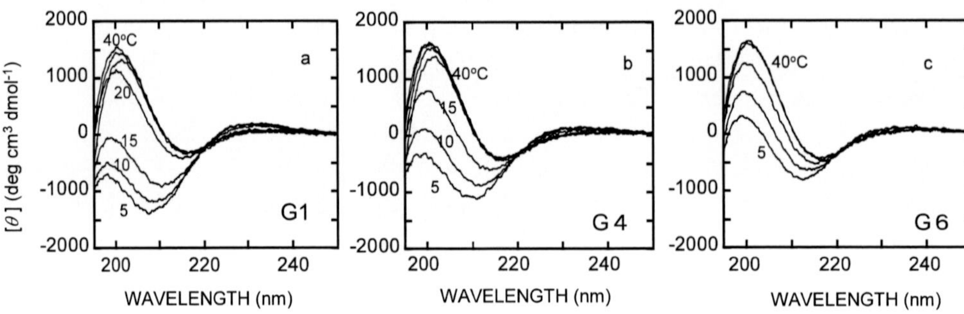

Figure 13. CD spectra of Na-gellan aqueous solutions. (a) G1, (b) G4, (c) G6.

The temperature dependence of the CD spectra for the 6 Na-gellan aqueous solutions (c_p= 0.25, 0.5, and 1.0 wt%) was measured. Typical examples are shown in Figures 13a-13c. The CD spectra for G1-G5 are similar and a peak was observed at around 201 nm which corresponds to the coil to double-helix conformational change of Na-gellan molecules. [18,33-36] The molecular ellipticity at 201 nm $[\theta]_{201}$ of Na-gellans is plotted against temperature. Typical examples are shown in Figures 14a and 14b. On lowering temperature, values of $[\theta]_{201}$ for G1-G5 drastically decreased below T_{ch}, suggesting that a coil to double-helix transition took place. On the other hand, as shown in Figures 14a and 14b, the values of $[\theta]_{201}$ for G6 showed a gradual decrease below T_{ch}, and T_{ch} for G6 was lower than those for G1-G5. It seems that the conformational behavior of the G6 sample in the temperature range 30 to 5 °C in aqueous solutions was different from that of G1-G5 samples, which are similar to the results obtained in aqueous solutions with 25 mmol NaCl. It is noted that the values of T_{ch} obtained in c_p= 0.5 wt% aqueous solutions are lower than the values obtained in c_p=1 wt% aqueous solutions.

In Figure 15, the values of T_{ch} for G1-G5 determined from the temperature dependence of $[\theta]_{201}$ are plotted against polymer concentration together with the values of T_{ch} obtained by viscometry in Figure 12.

In aqueous solutions without NaCl, T_{ch} obtained from $[\theta]_{201}$ and η_{sp}/c_p data were coincident, and these T_{ch} values increase with increasing polymer concentration. It is noted that the T_{ch} values for G1-G5 are the same, suggesting that these values do not depend on M_w when M_w is higher than 32×10^3. We investigated the molar mass dependence of T_{ch} for G1-G5 in aqueous solutions with 25 mmol NaCl. [33,35,36].

The results obtained in 25 mmol NaCl are also shown in Figure15. In aqueous solutions with NaCl, T_{ch} for G1-G5 did not depend on M_w in the same way as T_{ch} in aqueous solutions without NaCl. T_{ch} values for G1-G5 in aqueous solutions without NaCl are much lower than those in solutions with NaCl, and T_{ch} values in aqueous solutions without NaCl increase rapidly with increasing polymer concentration. It is considered that these results can be attributed to the fact that the polymer chains are expanded by the electrostatic repulsive interactions in dilute aqueous solutions without salt.

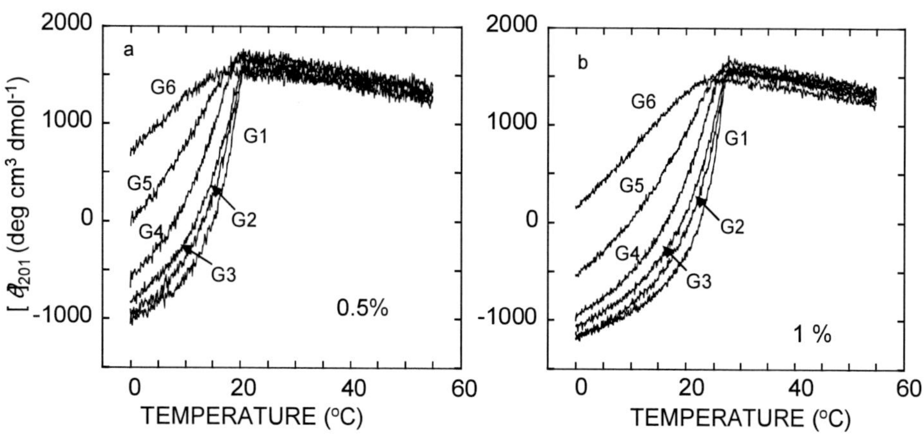

Figure 14. Temperature dependence of molar ellipticity at 201 nm, $[\theta]_{201}$, for Na-gellan aqueous solutions. (a) 0.5 wt%, (b) 1.0wt%.

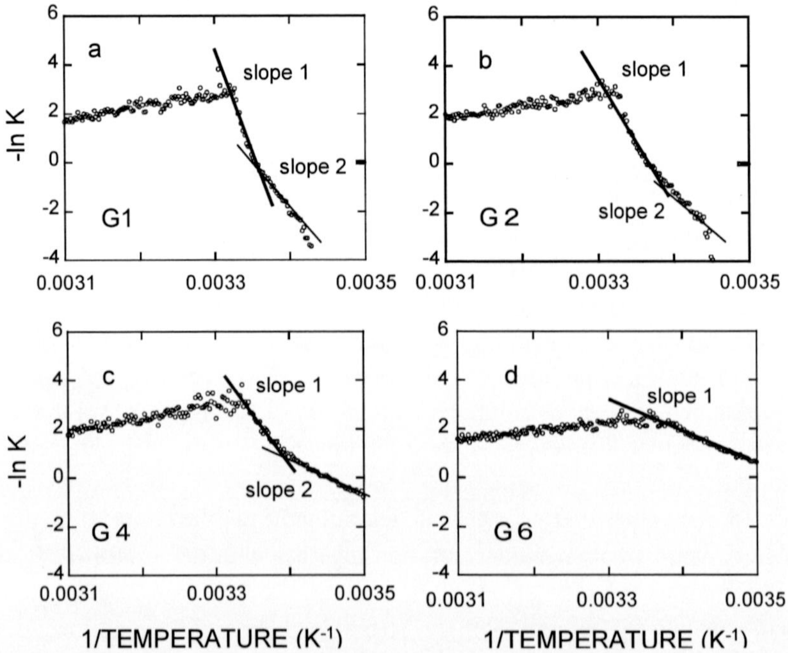

Figure 15. Polymer concentration dependence of coil to double-helix transition temperature, T_{ch}, for Na-gellan aqueous solutions with and without 25 mmol NaCl. T_{ch} obtained from CD data in aqueous solutions: G1( ), G2( ),G3( ), G4( ), G5( ), and G6( ). T_{ch} obtained from viscosity data in aqueous solutions: G1( ), G2( ),G3( ),G4( ), and G5( ). T_{ch} obtained from CD data in aqueous solutions with 25 mmol NaCl: G1( ), G2( ),G3( ), G4( ), G5( ), and G6( ). T_{ch} obtained from viscosity data in aqueous solutions with 25 mmol: G1( ), G2( ),G3( ), G4( ), and G5( ). Lines are experimental fit of G1-G5 data.

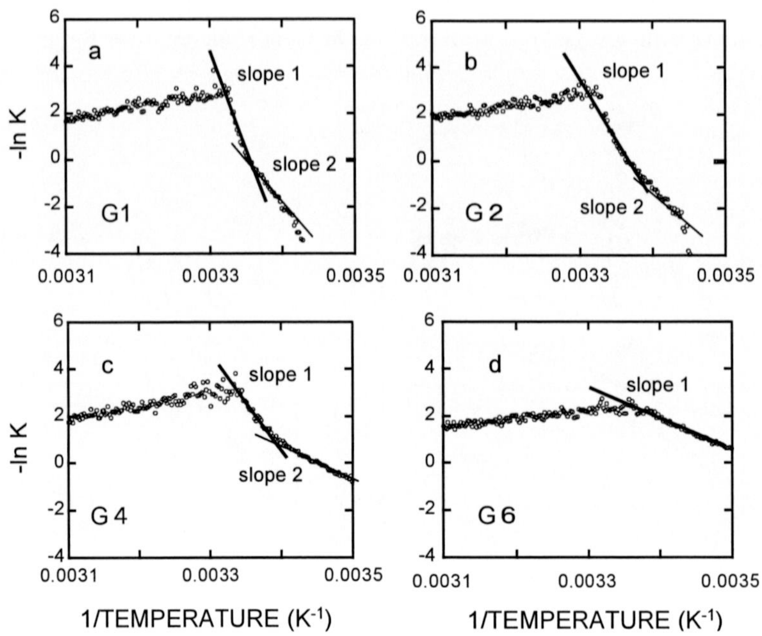

Figure 16. The van't Hoff plots of molar ellipticity, $[\theta]_{201}$, for Na-gellan aqueous solutions (c_p=1.0 wt%). (a) G1, (b)G2, (c)G4, (c)G6.

We analyzed the CD data using van't Hoff plots and calculated van't Hoff's transition enthalpy ΔH_{VH}. We concluded that the coil to double-helix transition for Na-gellan in aqueous solutions with 25 mmol NaCl is an all-or-none type transition [35]. Assuming that coils and double-helices coexist in aqueous solutions, double-helix contents in a chain f and equilibrium constant K can be calculated from Equations 1-3. Using the values of $[\theta]_{201c}$ and $[\theta]_{201h}$ obtained in 25 mmol NaCl [33,35], we calculated K values, and plotted them against $1/T$. Typical examples of van't Hoff plots are shown in Figures 16a-16d. ΔH_{vH} values for G1-G5 are determined from the slopes just below T_{ch} and are plotted against M_w in Figure 17. ΔH_{vH} values for G1-G5, which are nearly the same as the values obtained in aqueous NaCl solutions, are slightly dependent on the polymer concentration but depend markedly on M_w. These results suggest that, in the same way as the results obtained in aqueous solutions with NaCl, the coil to double-helix transitions in aqueous solutions without NaCl are accelerated with increasing M_w.

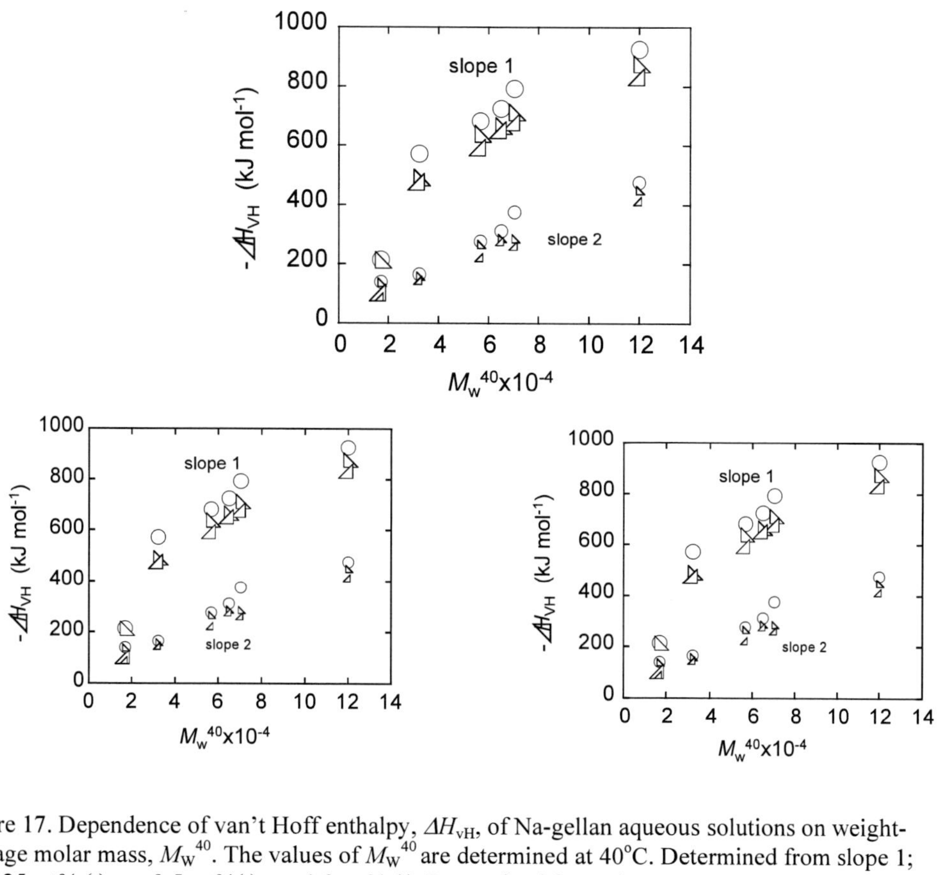

Figure 17. Dependence of van't Hoff enthalpy, ΔH_{vH}, of Na-gellan aqueous solutions on weight-average molar mass, M_W^{40}. The values of M_W^{40} are determined at 40°C. Determined from slope 1; c_p=0.25 wt% ( ), c_p=0.5 wt%( ), c_p=1.0 wt% ( ); Determined from slope 2; c_p=0.25 wt% ( ), c_p=0.5 wt%(), c_p=1.0 wt% ( ).

T_{ch} for G6 was lower than those for G1-G5 and ΔH_{vH} value for G6 was smaller than those of G1-G5. The conformational behavior of G6 molecule observed by viscometry and CD measurements in this study is undoubtedly different from that of G1-G5 molecules. It seems that for G6, the conformational transition such as the coil to double-helix transition did not occur in the temperature region of 30-10 °C.

CONCLUSION

Using 6 samples of well-purified Na-gellans (G1-G6, M_w=120x10^3-17x10^3 at 40 °C), effects of molar mass on the coil to double-helix transition in aqueous solutions with and without 25 mmol NaCl were studied by intrinsic viscosity, LS, CD, and DSC.

In aqueous solutions with 25 mmol NaCl, from the temperature dependence of $[\eta]$, M_w, $[\theta]_{201}$, and DSC exothermic curves, it was found that the coil to double-helix transitions for G1-G5 (M_w=120x10^3-32x10^3 at 40 °C) took place at the same temperatures, and the coil to double-helix transition is accelerated with increase of M_w. The G6 sample (M_w=17x10^3 at 40 °C) did not form a double-helix at 25 °C due to insufficient chain length. It is concluded that the lowest molar mass for double-helix formation at 25 °C lies between M_w=32x10^3 and M_w=17x10^3.

In aqueous solutions with 25 mmol NaCl, the $[\eta]$ and M_w for G1-G5 obtained in the temperature range of 40 to 25 °C can be explained by a simple coil/double-helix equilibrium model using the double-helix contents determined from CD data. The van't Hoff's transition enthalpy, ΔH_{vH} of Na-gellans depended on M_w. From these results, it is concluded that the coil to double-helix transitions are all-or-none type transitions and are accelerated with increasing M_w.

In aqueous solutions without NaCl, the coil to double-helix transitions for G1-G5 took place at the same temperatures, which are much lower than those in aqueous solutions with 25 mmol NaCl but depend markedly on polymer concentrations. ΔH_{vH} values for G1-G5 depend on M_w. From these results, it is concluded that, in the same way as the results obtained in aqueous solutions with 25 mmol NaCl, the coil to double-helix transition is accelerated with increasing M_w. Results of viscometry and CD measurements showed the conformational behavior of G6 was different from those of G1-G5.

REFERENCES

[1] O'Neill, M. A.; Selvendran, R. R.; Morris, V. J.1983, *Carbohyd. Res. 124*, 123-133.

[2] Jansson, P.; Lindberg, B.; Sandford, P. A. 1983, *Carbohyd. Res. 124*, 135-139.

[3] Sanderson, G. R. In *Food Gels*, Harris, P. Ed.; Elsevier Applied Science: London and New York, 1990, p201-232.

[4] Viebke, C.; Piculell, L.; Nilsson, S. 1994, *Macromolecules 27*, 4160-4166.

[5] Morris, E. R.; Rees, D. A.; Robinson, C. 1980, *J. Mol. Biol. 138*, 349-362.

[6] Grasdalen, H.; Smidsrød, O. 1981, *Macromolecules 14*, 1845-1850.

[7] Smidsrød, O.; Grasdalen, H. 1982, *Carbohyd. Polym. 2*, 270-275.

[8] Nijenhuis, K. te. In *Thermoreversible Networks*, Springer: Berlin, 1997, Chapter 12.

[9] Guenet, J. M. In *Thermoreversible Gelation of Polymers and Biopolymers*, Academic Press: London, 1992, Chapter 2, p110-112.

[10] Nishinari, K. 1997, *Colloid Polym. Sci. 275*, 1093-1107.

[11] Miyoshi, E.; Takaya, T.; Nishinari, K. 1996, *Carbohyd. Polym. 30*, 109-119.
[12] Miyoshi, E.; Nishinari, K. 1999, *Progr. Colloid Polym. Sci. 114*, 68-82.
[13] Moritaka, H.; Nishinari, K.; Nakahama, N.; Fukuba, H. 1992, *Biosci. Biotech. Biochem. 56*, 595-599.
[14] Yuguchi,Y.; Urakawa, H.; Kitamura, S.; Wataoka, I.; Kajiwara, K. 1999, *Prog. Colloid Polym. Sci. 114,* 41-47.
[15] Takahashi, R.; Akutu, M.; Kubota, K.; Nakamura, K. 1999, *Prog. Colloid Polym. Sci. 114*, 1-7.
[16] Ogawa, E. 1996, *Macromolecules 29*, 5178-5182.
[17] Ogawa, E. 1999, *Prog. Colloid Polym. Sci. 114*, 8-14.
[18] Ogawa, E.; Matsuzawa, H.; Iwahashi, M. 2002, *Food Hydrocolloids 16*, 1-9.
[19] Matsukawa, S.; Tang, Z.; Watanabe, T. 1999, *Prog. Colloid Polym. Sci. 114*, 15-24.
[20] Matsukawa, S.; Huang, Z.; Watanabe, T. 1999, *Prog. Colloid Polym. Sci. 114*, 92-97.
[21] Milas, M.; Rinaudo, M. 1996, *Carbohyd. Polym. 30*, 177-184.
[22] Bosco, M.; Miertus, S.; Dentini, M.; Segre, A. L. 2000, *Biopolymers 54*, 115-126.
[23] Annaka, M.; Honda, J. H.; Nakahira, T.; Seki, H.; Tokita, M. 1999, *Prog. Colloid Polym. Sci. 114*, 25-30.
[24] Milas, A.; Shi, X.; Rinoudo, M. 1990, *Biopolymers 30*, 451-454.
[25] Robinson, G.; Manning C. E.; Morris, E. R. In *Food Polymers,Gels and Colloids,* Dickinson, E. Ed; The Royal Society of Chemistry; Cambridge, 1991, p22.
[26] Nijenhuis, K.te. In *Thermoreversible Networks*; Springer: Berlin, 1997, Chapter 13.
[27] Atkin, N.; Abeysekera, R. M.; Kronestedt-Rombards, E. C.; Rombards, A. W. 2000, *Biopolymers 54*, 195-210.
[28] McIntire, T. M.; Brant, D. A. 1997, *Biopolymers 42*, 133-146.
[29] Gunning, A. P.; Kirby, A. R.; Ridout, M. J.; Brownsey, G. J.; Morris, V. J. 1996, *Macromolecules 29*, 6791-6796.
[30] Kirby, A. R.; Gunning, A. P.; Morris, V. J. 1996, *Biopolymers 38*, 355-366.
[31] Tanaka, F. 2003, *Macromolecules 36*, 5392-5405.
[32] Takahashi, R.; Tokunou, H.; Kubota, K.; Ogawa, E.; Oida, T.; Kawase, T.; Nishinari, K. 2004, *Biomacromolecules 5*, 516-523.
[33] Ogawa, E.; Takahashi, R.; Yajima, H.; Nishinari, K. 2003, *Trans. Ma. Re. Soc. Jap. 28*, 953-956.
[34] Ogawa, E.; Takahashi, R.; Yajima, H.; Nishinari, K. 2005, *Trans. Ma. Re. Soc. Jap. 30*, 851-854.
[35] Ogawa, E.; Takahashi, R.; Yajima, H.; Nishinari, K. 2005, *Biopolymers 79*, 207-217.
[36] Ogawa, E.; Takahashi, R.; Yajima, H.; Nishinari, K. 2006, *Food Hydrocolloids 20,* 378-385.
[37] Ogawa, E.; Nozawa, Y. 1998, *Rept. Prog. Polym. Phys. Japan 41*, 121-122.
[38] Ogawa, E. 2002, *Bull. Showagakuin Jr. College 38*, 67-74.
[39] Huggins, M. L. 1942, *J. Am. Chem. Soc. 64*, 2716-2718.
[40] Billmeyer, F. W. Jr. 1949, *J. Polym. Sci. 4*, 83-86.
[41] Mead, D. J.; Fuoss, R. M. 1942, *J. Am. Chem. Soc. 64*, 277-282.
[42] Yamakawa, H.; Fujii, M. 1974, *Macromolecules 7*, 128-139.
[43] Yamakawa, H.; Yoshizaki, T. 1980, *Macromolecules 13*, 633-643.
[44] Moritaka, H.; Nishinari, K.; Nakahama, N.; Fukuba, H. 1992, *Biosci. Biotech. Biochem. 56*, 595-599.

[45] Nakanishi, T.; Norisuye, T. 2003, *Biomacromolecules 4,* 736-742.
[46] Christensen, B.; E.; Knudsen, K. D.; Smidsrød, O.; Kitamura, S.; Takeo, K. 1993, *Biopolymers 33*,151-161.
[47] Gidley, M.; Bulpin, P. V. 1987, *Carbohyd. Re. 161,* 291-300.
[48] Chandrasekaran, R.; Millane, R. P.; Arnott, S.; Atkins, E. D. T. 1988, *Carbohyd. Re.175*, 1-15.
[49] Chandrasekaran, R.; Puigjaner, L. C.; Joyce, K.L.; Arnott, S. 1988, *Carbohyd. Re. 181*, 23-40.
[50] Okamoto, T.; Kubota, K.; Kuwahara, N. 1993, *Food Hydrocolloids 5*, 363-371.
[51] Nakanishi, T.; Norisuye, T. 2003, *Biomacromolecules 4*, 736-742.
[52] Kitamura, S.; Takeo, K.; Kuge, T.; Stokke, B. T. 1991, *Biopolymers 31*, 1243-1255.
[53] Stokke, B. T.; Knudsen, K. D.; Smidsrød, O.; Elgsaeter, A. 1991, *J. App. Polym. Sci. 42*, 2063-2071.
[54] Norton, I. T.; Goodall, D. M.; Frangou, S. A.; Morris, E. R.; Ree, D. A. 1984, *J. Mol. Biol. 175*, 371-394.
[55] Poland, D.; Scheraga, H. A. In *Theory of Helix-Coil Transitions in Biopolymers*; Academic Press: New York and London, 1970, Chapter 2.
[56] Okamoto, T.; Kubota, K.; Kuwahara, N. 1993, *Food Hydrocolloids 5*, 363-371.
[57] Schellman, J. A. 1995, *Compt. Rend. Trav. Lab. Carlsburg Ser. Chim. 29*, 230-259.
[58] Schellman, J. A. 1958, *J. Phys. Chem. 62*, 1485-1494.
[59] Applequist, J.; Damle, V. 1965, *J. Am. Chem. Soc. 87*, 1450-1459.

In: Biopolymer Research Trends
Editor: Tamas S. Nemeth, pp. 189-209

ISBN: 978-1-60021-983-2

Chapter 7

RAMAN SIGNATURES OF BIOPOLYMERS: DIAGNOSIS OF ORAL CANCERS AND INFLAMMATORY CONDITIONS

***C. Murali Krishna*[1*@], *V. B. Kartha*[1], *R. Malini*[1], *K. Venkatakrishna*[1], *Aparna Agarwal*[2], *Keertilata M. Pai*[2], *Betsy S. Thomas*[3], *Lakshmi Rao*[4], *Mohan Alexander*[5#] and *Jacob Kurein*[6]**

[1]Center for Laser Spectroscopy, Manipal Life Sciences Center,
[2]Department of Oral Pathology and Medicine, College of Dental Sciences,
[3]Department of Periodontics, College of Dental Sciences,
[4]Department of Pathology, Kasturba Medical College,
[5]Oral and Maxillofaical Surgery, College of Dental Sciences,
[6]Department of Surgical Oncology, Kasturba Medical College.
Manipal University, Manipal 576 104, Karnataka, India

ABSTRACT

Oral cancers are a serious health problem in developing as well as developed countries, and more so in India and other south Asian countries. Survival rate of these cancers, despite advances in treatment modalities are one of the poorest. This can be attributed to lack of reliable screening and early detection. Optical spectroscopy methods which are sensitive to biomolecular compositions of systems can be potential alternatives/ adjuvant diagnosis/screening approaches. Due to high sensitivity and

* Corresponding Author: Dr. C. Murali Krishna, Associate Professor, Center for Laser Spectroscopy, Manipal University, Manipal- 576104, INDIA. Telephone: +91-820-2571201 extn. 22526/22596. Fax: +91-820-2571931/2570061/62/63. E-mail: cm.krishna@manipal.edu, krishna.murali@univ-reims.fr

@ Present address: Dr. C. Murali Krishna, Scientifc officer E and Principal Investigator, Chilakapati Laboratory, Advanced Center for Treatment, Research and Education in Cancer, Tata Memorial Center, Sector 22, Kharghar, Navi Mumbai, 410 208, India. Ph. +91-22-2540 5039, mchilakapati@actrec.gov.in

simplicity of instrumentation, so far, autofluorescence has been the most popular among the optical diagnostic methods. Despite its inherently weak nature, other attributes of Raman spectroscopy such as *in vivo* applicability, rich information content through molecular finger print, easy extraction of data, and most importantly, use of less harmful Near Infrared (NIR) radiation with larger penetration depths for excitation, make this spectroscopy as an ideal choice. Presently, several data mining methods are available to spectroscopists to achieve objective discrimination which is a major advantage of optical spectroscopy methods over conventional approaches. We have demonstrated the efficacy of Raman spectroscopic discrimination of healthy and pathological oral tissues based on spectral signatures analyzed by PCA.

In the present chapter, we will provide a brief overview of oral cancers, spectroscopic approach for oral cancer diagnosis and basics of Raman spectroscopy. We also share our experiences on Raman spectroscopic discrimination of normal and disease conditions in oral tissues.

I. Introduction

I.1. Oral Cancer- Some Considerations for Early Detection

Malignant neoplasms anywhere in the human body tend to arouse fear in the general population. This is mainly because of the morbidity as well as mortality associated not only with the disease but also with its treatment. Malignant tumours of the oral cavity tend to cause a higher proportion of morbidity as it might cause disfigurement of the face as well as compromise of the masticatory apparatus which leads to decreased intake of food, thus affecting almost all the systems of the human body. This cancer starts as a sore that is not yet cancerous. Over a time, sometimes years, it can develop into a type of cancer that can spread to other parts of the body through the lymph systems and blood stream. Oral cancers can grow outwards also as a huge mass or they can be ulcers that invade inwardly. The longer oral cancers are allowed to grow untreated, the more likely they are to spread throughout the body.

World wide, oral cancer is the sixth most common cancer. But in the Indian subcontinent it accounts for almost 40% of all the cancers. In spite of great advances in the treatment of cancer, the five year disease free survival rates of oral cancer remains around 30-40%. The incidence of oral cancer varies markedly in various parts of the world. The highest incidence rates reported in 1985 were in the Indian subcontinent, Brazil, France and East European countries [1] .The age-standardized incidence rates per 1,00,000 population in South Asia were 25.1 for males and 14.9 for females as per this study. As per the National Cancer Registry Programme (1992) the most common site for oral cancer in the Indian population seems to be the tongue. In the west, almost 98% of the oral cancer patients are over the age of 40. But in South Asia, there are a large number of patients suffering from this disease who are less than 35 years old. Similarly, in the developed countries the incidence of oral cancer is two to three times more in men than in women, whereas the ratio is around one and a half times in South Asia. There seems to be an ethnic variation in the incidence of this disease. Indian migrants in Malaysia, South Africa etc., have been found to be having a much higher overall incidence of oral cancer compared to other populations. As mentioned earlier the survival rates globally have not shown much improvement in spite of great strides made in the field of surgery, radiotherapy, etc. These two modalities seem to be the main stay of oral cancer management with chemotherapy coming a distant third.

While considering oral cancer it is important that premalignant lesions and conditions (also known as potentially malignant lesions and conditions) are also included. Over 80% of the malignant lesions of the mouth are squamous cell carcinomas of the oral mucosa. The premalignant lesions tend to be more localized whereas the pre malignant conditions are more generalized and widespread, with systemic involvement. Examples of potentially malignant lesions include leukoplakia and erythroplakia. Potentially malignant conditions would include oral submucous fibrosis, erosive lichen planus, tertiary syphilis, sideropenic dysphagia, discoid lupus erythematosis, etc. As per the World Health Organization (WHO) definition leukoplakia is a white patch which cannot be rubbed off and cannot be characterized clinically or histologically as any other lesion. Biopsy of such a lesion may be advisable to establish diagnosis as well as for predicting malignant transformation chances, which is dependant on the degree of epithelial changes seen in the specimen. Various studies on the malignant transformation rate of leukoplakia have shown it to be around 5% with a range of 0.3-17.5%. There is no consensus as to whether to treat these premalignant lesions actively, since patients who underwent active treatment did not have a statistically significant decreased risk of malignant transformation. The malignant transformation rate of oral submucous fibrosis seems to be around 7% and this disease is predominantly found in patients from the Indian subcontinent.

The etiology of oral cancer, like other diseases, is multifactorial. There is not much evidence of genetic or familial predisposition, but studies on these are still in progress. The possible role of solar radiation in lip cancer has been noticed in some studies. But it is tobacco use and heavy alcohol consumption that may be the most important risk factors in oral cancer. Tobacco consumption may be in various forms. The smoking of tobacco in different forms like cigarettes, cigars, etc. is quite common. But much of the tobacco in the world is used in the form of smokeless tobacco. In India some of the various forms of oral smokeless tobacco use are paan, khaini, zarda, etc. These are kept in the mouth in contact with mucous membrane through which the nicotine is absorbed into the system. Another form of tobacco use is nasal snuff. There is also some evidence to indicate that betel quids without tobacco may also be associated with oral cancer. There are many confirmed carcinogens in tobacco belonging to the aromatic hydrocarbons like mono-, di-, and polycyclic aromatic hydrocarbons and aldehydes. Other carcinogens like N-nitroso compounds, and polycyclic aza-arenes, are also found in tobacco smoke. Alcohol is supposed to contribute to oral cancer in many ways [2]. Ethanol increases the permeability of oral mucosa to water itself and to many water-soluble molecules, and possibly also for many carcinogens. Acetaldehyde is the immediate metabolite of ethanol and this may cause damage to cells. In heavy drinking, there may be liver damage which in turn interferes with the detoxification of active carcinogens. Many of the heavy drinkers do not take adequate food and because of liver disease, metabolism also may be compromised. Viruses like the Human Papilloma Virus, Epstein-Barr virus, etc. may also contribute to the multi-step process of carcinogenesis in the oral cavity. Though denture wearing, dental fillings, etc. have not shown any increased risk to oral cancer, malignancy has been shown to arise in some areas where chronic trauma arising from a broken-down tooth, denture clasp, etc. has occurred.

Conventionally the method of diagnosis involves physical examination followed by a biopsy taken from suspected lesions which are subjected to histopatholgical examination which is the gold standard for diagnosis of malignancy. However, this method has been shown to have several shortcomings such as: (1) late detection as it depends on morphological changes which may be very late to occur; (2) visual decision making and subjective interpretation by pathologist; (3) time consuming in sample preparation and analysis and (4) poor prognosis support for clinician on premalignant lesions. The case in point oral cancers are one among the 10 leading cancers. Further, outcomes of these cancers, despite tremendous advances in treatment modalities, are one of the poorest which are attributed to late detection of the disease and lack of screening modalities. Typical survival-after- 5 years is reported to be about 55%. Twenty percent of mortality is reported to be due to direct consequence of malignancy and another 20% due to secondary cancers - even after apparent cure [3-6]. It is now recognized that for medical diagnosis, "the need has been created for reliable, rapid, inexpensive and sensitive tests that are non-invasive. Among the leading methods to perform such assays optically based methods hold the greatest promise" [7].

I.2. An Overview of Spectroscopic Approach in Cancer Diagnosis

Optical spectroscopy methods are uniquely poised to provide potential alternatives to conventional diagnosis. Advantages of these methods over the present methodology (pathological examination) are : (i) highly objective ; (ii) early diagnosis as they look at the molecular changes which occur at the onset of the disease ; (iii) less time consuming as virtually no sample pretreatments are needed and (iv) minimally invasive techniques can be adapted for *in situ/in vivo* applications by suitable fiberoptic probes. An important aspect of optical diagnosis is objectivity. As spectral data are amenable to multivariate statistical analysis, development of objective discrimination methods is facilitated. Multivariate analysis of data can be carried out by manually computing ratios of different selected peaks or by using algorithms of chemometrics, both supervised and unsupervised, which offer several possibilities of data mining. Principal Components Analysis (PCA), Discriminant Analysis (DA), and Hierarchical Cluster Analysis (HCA) are some of the widely employed multivariate tools in optical diagnosis.

Diagnosis of oral cancers by spectroscopic approach is well documented [8-53]. As evident from the literature, fluorescence and FTIR, because of higher sensitivity and simpler instrumentation, are more popular [8-41]. Several studies have demonstrated the efficacy of fluorescence spectroscopic diagnosis for oral cancer [8-34]. Diagnosis of oral and other cancers based on FTIR signals has also been reported [35-41]. In recent years, many studies using Raman or Raman micro-spectroscopy have shown clear discrimination between normal and pathological tissues [42-46]. Differentiation of healthy and diseased conditions by elastic scattering and proteomics based methods have also been reported in the literature [47-49]. As mentioned above, objective diagnosis is one of the major advantages of optical spectroscopy methods and spectroscopist utilizes multivariate tools to achieve this. Therefore, it is also necessary to explore several such tools in order to evaluate and identify most robust, user friendly methods of discrimination. Several groups have explored different discrimination methodologies such as PCA, ANN, KNN, MRDF on autofluorescence profiles in order to differentiate normal and pathological malignant tissues [43,50-53].

The emerging popularity of approaches based on Raman scattering compared to other spectroscopic techniques can be attributed to several unique features of this technique. Raman spectra are not influenced by presence of water like infrared spectra and it is not also influenced by sample thickness, which can influence infrared spectroscopy. Raman spectra do not depend on excitation wavelength since Raman scattering can be obtained by any photon, (exception being resonance and pre-resonance Raman). Information content of Raman spectra can be more easily exploited than in broad and featureless fluorescence spectra. Raman spectroscopy has some disadvantages like, as already mentioned, it is weak compared to fluorescence and Raman spectra of biological systems are therefore often swamped by parasitic fluorescence. However, latest developments in light sources (lasers) and detectors (CCDs) have made Raman spectroscopy of even weakly scattering samples like tissues and cells easy and good Raman spectra can be recorded without much effort. Use of near infrared photons, eg. 785, 830 or 850 nm, for excitation minimizes the associated fluorescence in biological samples. These wavelengths have an added edge of higher penetration, which could be advantageous in recording Raman spectra from lower depths in tissues. No or minimum sample preparation, applicability in *in vivo/in situ* measurements and use of harmless NIR sources make Raman spectroscopy an ideal tool in biomedical applications compared to infrared spectroscopy where sample needs to be dried, or to fluorescence where relatively harmful near UV or UV B radiation may have to be used for excitation.

I.3. Basics of Raman Spectroscopy

The interaction of photon promotes the molecule to some virtual state between the ground state and the lowest excited electronic state. Most of the time molecule returns to the original ground state where it originated. This leads to a phenomeneon known as Rayleigh scattering where the emitted light is of same wavelength as light used for excitation. In this case there is no change in the excitation energy which is a function of $h\upsilon$. However, a small fraction of light (approximately 1 in 10^7 photons) is scattered at frequencies different from the frequency of the incident photons (and usually lower than). The process leading to this inelastic scattering is termed 'Raman Effect'. When the molecule returns to the first vibrational level of the ground state, the change in energy (ΔE) causes the energy of scattered ight to be ($E - \Delta E$). Since the energy of emitted light is less, wavelength will be longer and this gives *Stokes lines*. However, if a molecule in the first vibrational level of the ground state is excited to a virtual state and returns to the lowest ground state then the emitted photon energy will be more than the energy of the initial photon by ΔE. In this case the wavelength of the emitted light will be shorter. These are called A*nti Stokes* lines. Figure 1 shows a schematic layout of these processes. From the discussion here and Figure 1 it can be seen that the difference in energy between the incident photon and the Raman scattered photon is equal to the energy of a vibration of the scattering molecule. A plot of intensity of scattered light versus energy difference, which is the 'Raman spectrum', thus gives information on the vibrational energy levels of the molecules contributing to the spectrum. The Raman spectrum is therefore highly characteristic of the composition of the sample.

The discovery of Raman effect in the year 1928 demonstrated that the analysis of inelastically scattered light from any molecule could provide a unique fingerprint of molecular structure [54,55]. In the last 75 years, popularity and versatility of Raman scattering spectroscopy have increased in many ways and a diverse family of Raman based techniques has been developed. More and more sensitive experimental approaches continue to be developed to explore the molecular mechanisms of complex biological phenomena. However, major drawback of this methodology is its inherent weak intensity. This is because Raman scattering is a non-linear and two-photon process. To improve sensitivity and make it more versatile, several improvisations have been developed over conventional Raman spectroscopy.

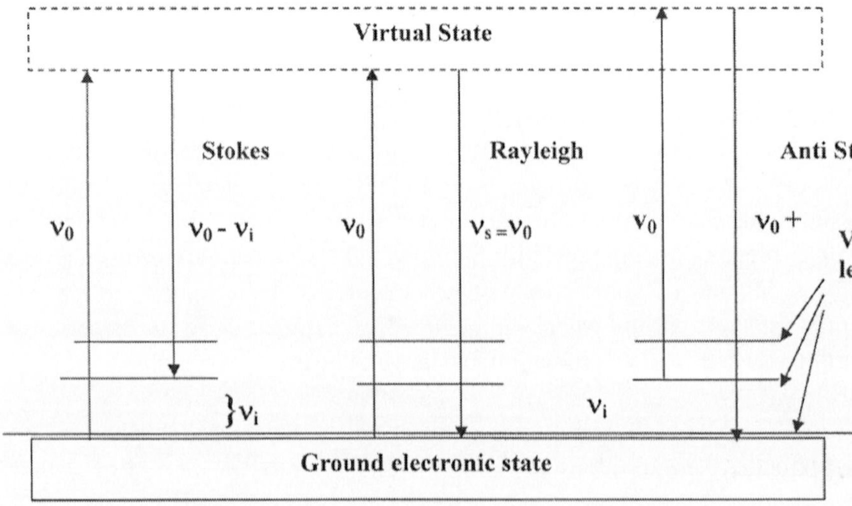

Figure 1. Schematic of Raman scattering processes.

Resonance Raman - increase in Raman scattering signal due to coupling of ground state vibration modes to excited vibronic levels [56,57], Surface Enhanced Raman Scattering (SERS) - enhancement in Raman signals due to attachment of molecules to nanometer sized metallic structures [58-64] and Raman microspectroscopy - Raman spectra recorded using microprobes, are widely used methods among the family of Raman based techniques.

Raman spectroscopy provides rapid and noninvasive characterization of physiological samples through its 'molecular finger print'. Raman spectroscopic differentiation of normal and cancerous tissues of oral [42-46], breast [65-72], cervix [73-76], colon [77-85], stomach [86-90] ovarian [91-93] and other forms of cancers have been reported in the literature. Reports of discrimination of cell types (multidrug resistant phenotypes of a cell type) have also appeared in the literature [94,95]. A recent study has demonstrated the feasibility of differentiating a particular cell type in a randomly distributed mixed cancer cell pellets [96]. In general, fresh tissues in saline are the ideal samples for *ex vivo* and *in vitro* optical spectroscopy studies. Because of the scarcity and other associated problems with fresh tissues and availability of *ex vivo* samples, it would be very useful to evaluate the suitability of fixed samples for optical pathology. Recently, Raman and combined Raman and FTIR studies have demonstrated the suitability of formalin-fixed tissues in optical pathology of oral, cervix, and ovarian cancers [44,91,92,97].

II. DISCRIMINATION OF ORAL TISSUES BY RAMAN SPECTROSCOPY – OUR EXPERIENCE

Although oral cancer is one of the leading cancers, and oral cavities are easily accessible, often cases are presented for treatment at fairly advanced stages. As pointed out earlier, despite very significant improvement in therapeutic methods, the mortality rate is very high. Formation of secondary cancer in other organs and metastases are also shown to be high for oral cancer. Early diagnosis and treatment only can change this trend. Hence, there is a need for development of new, rapid, reliable screening and early diagnosis methods which can be easily made available for the major section of susceptible population. It is well known that cancers involve multi-step processes which progress through precancerous conditions to eventual infiltrating malignant stage.

Raman spectroscopy is more molecular specific than other forms of spectroscopy like fluorescence as it is sensitive to vibrational modes of individual bonds. Therefore it can provide information on changes in composition of the tissue, leading to understanding of the fundamental biochemical processes taking place during onset, progression and regression of disease. Further, Raman spectroscopy needs no sample pretreatment, can be applied *in vivo*, is amenable to easy extraction of diagnostic parameters, and use less harmful NIR and thus provides an ideal method. In view of these considerations, development of Raman spectroscopic methods for discrimination of normal and malignant tissues was explored at Center for Laser Spectroscopy, Manipal University, Manipal, India. A Raman spectroscopic model for differentiation of normal and malignant tissues with more than 95% sensitivity and specificity was developed [46]. To the best of our knowledge this was the first investigation of this kind. These studies were further extended to pre-malignant and inflammatory situations. The aim of these investigations was to develop and evaluate the Raman spectroscopic diagnostic model for discrimination of normal and diseased tissues. In order to achieve this, our study was carried out in two steps. In the first step our earlier methodology [46] was revisited and tested by larger data set using certified samples. We have also extended analysis to premalignant and inflammatory conditions [42]. In the second step, blind samples, where spectroscopist was not aware of the nature of the sample, were used to test the models.

The certified samples, biopsy/surgical resection, were obtained from Dept. of Surgical Oncology, Sai Baba Cancer Hospital and Research Center and Dept. of Oral medicine and Radiology, College of Dental Sciences, MAHE, Manipal, India. Specimens from uninvolved areas from the same subjects were employed as control. In epithelial cancers, often a tissue specimen can have normal and anaplastic regions adjacent to each other. Further, in conventional pathological diagnosis also, several sites on a biopsy sample are examined and even if a single site is found pathological, the sample is treated as pathological. Moreover, in the case of '*in vivo*' screening or surgical demarcation by optical biopsy approach, it is necessary to identify the exact site where malignancy begins or terminates. For these reasons spectrum at several sites on each tissue were recorded and each spectrum was treated as a separate sample in data analysis. A total of 50 normal, 50 malignant, 10 inflammatory and 5 pre-malignant samples were available for these studies. Total 216 spectra were utilized for developing and testing the spectroscopic models. 24 blind samples were used to evaluate the models.

Raman spectra of both certified and blind samples were recorded using a Raman set up assembled in our laboratory. A schematic lay out of the Raman setup employed for the present studies is shown in Figure 2. This set up consists of a diode laser (SDL-8530, 785nm, 100 mW) as an excitation source, and HR 320 spectrograph (600 g/mm blazed at 900nm) and Spectrum One liquid N_2 cooled CCD for dispersion and detection of Raman signals. A holographic filter (HLBF-785.0, Kaiser Optics) was used to filter out unwanted lines from the excitation source. A notch filter (HSPF-5812, Kaiser Optics) was used to reject the Rayleigh scattering from the Raman signals. For the reasons explained above, depending on dimension of tissue, 5 or more spectra at different sites were recorded on each tissue.

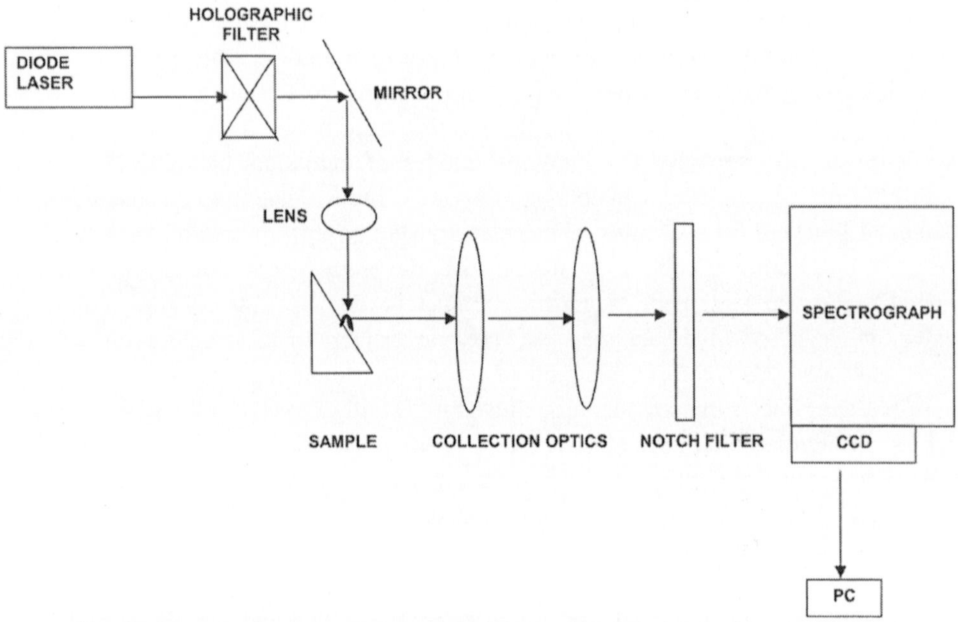

Figure 2. Lay out of Raman spectrometer used in the investigations.

Each spectrum was integrated for 30 seconds and averaged over 20 accumulations. The samples were kept moist with saline during the experiment. We have verified that under these conditions the Raman spectra remained unchanged for several hours. The recorded spectra were wavenumber calibrated with a cubic fit to known frequencies of Tylenol (4-acetamidophenol).

The mean Raman spectra of normal, inflammatory, pre-malignant, and malignant oral tissues are shown in Figure 3. As seen from the figure, the spectra show noticeable differences from each other. The differences between normal and malignant spectra are more pronounced in the 1200-1800 cm^{-1} region, while differences between malignant, pre-malignant and inflammatory conditions, though relatively less prominent, seem to be mainly in the 900-1400 cm^{-1} region. The features of normal spectra -weak, sharp C=O peak at 1750 cm^{-1}, sharp, weak C=C at 1650 cm^{-1}, strong CH_2 bend at 1440 cm^{-1}, two sharp peaks at 1330 cm^{-1} and 1270 cm^{-1}, and the broad peak at 1080 cm^{-1}- can be attributed to typical phospholipids, with practically no contribution from proteins. Malignant spectral features include broad and strong amide I at 1650 cm^{-1}, CH_2 bend at 1450 cm^{-1}, broad peaks in the amide III 1200-1350 cm^{-1} region, and the sharp phenyl alanine peak at 1000 cm^{-1}, all of which indicate a major contribution from proteins. In the case of benign and inflammatory conditions, spectra suggest variations in protein secondary structure as well as protein composition as indicated by features in amide I and III regions.

Both the clinician and pathologist rely on visual examination to make a decision. The decision is thus highly subjective, and will depend on their experience based on which they make the judgment. The fatigue factor in the examination of large number of samples and inexperience are often sited as reasons for the high error rate in conventional diagnosis for cancer [98]. These problems are removed in spectroscopic diagnosis. Here mathematical parameters are derived for a test spectrum, and these parameters are checked to see whether they fall into a given class within a desired range of standard deviation. No visual decision making is involved and the system (computer) is COMPLETELY BLIND as to what sample is given for analysis.

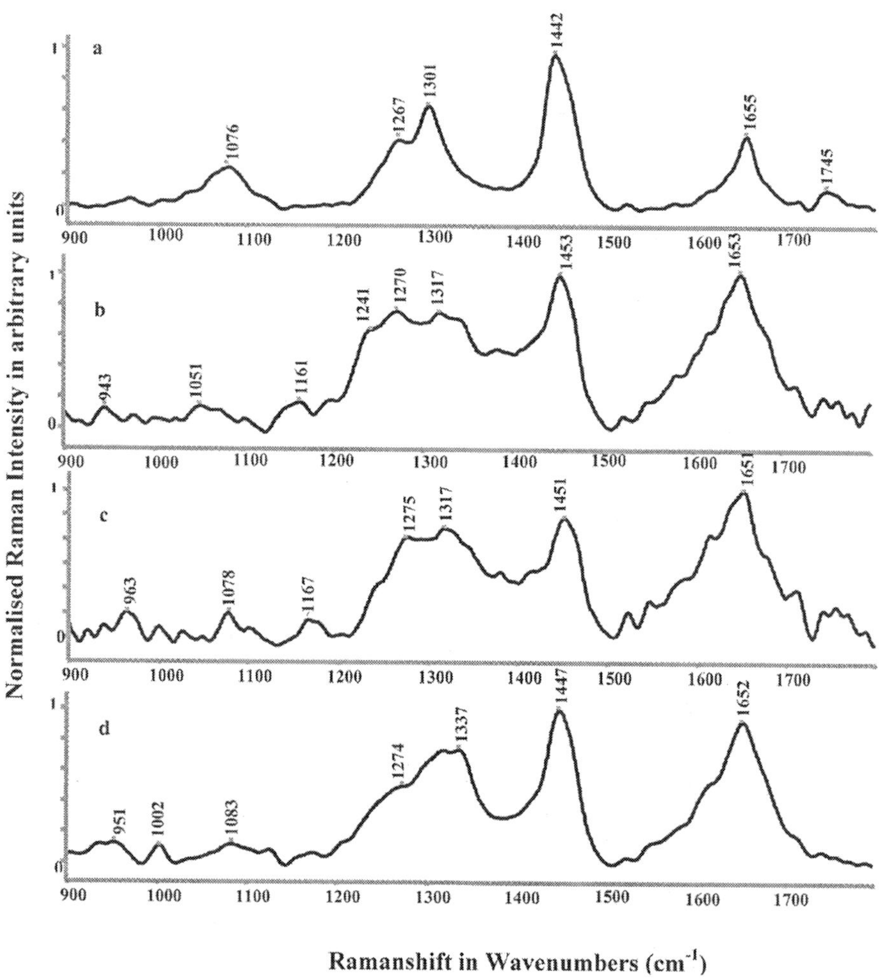

Figure 3. Mean Raman spectra of a.normal, b. malignant, c. inflammatory and d. benign oral tissues.

Even though many statistical tools such as PCA, ANN, SIMCA, HCA etc. are available for spectral discrimination of biological samples, PCA is found to be more popular. PCA is a classical data regression method where large spectral data are decomposed into small number of independent variations known as Factors or Principal Components and contributions of these factors to each spectrum are called Scores. Besides scores of factors, which are widely used as discriminating parameters, as will be discussed later, PCA provides other methodologies to bring out objective discrimination between samples in study. In our method of PCA, the mean of all samples in the data set is first formed. The differences of this mean from each sample is calculated to give the variations of each sample from the mean. With in samples, each having p data points we thus get an [n x p] matrix of these variations. Because all the samples contain more or less the same components (eg. Lipids, proteins, collagen etc) the large amount of data can be represented by a much smaller set of components, their contributions to each spectrum depending on their concentrations. In matrix language this implies that the [n x p] matrix of variations discussed above is highly redundant. It will have only a few eigen vectors (principal components) and the eigen values of these will rapidly come down to almost zero after the first few. Solving the eigen value - eigen vector problem will give us the principal components (factors), % variance (contribution of the factors to the variations in the data set), and scores of factors for each sample. The scores for a given sample correspond to the contribution of each principal component to the variation of that sample from the mean. It is therefore possible to simulate the observed spectrum of any sample by multiplying the eigen vectors with their respective scores for that sample and adding these products to the mean of the data set.

The preprocessed spectra (baseline corrected, smoothened, calibrated and normalized to δCH_2 peak) were used for analysis. In the case of certified sample spectra, analysis was carried out in two different approaches. In the first approach, spectra from normal, premalignant, inflammatory and malignant tissues were combined and analyzed by unsupervised PCA classification. Score of factor 1 was employed for cluster analysis, as shown in Figure 4. It is seen that almost all normal samples have a positive score, whereas pathological samples have a negative score value for the first factor. The use of the scores alone thus gives good discrimination between the normal and the various pathological conditions. But it fails to bring out discrimination between the different pathological conditions, i.e., premalignant, inflammatory, and malignant.

Further this approach of classification is somewhat cumbersome and tedious because, diagnosis of a sample needs entire analysis to be repeated along with new spectra. Moreover, it may be of limited practical utility for the end users, clinicians, since a visual decision making is involved in the case of border line samples. In view of these considerations, we have developed a second method using multiple discriminating parameters to give a better and objective diagnosis.

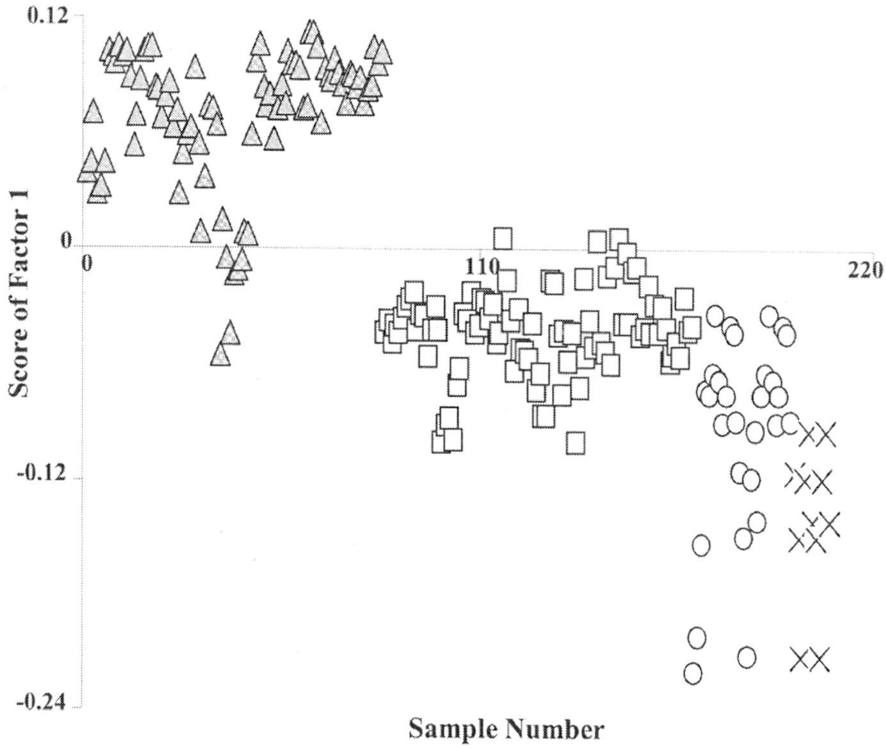

Figure 4. PCA of oral Raman spectra : A plot of score of factor 1 VS sample number. (▲: Normal □: Malignant O: Gingival, X: Pre malignant)

For this, like in any Analytical Technique where Standards with calibration curves are used for routine analysis, spectra of a set of Clinically/ Pathologically diagnosed samples can be used as a STANDARD CALIBRATION SET. This standard calibration set can be subjected to PCA to derive parameters which will be highly characteristic for any sample of that type. Any TEST SAMPLE can then be included in the set and the corresponding parameters for the test sample can be compared to the mean parameters for the set to decide whether the test sample belongs to that set, and if so, with what statistical probability. We have thus several statistical parameters available for decision-making in PCA, especially when standard calibration sets are used. These include scores, the spectral residuals, and the Mahalanobis Distance. The spectral residual is given by $Rs = \sum^{p}_{i=1}(S_o^i - S_s^i)^2$, calculated as the sum of the squares of the differences between observed and simulated intensities across the p spectral points. The Mahalanobis Distance, another well-known statistical parameter, which is computed in the present study by the equation $D^2 = (S_{unk}) M^{-1} (S_{unk})'$ where S_{unk} is the vector of scores and spectral residual for the unknown sample. M is the Mahalanobis matrix, given by M= S'S/ (n-1) where S is the corresponding (n X f) matrix of the standard calibration set of n standards. The Mahalanobis distance is already in units of standard deviation. When a test spectrum and any standard set are of same class then the values for the parameters for the test sample will be low and fall in the range for that of standard class, otherwise they will be very different.

In our second approach, therefore, 33 normal, 31 malignant, and 15 inflammatory tissue spectra, randomly selected from their respective classes were selected as standard sets for respective conditions. These standard sets were then checked for internal consistency by rotating out each spectrum from the standard set and comparing the three discriminating parameters, scores, Mahalanobis distance and spectral residual. As discussed above, these values will be very low if test spectrum and standard set belong to same class. We show the example of verification of standard set against malignant standard set, Figure 5A.

As one can expect, all malignant spectra showed low values whereas others yielded much higher values. Therefore good classification of all tissue types could be achieved. These results also suggest that standard sets are exclusive and representative of their respective classes. All other spectra were also tested against the standard sets, once again by computing same parameters. In this case, results obtained against normal standard set are given as example. Once again good classification was achieved by this methodology, (Figure 5B).

Once the validity of formation of highly differentiated standard sets of spectra of pathologically certified samples is established, any "blind" test sample spectrum can be diagnosed as belonging to a given class, by matching it with each standard set. We have made use of this fact and explored the user friendly limit test approach for routine diagnosis. This methodology is based on match/mismatch status of spectra against standard set. In this analysis spectra are matched against all available standard sets with inclusion/exclusion criterion for all 3 discriminating parameters, scores of factors, Mahalanobis distance and spectral residual. If spectra yield/do not yield values within limits fixed for a given standard set, then spectra are labeled as YES (PASS) or NO (FAIL). Typical results of this analysis against the inflammatory standard set is provided here as an example (Table 1).

As can be seen in Table I sample numbers 1-79, which are all normal, "PASS" against this standard set of 33 normal sample spectra. All other spectra "FAIL", against this standard set. The complete set of normal spectra fails against all other standard sets (data not shown). Thus classification/discrimination of a sample not only depends on matching against a particular standard set but also not matching against other sets. Thus this methodology provides unambiguous, objective discrimination.

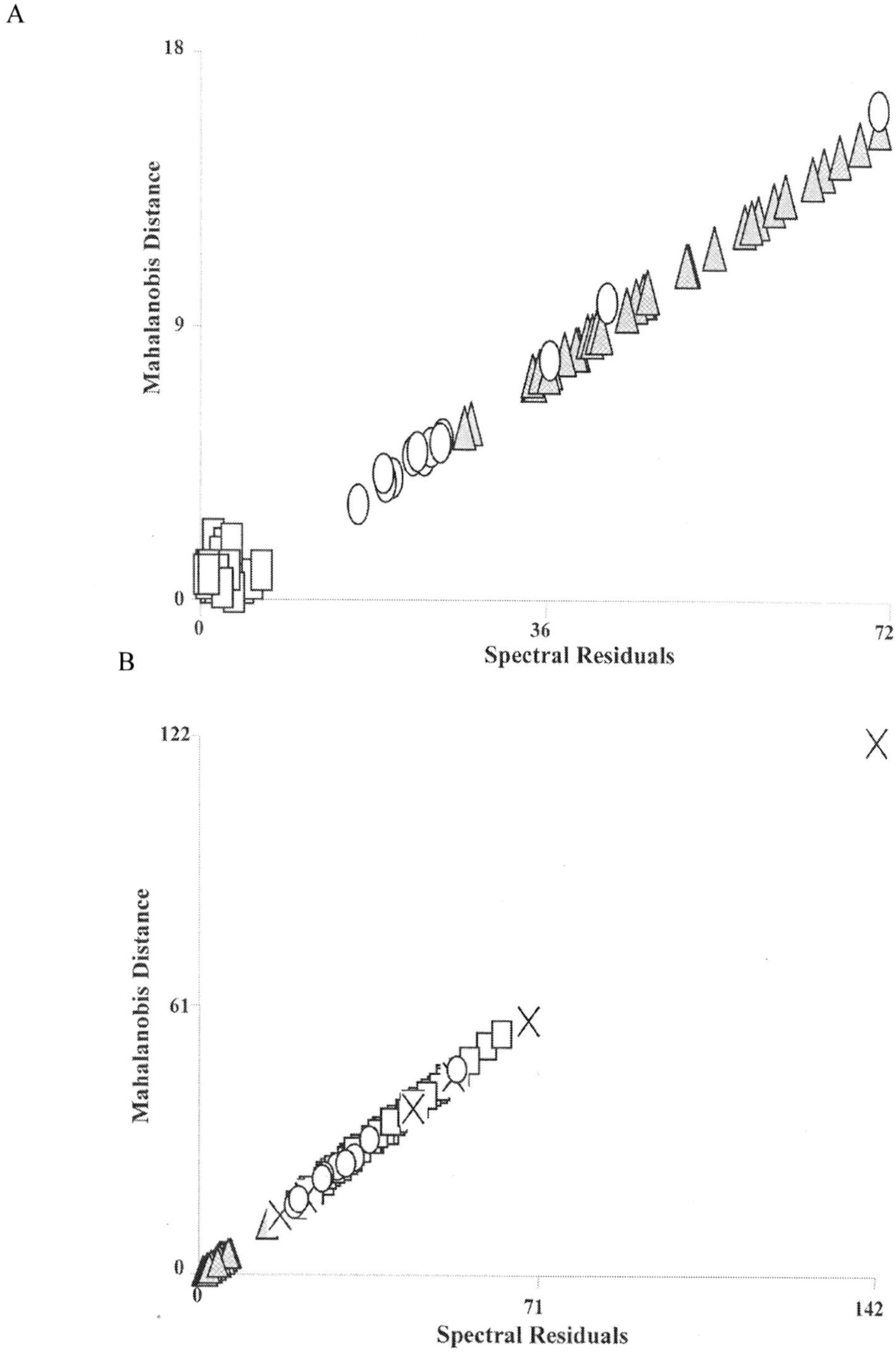

Figure 5. Discrimination of oral tissues using Mahalonbis distance and spectral residuals. A. Verification of standard sets (against malignant standard set). B. Testing of standard sets (against normal standard set). (▲: Normal □: Malignant, O: Gingival, X: Pre malignant)

Table 1. Limit test approach for discrimination of oral tissues (spectra were matched against normal standard set)

Spectrum No.	Match	Limit Test
1	YES	PASS (PPP#)
2	YES	PASS (PPP#)
3	YES	PASS (PPP#)
4	YES	PASS (PPP#)
5	YES	PASS (PPP#)
6-79	YES	PASS (PPP#)
80	NO	FAIL (FFF#)
81	NO	FAIL (FFF#)
82	NO	FAIL (FFF#)
83	NO	FAIL (FFF#)
84	*NO*	*FAIL (FFF#*
85-169	*NO*	*FAIL (FFF#)*
170	*NO*	*FAIL (FFF#)*
171	*NO*	*FAIL (FFF#)*
172	*NO*	*FAIL (FFF#)*
173	NO	FAIL (FFF#)
174	NO	FAIL (FFF#)
175-206	NO	FAIL (FFF#)
207	NO	FAIL (FFF#)
208	NO	FAIL (FFF#)
209	NO	FAIL (FFF#)
210	NO	FAIL (FFF#)
211	NO	FAIL (FFF#)
212-216	NO	FAIL (FFF#)

Table 2. match/mismatch status of blinded tissues against standard sets by 'Limit test' approach

S. No	Spectra analyzed	Normal standard set	Malignant standard set	Inflammatory standard set	Histopathology report	Raman spectroscopy report
1	a,b,c,d,e,f (6)	N,N,N,N,N,N	P,P,P,P,*N*,P	N,N,N,N,N,N	Malignant	Malignant
2	a,b,c,d,e (5)	N,N,N,N,N	N,N,N,N,N	P,P,P,P,P	*Inflammatory*	*Inflammatory*
3	a,b,c,d,e,f (6)	N,N,N,N,N,N	P,P,P,P,P,P	N,*P*,N,N,N,N	Malignant	Malignant
4	a,b,c,d,e,f (6)	N,N,N,N,*P*,*P*	N,N,N,N,N,N	N,N,N,N,N,N	Normal	Normal???
5	a,b,c,d,e,f (6)	N,N,N,N,N,N	P,P,P,P,P,P	*P*,N,N,N,N,N	Malignant	Malignant
6	a,b,c,d,e,f,g (7)	N,N,N,N,N,N,N	P,P,P,P,P,P,P	N,N,N,N,N,N,N	Malignant	Malignant
7	a,b,c,d,e,f (6)	N,N,N,N,N,N	N,*P*,N,N,P,P	N,N,N,N,N,N	Malignant	Malignant
8	a,b,c,d,e,f (6)	N,N,N,N,N,N	N,N,N,N,N	N,N,N,*P*,*P*,*P*	*Inflammatory*	*Inflammatory*
9	a,b,c,d,e (5)	N,N,N,N,N	N,N,*P*,*P*,N	N,N,N,N,N	Malignant	Malignant
10	a,b,c,d,e,f (6)	N,N,N,N,N	P,P,P,*N*,P,P	N,N,N,N,N,N	Malignant	Malignant
11	a,b,c,d,e,f (6)	N,N,N,N,N,N	*P*,*P*,N,N,*Y*,N	N,N,N,N,N,N	Malignant	Malignant
12	a,b,c,d,e,f (6)	N,N,N,N,N,N	N,N,N,N,*P*,*P*	N,N,N,N,N,N	Malignant	Malignant
13	a,b,c,d,e,f (6)	N,N,N,N,N,N	P,P,P,P,N,N	N,N,N,N,N,N	Malignant	Malignant
14	a,b,c,d,e,f (6)	P,P,P,P,P,P	N,N,N,N,N,N	N,N,N,N,N,N	Normal	Normal
15	a,b,c,d,e,f (6)	N,N,N,N,N,N	P,Y,Y,P,Y,P	N,N,N,N,N,N	Malignant	Malignant
16	a,b,c,d,e (5)	N,N,N,N,N	Y,P,Y,*N*,*N*	N,N,N,N,N	Malignant	Malignant
17	a,b,c,d,e,f (6)	N,N,N,N,N,N	P,P,*N*,P,P,P	N,N,N,N,N,N	Malignant	Malignant
18	a,b,c,d,e,f (6)	N,N,N,N,N,N	N,N,N,N,N,N	N,N,N,N,N,N	Normal (Gingival)	DNM
19	a,b,c,d,e (5)	N,N,N,N,N	N,N,N,N,N	N,N,N,N,N	Normal (Gingival)	DNM
20	a,b,c,d,e,f (6)	N,N,N,N,N,N	P,P,P,P,P,P	N,N,N,N,N,N	Malignant	Malignant
21	a,b,c,d,e,f (6)	N,N,N,N,N,N	*N*,*N*,P,P,P,P	N,N,N,N,N,N	Malignant	Malignant
22	a,b,c,d,e,f (6)	N,N,N,N,N,N	N,N,N,N,N,N	N,N,N,N,N,*P*	*Inflammatory*	*Inflammatory*
23	a,b,c,d,e,f (6)	N,N,N,N,N,N	N,N,N,N,N,*P*	N,N,N,N,N,N	Malignant	Malignant
24	a,b,c,d,e,f (6)	N,N,N,N,N,N	P,P,P,P,P,P	N,N,N,N,N,N	Malignant	Malignant

N = NO (no match), Y = YES (match), P = POSSIBLE, DNM = decision not made.

Validation of diagnostic models is prerequisite before it is routinely implemented. This is more important in the case of clinical practice. Hence, we have tested our models by blinded samples where we were not provided information about the samples. In this analysis 24 samples (spectra) were recruited. Spectra were matched against all available standard sets in 'limit test' approach. Based on match/mismatch status of spectra recorded from each tissue final diagnostic results were given. Raman spectroscopic diagnosis was correlated with the histopathological observations as shown in Table II. For example samples were diagnosed as normal (14), malignant (3,5,6,15,20,24) and inflammatory (2), respectively, as all the spectra recorded from these tissues have matched with only one of the sets and mismatched against all other standard sets. In other cases, sample numbers 1,7,9,10,11,12,13,16,17,21 and 23 were diagnosed as malignant though some of the spectra from these samples did not match with malignant set, because at least one from each sample matched with malignant set. In these cases matching of rest of the spectra of the same tissues with either normal or inflammatory were not taken into consideration. This is in tune with conventional histopathological analysis. As already been discussed earlier, pathologists make several sections from any single sample and observe several sites on each section. Even when one of the sites is pathological, the sample is considered as a pathological specimen. On the same token, sample number 8 and 22 were treated as inflammatory. However, there are few cases (sample numbers 18 and 19), where decision could not be reached. In these cases, spectra have not matched with any of the available standard sets. The samples in point are healthy but harvested from gingival and we have normal set for buccal tissues and inflammatory tissues of gingival origin. Therefore it is natural that spectra of these samples have not matched with either of these standard sets. The most important advantage of the spectroscopic method is that in such cases the subject can be called in for a second (repeat) examination immediately or later, since no biopsy is needed. The present results demonstrate the great usefulness of Raman spectroscopic models for diagnosis of oral pathological conditions.

III. CONCUSION

Oral cancers are one of the leading malignancies in developing countries, with poor survival rates. Presently there are no prescribed routine screening or clinical diagnosis methods. Optical spectroscopic approaches, mostly autofluorescence, are being shown as potential screening/diagnosis methodologies. Several groups are working on *in vivo* instrumentation for screening and early clinical diagnosis of oral cancers. Raman spectroscopy facilitates *in vivo* analysis by relatively less harmful near infrared radiations. Our experience of Raman spectroscopic discrimination of oral tissues has been very encouraging. The discrimination methodologies employed by us have not only provided objective discrimination but also are user friendly for clinical set up. However, the models developed by us needs further rigorous evaluation by double-blinded study on a larger sample size before being implemented into routine practice. And it also requires extending this approach (building up of models) to other oral pathologies like different categories in premalignant, inflammatory, and malignant conditions. Prospective development of models encompassing more pathological conditions as well as suitable fiberoptic probes will facilitate noninvasive diagnosis of oral pathologies based on Raman signatures.

ACKNOWLEDGMENT

We are thankful to Department of Science Technology, Government of India for the financial support ("Raman Spectroscopy Studies for Early Diagnosis in Oral Malignancy" No. SP/S2/L04/2001 DST). Ms Keerti and Mr. Chetan Anand are gratefully acknowledged for their support in sample handling and acquiring Raman data, respectively. Authors sincerely thank Mr. K. Kalyan Kumar for his help in preparing the MS. Our special thanks to our colleague Dr. Sajan. D. George for his critical comments and suggestions.

REFERENCES

[1] Parkin, DM; Pisani, P; Feslay, J. *Int. J. Cancer*, 1993, 54, 594-606.

[2] International Agency for Research on Cancer. Alcohol drinking. *IARC monographs on the evaluation of carcinogenic risks to humans*. Vol.44.Lyon:IARC.1988.

[3] Pingo, WA; Tong, T; Bolden, S. CA cancer *J. Clin.*, 1995, 45, 8-30.

[4] Partridge, M. *Dent Update,* 2000, 27, 288-294.

[5] Saranath, D. In *Carcinogenicity testing, predicting and interpreting chemical effect* (ed: K.T. Kitchin), Marcel Dekker Inc, NewYork. 1995, 653-675.

[6] Notani, PN. *Curr. Sci.*, 2001, 81, 465-474.

[7] Benaron, DA; Sevick-Muraca, EM; Florida, O. *Trends in Optics and Photonics*, March, 1996.

[8] Poh, CF; Ng, SP; Williams, PM; Zhang, L; Laronde, DM; Lane, P; Macaulay, C; Rosin, MP. *Head Neck*, 2007, 29, 71-76.

[9] Hsu, ER; Gillenwater, AM; Richards-Kortum, RR. *Appl. Spectrosc.*, 2005, 59, 1166-1173.

[10] Chen, HM; Chiang, CP; You, C; Hsiao, TC; Wang, CY. *Lasers Surg. Med.*, 2005, 37, 37-45.

[11] de Veld, DC; Bakker Schut, TC; Skurichina, M; Witjes, MJ; Van der Wal, JE; Roodenburg, JL; Sterenborg, HJ. *Lasers Med. Sci.*, 2005, 19, 203-209.

[12] Svistun, E; Alizadeh-Naderi, R; El-Naggar, A; Jacob, R; Gillenwater, A; Richards-Kortum, R. *Head Neck*, 2004, 26, 205-215.

[13] Skala, MC; Palmer, GM; Zhu, C; Liu, Q; Vrotsos, KM; Marshek-Stone, CL; Gendron-Fitzpatrick, A; Ramanujam, N. *Lasers Surg. Med.*, 2004, 34, 25-38.

[14] Manjunath, BK; Kurein, J; Rao, L; Murali Krishna, C; Chidananda, MS; Venkatakrishna, K; Kartha, VB. *J. Photochem. Photobiol B:Biol.* 2004, 73, 49-58.

[15] Majumder, SK; Ghosh, N; Kataria, S; Gupta, PK. *Lasers Surg Med.*, 2003, 33, 48-56.

[16] Wang, CY; Tsai, T; Chen, HM; Chen, CT; Chiang, CP. *Lasers Surg Med.*, 2003, 32,318-326.

[17] Tsai, T; Chen, HM; Wang, CY; Tsai, JC; Chen, CT; Chiang, CP. *Lasers Surg Med.*, 2003, 33, 40-47.

[18] Song, JM; Jagannathan, R; Stokes, DL; Kasili, PM; Panjehpour, M; Phan, MN; Overholt, BF; DeNovo, RC; Pan, X; Lee, RJ; Vo-Dinh ,T. Technol. *Cancer Res. Treat.*, 2003, 2, 515-523.

[19] Müller, MG; Valdez, TA; Georgakoudi, I; Backman, V; Fuentes, C; Kabani, S; Laver, N; Wang, Z; Boone, CW; Dasari, RR; Shapshay, SM; Feld, MS. *Cancer*, 2003, 97, 1681-1692.

[20] de Veld, DC; Skurichina, M; Witjes, MJ; Duin, RP; Sterenborg, DJ; Star, WM; Roodenburg, JL. *Lasers Surg. Med.*, 2003, 32, 367-376.

[21] Betz, CS; Stepp, H; Janda, P; Arbogast, S; Grevers, G; Baumgartner, R; Leunig, A. *Int. J. Cancer,* 2002, 97, 245-252.

[22] Miller, MG; Valdez, TA; Gerogakoudi, I; Backman, V; Fluentes, C; Kabani, S; Wang, Z; Boone, CW; Dasari, RR; Shapshay, SM; Feld, MS. *Cancer*, 2003, 97, 1681-1692.

[23] Palmer, GM; Marshek, CL; Vrotsos, KM; Ramanujam, N. *Las. Surg. Med.*, 2002, 30, 191-200.

[24] Heintzelman, DL; Utzinger, U; Fuchs, H; Zuluaga, A; Gossage, K; Gillenwater, AM; Jacob, R; Kemp, B; Richards-Kortum, RR. *Photochem. Photobiol.*, 2000, 72, 103-113.

[25] Majumder, SK; Gupta, PK. *Lasers in Life Sci.* 2000, 9, 143-152.

[26] Gillenwater, A; Jacob, R; Ganeshappa, R; Kem, B; El-Naggar, AK; Palmer, JL; Clayman, G; Mitchell, MF; Richards-Kortum, R. *Archives of Otolaryngology - Head and Neck Surgery*, 1998, 124, 1251-1258.

[27] Wang, CY; Chen, CT; Chiang, CP; Young, ST; Chow, SN; Chiang, HK. *Photochem. Photobiol.*, 1999, 69, 471-477.

[28] Zulwaga, AF; Utzinger, U; Durkin, A; Fuchs, H; Gillenwater, A; Jacob, R; Kemp, B; Fan, J; Richards – Kortum, R. *Appl. Spectrosc.*, 1999, 53, 302-311.

[29] Majumder, SK; Gupta, PK; Uppal, A. *Lasers in the Life Sci.*, 1999, 8, 211-277.

[30] Chen, CT; Chiang, HK; Chow, SN; Wang, CY; Lee, YS; Tsai, JC; Chiang, CP. *J. Oral Pathology and Medicine*, 1998, 64, 470-474.

[31] Chen, CT; Wang, CY; Kuo, YS; Chiang, HH; Chow, SN; Hsiao, IY; Chiang, CP. *Proc. Natl. Sci. Counc.* Repub. China B, 1996, 20, 123-130.

[32] Dhingra, JKD; Perrault, DF; McMillan, K; Rebeiz, EE; Kabani, S; Manoharan, R; Itzkan, I; Feld, MS; Shapshay, SM. *Archives of Otolaryngology - Head and Neck Surg.*, 1996, 122, 1181-1186.

[33] Kolli, VR; Shaha, AR; Savage, HE; Sacks, PG; Calash, MA; Schantz, SP. *American J. Surgery,* 1995, 170, 495-498.

[34] Yang, YL; Ye, YM; Li, FM; Ma, PZ. Lasers Surg. Med., 1987, 7, 528-532.

[35] Tobin, MJ; Chesters, MA; Chalmers, JM; Rutten, FJ; Fisher, SE; Symonds, IM; Hitchcock, A; Allibone, R; Dias-Gunasekara, S. *Faraday Discuss*, 2004, 26, 27-39.

[36] Wu, JG; Xu, YZ; Sun, CW; Soloway, RD; Xu, DF; Wu, QG; Sun, KH; Weng, SF; Xu, GX. *Biopolymers*, 2001, 62, 185-192.

[37] Jackson, M; Mantch, HH. *Infrared spectroscopy of biomolecules*, D.Chapmann eds, New york, Wiley-Liss, 1996, 311-340.

[38] Wu, JG; Xu, YZ; Sun, CW; Soloway, RD; Xu, DF; Wu, QG; Sun, KH; Weng, SH; Xu, GX. *Biopolymers* (Biospectroscopy), 2001, 62, 185-192.

[39] Dumas, P; Tobin, MJ. *Spectroscopy Europe*, 2003, 15, 17-23.

[40] Nanba, T. *Sci. Instrum.*, 1989, 60, 1680-1685.

[41] Schultz, CP; Liu, KZ; Kerr, PD; Manch, HH. *Oncol. Res.*, 1998, 277-286.

[42] Malini, R; Venkatakrishna K; Kurien, J; Keethilatha Pai, M; Kartha, VB; Murali Krishna, C. *Biopolymers*, 2006, 81, 179-193.

[43] Oliveira, AP; Bitar, RA; Silveira, L; Zângaro, RA; Martin, AA. Photomed. *Laser Surg.*, 2006, 24, 348-353.

[44] Murali Krishna, C; Sockalingum, GD; Jacob Kurien; Lakshmi Rao; Venteo, L; Pluot, M; Manfait, M; Kartha, VB. *Appl. Spectrosc.*, 2004, 58, 107-114.

[45] Kartha, VB; Jyothi Lakshmi, R; Krishna Kishore Mahato; Murali Krishna, C; Santhosh, C; Venkata Krishna, K. *Indian J. Phys.*, 2003, 77B, 113-123.

[46] Venkatakrishna, K; Jacob Kurien; Keerthilatha Pai, M; Murali Krishna, C; Ullas, G; Kartha, VB. *Curr. Sci.,* 2001, 80, 101-105.

[47] Sharwani, A; Jerjes, W; Salih, V; Swinson, B; Bigio, IJ; El-Maaytah, M; Hopper, C. *Oral Oncol.*, 2006, 42, 343-349.

[48] He, QY; Chen, J; Kung, HF; Yuen, AP; Chiu, JF. *Proteomics*, 2004, 4, 271-278.

[49] Venkatakrishna, K; Keerthilatha Pai, M; Ravikiran, O; Jacob Kurien; Mohan Alexander; Murali Krishna, C; Kartha, VB. *Curr. Sci.*, 2003, 84, 551-557.

[50] Kamath, SD; Mahato, KK. *J. Biomed. Opt.*, 2007, 12, 014028.

[51] Majumder, SK; Gupta, A; Gupta, S; Ghosh, N; Gupta, PK. *J. Photochem. Photobiol B.* 2006, 85, 109-117.

[52] Nayak, GS; Kamath, S; Pai, KM; Sarkar, A; Ray, S; Jacob Kurien; D'Almeida, L; Krishnanand, BR; Santhosh, C; Kartha, VB; Mahato, KK. *Biopolymers*, 2006, 82, 152-166.

[53] Wang, CY; Tsai, T; Chen, HM; Chen, CT; Chiang, CP. *Laser Surg. Med.*, 2003, 32, 318-326.

[54] Raman, CV; Krishnan KS. *Ind. J. Phys., Part I*, 1928, 2, 399-419.

[55] Raman, CV; Krishnan KS. *Nature*, 1928, 121, 501 - 502.

[56] Greve, J; Puppels, GJ. *Biomolecular Spectroscopy Part* A, RJH. Clark and RF. Hester, Eds. (John Wiley and Sons/Chichester.p231, 1993.

[57] Schweitzer-Stenner, R. J. Raman. *Spectrosco.*, 2001, 32 ,711-732.

[58] Fleischmann, M; Hendra, PJ; McQuillan, AJ. *Chem. Phys. Lett.*, 1974, 26,163.

[59] Moskovits, M. *Rev. Mod. Phys.*, 1985, 57, 783.

[60] Kneipp, K; Kneipp, H; Itzkan, I; Dasari, RR; Feld, MS. *Chem. Rev.*, 1999, 99, 2957.

[61] Vo-Dinh, T; Stokes, DL; Griffin, GD; Volkan, M; Kim VJ; Simon, MI. *J. Raman Spectrosc.*, 1999,30, 785.

[62] Kneipp, K; Kneipp, H; Itzkan, I; Dasari, RR; Feld, MS. *Curr. Sci.*, 1999,77, 915.

[63] Kneipp, K; Kneipp, H; Kartha, VB; Manoharan, R; Deinum, G; Itzkan, I. *Phys. Rev.* 1998, E, 57.

[64] Lee, NS; Hsieh, YZ; Paisley RF; Morris, MD. *Anal. Chem.*, 1990, 60, 442.

[65] Chowdary, MVP; Kalyan Kumar, K; Keerthi, Anand, R; Kurien, J; Murali Krishna, C; Mathew S; *Biopolymers*. 2006 83, 556-69.

[66] Haka, AS; Volynskaya, Z; Gardecki, JA; Nazemi, J; Hicks, D; Fitzumaurice, M; Dasari, RR; Crowe, JP; Feld, MS. *Cancer Research*, 2006, 66, 3317-3322.

[67] Abigail, SH; Shafer-Peltier, KE; Maryann, F; Joseph, C; Dasari, RR; Feld, MS. *Proc. Natl. Acad. Sci.,* 2005, 102, 12371-12376.

[68] Kneipp, J; Tom, BS; Kliffen, M; Marian, MP; Puppels, G. *Vibrational spectroscopy* 2003, 32, 67–74.

[69] Abigail, SH; Shafer-Peltier, KE; Maryann, F; Crowe, J; Dasari, RR; Feld, MS. *Cancer Research*, 2002, 62, 5375-5380.

[70] Shafer-Peltier, KE; Haka, AS; Fitzumaurice, M; Crowe, J; Myles, J; Dasari, RR; Feld, MS. *J. Raman spectroscopy*, 2002, 33, 552-563.
[71] Shafer-Peltier, KE; Haka, AS; Motz, JT; Fitzumaurice, M; Dasari, RR; Feld, MS. *J. of Cellular Biochem.*, 2002, 87, S125-S137.
[72] Manoharan, R; Shafer, K; Perelman, I; Wu, J; Chen, K; Deinum, G; Fitzumaurice, M; Myles, J; Crowe, J; Dasari, RR; Feld, MS. *Photochem. Photobiol.*, 1998, 67, 15-22.
[73] Murali Krishna, C; Prathima, NB; Malini, R; Vadhiraja, BM; Rani Bhat, A; Donald, JF; Pralhad, K; Vidyasagar, MS; Kartha, VB. *Vib. Spectrosc.* 2006, 41,136 -141.
[74] Utzinger, U; Douglas, L; Heintzelman, A; Mahadevan-Jansen, A; Michelle, MF; Richards-Kortum, R. *Applied Spectroscopy*, 2001, 55, 955-959.
[75] Mahadevan-Jansen, A; Mitchell, MF; Ramanujam, N; Utzinger, U; Richards-Kortum, R. *Photochem. Photobiol.*, 1998, 68, 427-431.
[76] Mahadevan-Jansen, A; Mitchell, MF; Ramanujam, N; Malpica, A; Thomsen, S; Utzinger, U; Richards-Kortum, R. *Photochem, Photobiol.*, 1998, 68,123-132.
[77] Chowdary, MVP; Kalyan Kumar, K; Keerthi, Anand, A; Kurien, J; Murali Krishna, C; Mathew, S; *Photomed. and Laser Surg.* (In Press), 2007.
[78] Andrade, PO; Bitar, RA; Yassoyama, K; Matinho, H; Santo, AM; Bruno, PM; Martin, AA. Anal. Bioanal. chem., 2007,387,1643-1648.
[79] Li, X; Lei, M; Wang, D; Lin, J. *Proceedings of the 2005 IEEE Engineering in Medicine and Biology 27th Annual Conference*, Shanghai, China. 2005, 6961-6964.
[80] Molckovsky, A; Song, LM; Shim, MG; Macron, N.E; Wilson, BC. *Gastrointest. Endoscopy.*, 2003, 57, 396-402.
[81] Shim, MG; Song, LM; Marcon, NE; Wilson, BC. *Photochem. Photobiol.*, 2000, 72, 146-150.
[82] Boustany, N; Crawford, JM; Manoharan, R; Dasari, RR; Feld, MS. *Lab. Invest.*, 1999, 79, 1201-1214.
[83] Manoharan, R; Wang, Y; Dasari, RR; Rava, R; Singer, SS; Feld, MS. *Lasers in Life Science*. 1995, 6, 217-227.
[84] Feld, MS; Manoharan, R; Salenius, J. In: *Advances in Fluorescence Sensing Technology II.* (Lakowicz, J.R.), SPIE, 1995, 2388, 99-104.
[85] Cellmer, E. *Lasers Life sci.*, 1991, 4, 23.
[86] Kalyan Kumar K; Anand R; Chowdary MVP; Keerthi; Kurien, J; Murali Krishna, C; Mathew S; Vib. *Spectrosc*. 2007; 44: 382-387.
[87] Yan, XL; Dong, RX; Zhang, L;. Zhang, ZW. *World J. Gastroenterol.* 2005,11, 3290-3292.
[88] Shen, AG; Ye, Y; Zhang, JW; Wang, XH; Hu, JM; Xie W; Shen, *J. Vib. Spectrosc.*, 2005, 37, 225.
[89] Yang, J; Guo, J; Wu, L; Sun, Z; Cai, W; Wang, Z. *Chinese Optics Letters*, 2005, 3, 705-707.
[90] Li, X; jin, H; Ding, C; Lin, J. *Proc. of the 4th Annual IEEE conf. on information technology Applications in Biomedicine*, UK, 2003, 250.
[91] Murali Krishna, C; Sockalingum, GD; Rani Bhat, A; Venteo, L; Pralhad, K; Pluto, M; Manfait, M. *Analytical and Bioanalytical Chemistry*, 2007, 387, 1649-1656.
[92] Murali Krishna, C; Sockalingum, GD; Venteo, L; Rani Bhat, A; Pralhad, K; Pluto, M; Manfait, M. *Biopolymers*, 2005, 79, 269-276.

[93] Chad Lieber, A; Molpus, K; Brader, K; Mahadevan Jansen, A. *Proceedings of SPIE*, 2000, 3918, 129-134.

[94] Murali Krishna, C; Kegelaer, G; Adt, I; Rubin, S; Kartha, VB; Manfait, M; Sockalingum, GD. *Biopolymers*,, 2006, 82, 462-470.

[95] Murali Krishna, C; Kegelaer, G; Adt, I; Rubin, S; Kartha, VB; Manfait, M; Sockalingum, GD. *Biochem. Biophys. Acta.*, 2005, 1726, 160-167.

[96] Murali Krishna, C; Sockalingum, GD; Kegelaer, G; Rubin, S; Kartha, VB; Manfait, M. *Vibrational Spectroscopy*, 2005, 38, 95-100.

[97] Murali Krishna, C; Sockalingum, GD; Vadhiraja, BM; Maheedhar, K; Anuradha Rao, CK; Rao, L; Venteo, L; Pluot, M; Donald, JF; Vidyasagar, MS; Kartha, VB; Manfait, M. *Biopolymers*, 2007, 85, 214-221.

[98] Koss LG; *J. Am. Med. Assoc.* 1989, 261, 737-743.

INDEX

A

D

E

F

G

H

I

K

L

O

P

Q

R